EL NIÑO
SOUTHERN OSCILLATION
& CLIMATIC VARIABILITY

Rob Allan

Climate Impact Group,
CSIRO Division of
Atmospheric Research,
Mordialloc, Victoria,
AUSTRALIA.

Janette Lindesay

Department of Geography,
Faculty of Science,
Australian National
University,
Canberra, ACT,
AUSTRALIA.

David Parker

Hadley Centre for Climate
Prediction and Research,
United Kingdom
Meteorological Office
(UKMO),
Bracknell, Berkshire,
UNITED KINGDOM.

CSIRO
AUSTRALIA

National Library of Australia Cataloguing-in-Publication entry

Allan, Rob
El Niño Southern Oscillation and Climatic Variability

Includes bibliography
ISBN 0 643 05803 6

1. El Niño Current. 2. Ocean–atmosphere interaction.
3. Southern Oscillation. 4. Climatic changes.
I. Lindesay, Janette. II. Parker, David. III CSIRO.
IV. Title.

551.5

This book is available from:
 CSIRO Publishing
 PO Box 1139 (150 Oxford Street)
 Collingwood, VIC 3066
 Australia

Tel. (03) 9662 7666 Int:+(613) 9662 7666
Fax (03) 9662 7555 Int:+(613) 9662 7555
Email: sales@publish.csiro.au
World Wide Web: http://www.publish.csiro.au

Cover photo: *Thunderstorm over Darwin*,
Peter Jarver, Thunderhead Photographics

Editor: Marta Veroni
Cover and layout design: Melissa Spencer
Typesetting: Melissa Spencer
Production Manager: Jim Quinlan

FOREWORD

El Niño Southern Oscillation and Climatic Variability, in book and CD-ROM versions, is based on a unique record of more than a century of scientific observations of our oceans and atmosphere. The observations of pressure and sea surface temperature are reinterpreted in the light of the relatively recent comprehension of the El Niño Southern Oscillation (ENSO) as an interaction of the ocean and atmosphere. ENSO is now understood to play a critical role in many of the extremes of climate, and the atlas is interspersed with examinations of the physical impacts of major ENSO phases (El Niño and La Niño). For example, an all too topical reminder of the impacts of El Niño on the economic and rural sectors is the Australian drought experience of the 1990s, for many the worst in over a century of records.

This atlas will be a valuable heritage, enriching the perspective of researchers and students, and linking wider fields of science with climatology and oceanography. These fields include a new and expanding audience studying the economic and social impacts of climate variability and how improved predictions of ENSO behaviour can reduce impacts. ENSO is now accepted as the major climate pattern after the underlying seasonal rhythm. This book makes a major contribution in documenting ENSO behaviour as defined, for example, by the patterns of propagation.

The completion of the atlas illustrates the dependence of science on instrumental records built up from dedicated and often unsung observers. Their records, whether from ships at sea or from isolated islands, have made a major international achievement possible. The authors have themselves been involved in the often painstaking collection and reconstruction of records dating back to the 1870s.

The authors have previously made significant contributions to the understanding of ENSO in Australia, southern Africa and globally, and this atlas will be a major resource supporting recent efforts to make ENSO knowledge available internationally. Initiatives include Climate Information and Prediction Services (CLIPS) by the World Meteorological Organisation and the International Research Institute developing applications in seasonal to interannual climate prediction in the United States of America.

The 'Managing with Climate Variability' Research Program has been pleased to join with CSIRO, the Australian National University and the Hadley Centre for Climate Prediction and Research at the United Kingdom Meteorological Office in a major international and institutional cooperative venture to bring this initiative to a wider audience. The Program is funded under National Drought Policy by the Australian Government and, with other R&D Corporations, funds a range of collaborative projects on improved management of climate risks in agriculture and natural resource management.

Barry White
National Coordinator
'Managing with Climate Variability' R&D Program
Land and Water Resources Research and Development Corporation

CONTENTS

ACKNOWLEDGEMENTS

The participation of Rob Allan in this publication was made possible under the CSIRO Climate Change Research Program, funded in part by the Australian Federal Government's Commonwealth Department of Environment, Sport and Territories, and the State Governments of the Northern Territory, Queensland and Western Australia, and the CSIRO Multidivisional Program on Climate Variability and Impacts. David Parker's contribution to this atlas is © Copyright, Controller, Her Majesty's Stationery Office, Norwich, England, 1996.

Comments on all or part of the manuscript were kindly provided by Drs Stuart Godfrey (CSIRO Division of Oceanography), George Kiladis (National Oceanic and Atmospheric Administration, Environmental Research Laboratory, USA), Neville Nicholls (Bureau of Meteorology Research Centre), Chris Reason (School of Earth Sciences, University of Melbourne), Ian Smith (CSIRO Division of Atmospheric Research), Peter Whetton (CSIRO Division of Atmospheric Research) and Steve Wilson (CSIRO Division of Atmospheric Research).

The drafting and construction of digital images and figures was undertaken by Louise Carr (CSIRO Division of Atmospheric Research), Joanne Richmond (CSIRO Division of Atmospheric Research), Val Lyons (Geography Department, Australian National University), and Gail Willetts (Bureau of Meteorology, National Climate Centre). Photographic assistance was provided by Mr David Whillas (CSIRO Division of Atmospheric Research).

Data processing for this book was performed by Ms Tracy Basnett (Hadley Centre for Climate Prediction and Research, United Kingdom Meteorological Office), Mr Malcolm Haylock (CSIRO Division of Atmospheric Research), Mr Jason Li (CSIRO Division of Atmospheric Research), Dr Xingren Wu (CSIRO Division of Atmospheric Research) and Mr Clive Elsum (CSIRO Division of Atmospheric Research).

The front cover photograph was supplied by Peter Jarver (Thunderhead Photographics).

Data sources were embellished significantly by contributions from Drs Aida Jose (Climate Data Section, Philippine Atmospheric, Geophysical and Astronomical Services Administration), Volker Wagner (Deutsher Wetterdienst Seewetteramt, Maritime Meteorology Data Division), Phil Jones (Climatic Research Unit, University of East Anglia), Kenneth Young (University of Arizona), Jim Salinger (National Institute of Water and Atmospheric Research, New Zealand), Nicola Rayner (Hadley Centre for Climate Prediction and Research, UK), and staff at the National Climate Centre and the Queensland Regional Office of the Australian Bureau of Meteorology (in particular Ms Julienne Morrison and Mr Ian Grantham).

Specific subject information was generously provided by Mrs Patricia and Felicity Troup (Alexander James Troup), Dr Gene Rasmusson (Niño regions genesis) and Dr Luc Ortlieb (ENSO research in South America).

Library and reference assistance was kindly made available by Mrs Liz Davy (CSIRO Division of Atmospheric Research Library), Ms Clare Body, Mrs Jill Nicholls and Mr Andrew Hollis (Bureau of Meteorology Library), Ms Anne Barrett (Archivist, Imperial College London), Ms Mary Spence (Royal Meteorological Society, Reading, UK), Ms Sarah Connolly (Science and Society Picture Library), Mr Maurice Crewe (United Kingdom Meteorological Office Library), Ms Kate Ellis (The Royal Society of Edinburgh) and Ms Barbara Morris (Royal Scottish Geographic Society).

The authors are grateful to The American Meteorological Society, CSIRO Division of Atmospheric Research, Koninklijk Nederlands Meteorologisch Instituut (KNMI) Library, India Meteorology Department, Royal Meteorological Society, Macmillan Distributors, Science and Society Picture Library, The Royal Society of Edinburgh and the Royal Scottish Geographical Society for permission to reproduce, and provision of, black and white photographs used in this publication.

Use of specific diagrams was only possible through the provision of such material by Drs Gary Meyers (CSIRO Division of Oceanography) (Figure 1), Janice Lough and David Barnes (Australian Institute of Marine Science) (Figure 42), Peter Webster (University of Colorado) (Figure 28), Don Chambers (University of Texas at Austin) (Figure 38) and the NOAA Pacific Marine Environmental Laboratory (PMEL) (Figures 23 and 27).

The authors are indebted to the Land and Water Resources Research and Development Corporation (LWRRDC) for their generous funding support for this project.

Land & Water
Resources
Research &
Development
Corporation

That these prolonged and widely extended oscillations of pressure exercise an important influence on the meteorology of the region, can scarcely be doubted; since, although originating apparently in the higher strata of the atmosphere, they necessarily communicate this excess to the lower strata, and must tend to produce an outflow of air from the axis of maximum intensity; and, since this appears to lie over the continental masses of Australia and Asia, to perpetuate dry land winds, and bring about a certain diminution of the rainfall.

Henry Francis Blanford (1880b, p. 58), the first Imperial Meteorological Reporter to the Government of India from 1875 to 1887, describing climatic conditions in the Indo–Australasian region during what we now know as the 1877–1878 El Niño event.

I think that the relationships of world weather are so complex that our only chance of explaining them is to accumulate the facts empirically; we know now that it was impossible to explain cyclones (lows) until data of the upper air conditions were available, and there is a strong presumption that when we have data of the pressure and temperature at 10 and 20 km, we shall find a number of new relations that are of vital importance.

Montgomery (1940, p. 9) quoting Sir Gilbert Thomas Walker, the second Director-General of Observatories in India from 1904 to 1924.

CONCEPTS AND
SCENE SETTING

1. Introduction

Since the mid 1970s, the El Niño Southern Oscillation (ENSO) phenomenon has received considerable attention worldwide. Interest in ENSO has spread to encompass not only scientists from many fields, but also primary producers, policy analysts, management specialists and the wider community. This has resulted primarily because of the severe global impacts of the 1972–73 and 1982–83 episodes, and the 'persistent' El Niño or climatic anomaly sequence that dominated the climate in the first half of the 1990s. In many countries, the phenomenon is now described regularly in the print media and even on television news and weather segments, and it is thus becoming increasingly familiar to the general public.

Although there is no overall theory that can explain all aspects of the ENSO phenomenon, research efforts over many decades have resolved some of its important physical and dynamical elements. ENSO is known to be one of the largest sources of natural variability in the climate system, and involves two opposite but closely interacting phases (often termed 'warm' or El Niño and 'cold' or La Niña events). In fact, the phenomenon is part of a continuum, with El Niño and La Niña events representing the tendency for the ocean–atmosphere system to be in one extreme or the other. These events occur aperiodically (usually in the 2–10 year time frame) but, once established, there is a tendency for the phenomenon to exhibit quasi-biennial characteristics (to change from one phase to the other during the 'life cycle' of the phenomenon). Climatologists consider the ENSO phenomenon to be a large-scale, natural fluctuation in the global climate system that occurs irregularly and involves a close coupling of the oceans and atmosphere (Figure 1).

Considerable marine and terrestrial impacts are related to ENSO, particularly across its dynamical core region in the Indo–Pacific region and as a result of interactions with the planetary monsoon system. However, it has become apparent that major near-global climatic impacts result from ENSO extremes, and that they tend to become more extensive during strong events. This occurs via perturbations to the climate system that develop as significant ocean–atmosphere redistributions in the Indo–Pacific basin become established during the ENSO cycle. These organised perturbations, or teleconnection patterns, extend to higher latitudes in both hemispheres and vary seasonally. In some regions, teleconnections show a degree of persistence that can extend for a number of months or seasons, and this has formed the basis of attempts to forecast climatic impacts such as rainfall and air temperature.

Concerns about the stability of ENSO effects on the environment in the longer term have led to an increasing focus on possible modulations of the phenomenon by natural fluctuations in the climate system. These climatic variability and ENSO initiatives have occurred as new perspectives on the nature and characteristics of ENSO and its teleconnections are emerging, a better understanding of the physical mechanisms driving ENSO phases is available, and there are continuing improvements in numerical modelling of ENSO.

This book uses long-term gridded fields of monthly atmospheric mean sea level pressure (MSLP) and sea surface temperature (SST) over the globe to provide the basis for a synthesis of past ENSO phases, and the extension of our understanding of the phenomenon and its relationship to natural climatic variability. In order to appreciate fully the scope of the ENSO phenomenon presented in this publication, it is important to examine something of the historical context in which scientific understanding of ENSO has evolved.

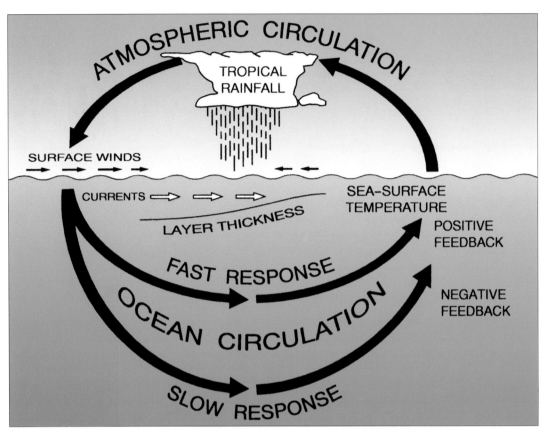

Figure 1: Schematic diagram of oceanic and atmospheric variable interactions intrinsic to the ENSO phenomenon in the tropical Pacific Ocean. The fast and slow responses in the ocean point to current concepts of the different feedbacks caused by physical processes involving wave dynamics (see discussions in 'Ocean–Atmosphere Interaction' p. 23) (Meyers, 1994 pers. comm.).

El Niño and La Niña, Southern Oscillation and SOI, ENSO

The evolution in understanding and thinking that has led to the term ENSO being used to describe this large-scale, ocean–atmosphere phenomenon had, until 30 years ago, involved separate research histories in oceanic and atmospheric fields. Our current perspective on ENSO has been a relatively recent development, and many aspects of the phenomenon are still to be resolved. It is only in the last 10 years that there has been a more balanced focus on both El Niño and La Niña extremes of the ENSO phenomenon.

The Oceans: El Niño and La Niña

The Spanish term El Niño has been used for centuries by South American mariners and fishermen to define the annual occurrence of warm, southward-flowing oceanic current waters off the coast of Ecuador and Peru around Christmas time each year (Carrillo, 1892). A study of South American coastal fishermen's knowledge of changing ocean currents and conditions by Suzuki (1973), highlights the geographical experience of both the El Niño and the cold, northward-flowing Humboldt Currents (Figure 2). El Niño translates as 'the child', and in this case specifically 'the Christ child'. According to Hisard (1992), the earliest documented scientific encounters with anomalous ocean-current conditions off the western coast of South America indicative of the El Niño phenomenon occurred in the years 1791 and 1804. Although documentary evidence of local, more regular El Niño influence has been compiled from texts back to the early 1500s (Quinn *et al.*, 1987; Hocquenghem and Ortlieb, 1992; Quinn, 1992, 1993; Quinn and Neal, 1992; Ortlieb and Machare, 1993; Ortlieb, 1994), it was the studies of Carranza (1891), Carrillo (1892) and Eguiguren (1894), all inspired by the 1891 El Niño event, that were amongst the first to document such warm current episodes and suggest that they were a strong influence on regional climatic patterns, especially Peruvian coastal rainfall. Wider scientific exposure to El Niño was first reported around 100 years ago in an address to the 6th International Geographical Congress in London in 1895 (Pezet, 1896). Interest in this oceanic feature during this century was expanded primarily through papers by Murphy (1926, 1932), Schott (1931), Gunther (1936), Lobell (1942), Mears (1943, 1944), Schweigger (1945, 1961), Wooster (1960) and Bjerknes (1961). Such studies showed strong links with local flooding rains in Peru during the 1925 and 1941 warm current episodes, and suggested local variations in oceanic advection or upwelling due to changes in prevailing winds as causes for the phenomenon. Schott was the first to use the term El Niño for the wider oceanic occurrence of warm waters advected southward from around the Galapagos Islands. It was the International Geophysical Year (IGY) of 1957–58 that provided, for the first time, observations of large-scale oceanic warming extending across the equatorial Pacific beyond the dateline in association with an El Niño event during that period. By the latter part of 1960, newspapers in Lima, Peru, carried a forecast of El Niño conditions in the austral summer of 1961

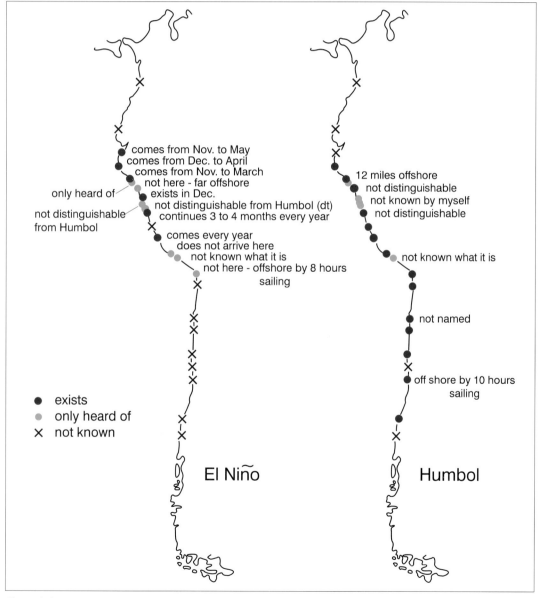

Figure 2: South American west coast fishermen's knowledge of the El Niño and Humboldt (Humbol) Currents (after Suzuki, 1973).

made by the Dutch meteorologist Hendrik Berlage. Although an event did not occur, Berlage had provided the impetus for a more holistic view of the El Niño phenomenon. It was Jacob Bjerknes, from the University of California-Los Angeles, who was the first to argue concisely that the wider spatial pattern of warming and observed atmospheric circulation fluctuations across the North Pacific to North American sector were linked, and had occurred during previous El Niño events. In a series of papers, Bjerknes (1966, 1969, 1972) linked El Niño episodes with what meteorologists termed the Southern Oscillation, and postulated that these irregular warmings were the result of large-scale, ocean–atmosphere interactions across the Pacific basin. A resurgence in scientific interest ensued, and has continued up to the present.

One of the most significant contributions to the longer term record of El Niño events was made during the 1970s–1980s through the efforts of William Quinn, an oceanographer at Oregon State University, who produced a series of papers using historical evidence from the South American region back to the 1500s (Quinn et al., 1978; Quinn and Neal, 1983; Quinn et al., 1987; Quinn and Neal, 1992). However, as research has shown, these events need not always be representative of the wider ENSO phenomenon across the Indo–Pacific basin. As a consequence, Quinn (1992, 1993) attempted to look at evidence for responses to large-scale ENSO events, such as droughts in India, Australia and Indonesia, and weak Nile River floods. These later papers dealt with only the 'warm' or El Niño phase of the phenomenon when referring to ENSO, and the only similar compilation of the 'cold' or La Niña phase through time is in a study by Whetton and Rutherfurd (1994). Recent papers by Hocquenghem and Ortlieb (1992), Mabres et al. (1993), Ortlieb and Machare (1993), Ortlieb (1994) and Whetton et al. (1996) have questioned aspects of the most recent South American regional index given in Quinn (1992, 1993). This latest Quinn sequence is shown to be influenced by wider global indications of ENSO warm phases. Most affected are some of the dates, magnitudes and reliability of events in the earliest centuries of the Quinn (1992, 1993) regional series. Hocquenghem and Ortlieb (1992), Ortlieb and Machare (1993) and Ortlieb (1994) indicate that they are undertaking revisions of the Quinn regional El Niño series. A detailed bibliography of publications on the El Niño phenomenon up to 1985 has been produced by Mariategui et al. (1985), and recent broad research programs/projects in South America are detailed in Ortlieb and Machare (1992) and Machare and Ortlieb (1993).

Direct instrumental measurements of the equatorial Pacific SST warming signature of El Niño events have provided the basis for simple indices of the phenomenon. Initially, these measurements centred on in situ SST data from individual South American west coast ports such as Puerto Chicama and Chimbote in Peru (Berlage, 1966; Ramage, 1975; Wright, 1977; Julian and Chervin, 1978). In the late 1970s, Wright (1977) used historical SST data in the region 10°N–10°S, 90°W–180°W from the United Kingdom Meteorological Office (UKMO) for the period 1949–1969 to construct an equatorial Pacific SST anomaly index. Wright (1984) built on his earlier index by using historical and contemporary SST data and maximising spatial SST correlations in the above region to create a spatially averaged SST anomaly index for the period 1950–1983. This approach was expanded further in Wright (1985, 1986, 1989) to produce a monthly SST anomaly index from 1872 to 1986. Other analyses of relationships between observations of equatorial Pacific SST along various ship tracks and other Indo–Pacific and global signatures of ENSO have led to an objective derivation of oceanic regions that more completely measure the evolution of the phenomenon (Newell et al., 1982; Rasmusson and Carpenter, 1982; Wright et al., 1985, 1988; Kiladis and Diaz, 1989; Diaz and Kiladis, 1992). The most widely used indices of Pacific SST fluctuations due to ENSO are the Niño 1, 2, 3 and 4 regions devised by Gene Rasmusson at the then Climate Analysis Center (CAC) in the early 1980s, and detailed in the regular monthly Climate Diagnostics Bulletin of the now Climate Prediction Center (CPC) in Washington, D.C. (Figure 3). Expansion of observational networks and improvements in technology have resulted in efforts to develop high resolution, real time measurements of other oceanic variables to monitor ENSO phases. Most notable are in situ sea level measurements at Indo–Pacific island locations pioneered by Wyrtki (1973, 1974, 1975a, b, 1976, 1977), with details of the array in Wyrtki et al. (1988a, b); sea level data on a Pacific basin-wide scale are available from satellite observations alone, or are blended with Pacific tide gauge data (Cheney et al., 1987, 1989, 1991; Delcroix et al., 1991; Miller et al., 1993a). Increasing emphasis has also been placed on obtaining subsurface data on parameters such as oceanic temperature and salinity using expendable bathythermographs (XBTs) (Levitus, 1982; Smith, 1993; Donguy, 1994).

Figure 6: Henry Francis Blanford (India Meteorological Department, 1976).

Figure 7: Sir Charles Todd (Bureau of Meteorology, 1996 pers. comm.).

as Zi-ka-wei [Shanghai] in Northern China, and Sydney in Australia; but these stations were probably near the limits of its influence in that direction. It was less intense at Ceylon [Sri Lanka] than at the Nicobars and in the Western Himalaya than in Sikkim. Superimposed on the principal wave of pressure, were other and possibly independent oscillations, which appear to have reached their maxima, in each hemisphere alternatively, in the winter [boreal] or early spring [boreal] season, and to have extended with diminished influence into the opposite hemisphere. But there are geographical variations in these subordinate oscillations which cannot at present be reduced to rule; and it is noteworthy that, at Ekatarinenburg [Sverdlovsk] in Western Siberia, the oscillations, up to June 1877, correspond closely to those of South Australia.

One of the respondents to Blanford's request, who provided monthly pressure data for observing stations in Mauritius, Australia and New Zealand, was Sir Charles Todd (Figure 7), the then Deputy Postmaster-General of the British Commonwealth, Government Astronomer and Meteorologist in South Australia. Todd (1888, p. 1456) was later to remark:

> Comparing our records [Australia] with those of India, I find a close correspondence or similarity of seasons with regard to the prevalence of drought, and there can be little or no doubt that severe droughts occur as a rule simultaneously over the two countries.

During his term in India (1875–1887), Blanford examined climatic conditions over the subcontinent in an attempt to forecast monsoonal fluctuations. One forecasting approach that he had considered in 1881, and discounted as unsubstantiated but worth further examination, was Lockyer's earlier relationship between the solar cycle and Indian monsoonal rainfall. By 1882, climatic data of another type had become available following the Indian Government's initiation of regular monthly reports of Himalayan snowfall from regional authorities. These snowfall reports, together with wind and pressure distributions over the subcontinent, provided the basis of the experimental monsoonal forecasts that were prepared by Blanford during the period 1881–1884, but never issued. In 1885–86, the India Meteorological Department began publishing forecasts of Indian summer monsoon rainfall using the previous January to May snowfall data over the Himalayas and the general nature of regional climatic variables.

Blanford's successor, Sir John Eliot (Figure 8), was Imperial Meteorological Reporter, then Director-General of Observatories, in the India Meteorological Department from 1887 until 1903, during which time he looked at past analogue conditions over India and widened the scope of precursors used to forecast the Indian monsoon to include variations in the south-east trade winds over the Indian Ocean, the strength of the South Indian Ocean anticyclone, Nile floods and data from southern Australia and South Africa (Eliot, 1896). Forecasts provided more and more information, and during the 1890s drew high praises.

> The importance to the farmer, the horticulturist, and pastoralist of knowing beforehand the probabilities of dry or wet winter seasons, and whether the rains will be early or late, or both, has naturally led to a desire for seasonal forecasts. They have them, it is said, in India; why not in Australia? (Todd, 1893, p. 267)

> In days when the cui bono of everything connected with scientific research is subjected to the glare of criticism by a public which is frequently too busy to analyse or understand the laborious methods by which accurate knowledge is attained, the Meteorological Service of India poses as a happy exception to that of many other scientific departments in being able to demonstrate its practical utility by the success, not merely of its everyday routine forecasts, but by its unique initiation and development of seasonal or long-period forecasts of the alternate monsoons. (Archibald, 1896, p. 85)

However, within the India Meteorological Department, a change in the nature of the relationships between monsoonal rainfall and Himalayan snowfall, and South East Asian–Indian climatic characteristics in general, had been known since 1892 (Eliot, 1897), and the resulting inaccuracy of monsoonal forecasts caused wider consternation. Newspaper criticisms of Eliot's failed forecasts in 1899 and 1901 led to the monsoon forecasts from 1902 to 1905 being made confidentially to the Indian Government (India Meteorological Department, 1976). It was also during this time that more academic discussions of solar–weather relationships had again emerged (Blanford, 1891), and Sir Norman Lockyer and his son William (Figure 9) re-entered the growing meteorological forecasting field, claiming a new relationship between sunspots (solar temperature) and rainfall 'pulses' in countries bordering the Indian Ocean (Lockyer and Lockyer, 1900a, b). This work captured Eliot's interest, and was reinforced when he met the Lockyers and the Astronomer Royal south of Bombay during observations of the total eclipse of 1898. They subsequently undertook an inspection of all of the

Figure 8: Sir John Eliot (India Meteorological Dept, 1976).   **Figure 9**: William James Stewart Lockyer (Meadows, 1972).

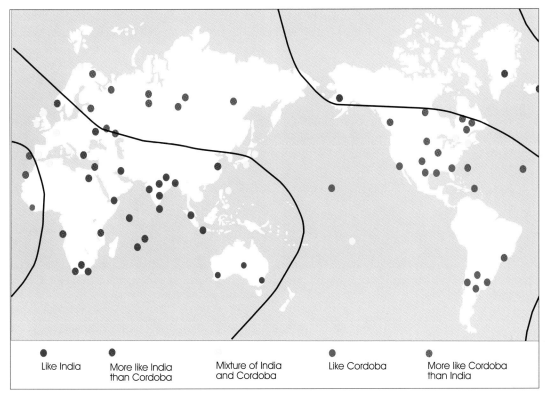

Figure 10: Distribution of stations that display varying associations with atmospheric pressure fluctuations in India and at Cordoba in South America (after Lockyer and Lockyer, 1904).

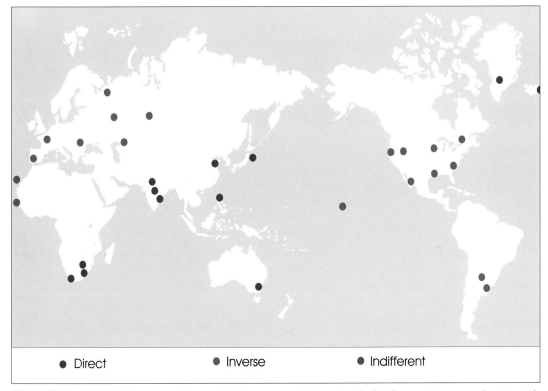

Figure 11: Distribution of stations with atmospheric pressure variations that display direct or inverse synchronism with fluctuations of solar prominences, or are of indifferent type (after Bigelow, 1903).

Figure 12: Sir Gilbert Thomas Walker (Quart. J. Roy. Meteor. Soc., **54**, 1928).

Figure 13: Robert Cockburn Mossman (Royal Scottish Geographical Society, pers. comm.).

observatories in India, with Sir Norman Lockyer and the Astronomer Royal asked to provide an assessment of their operations by the India Office (Meadows, 1972). All of these factors appear to have given Eliot the motivation to write a paper on the 'The meteorology of the Empire during the unique period, 1892–1902' (Eliot, 1904), perhaps the first work concerning a period of reduced statistical forecasting capacity in the broader climate system. Eliot's successor (Walker, 1910a, p. 6) was later to suggest that such findings in the region of the Indian subcontinent were due to '...something abnormal in the larger movements of the atmosphere and not due to human agency in India...'

As a result of their Indian experience, the Lockyers went on to publish several pieces of work that examined solar prominence and sunspot numbers and variations in atmospheric temperature and pressure (Lockyer and Lockyer, 1900a, b, 1902a, b, 1904; Lockyer, 1904). This research found support in similar studies of solar prominences and global atmospheric pressure and temperature by Bigelow (1903). Using observations of atmospheric pressure around the globe, the Lockyers discovered that relationships of solar prominences to atmospheric pressure in India were the opposite to those in South America (Figure 10). When all available data were analysed, they found broad regions of the eastern and western hemispheres where atmospheric pressure records tended to be either in- or out-of-phase with those at Indian or South American stations. This result had close similarities to the pattern of direct, inverse and indifferent synchronism between fluctuations of solar prominences and atmospheric pressure defined by Bigelow (Figure 11). Refining earlier studies of regions of similar meteorological responses, such as the out-of-phase relationship in pressure between Sydney, Australia, and Buenos Aires, Argentina, reported by Hildebrandsson (1897), the Lockyer's talked of a barometric 'see-saw' set up by solar influences between the hemispheres (Lockyer, 1906). In addition, Lockyer (1910, p. 178) remarked:

> ...meteorological elements not only oscillate, about a mean value, rapidly and to a considerable extent in magnitude, in a short period of time of about three or four years, but that, in addition to these variations, there is another oscillation, or perhaps several others, occurring over very much longer periods.

Sir Norman and William Lockyer in the United Kingdom and Frank Bigelow in the United States had observed the Southern Oscillation at work.

However, it was Sir Gilbert Walker (Figure 12), the second Director-General of Observatories in India from 1904 to 1924, who, through statistical attempts to assess, objectively, relationships between potential Indian monsoon precursors, made one of the most significant contributions to studies of global climate variability. Beginning this work in India at a time of general discontent with the type of forecasts Eliot had made, Walker used his mathematical background to work towards more objective forecast methods. His work was aided by the findings of Hildebrandsson (1897), Lockyer and Lockyer (1902a, b, 1904), Bigelow (1903), Lockyer (1904, 1906) and Mossman (1913) (Figure 13) , whom he acknowledged as finding the first evidence for the presence of a strong atmospheric pressure fluctuation between the Indo–Australasian and South American regions. Such research had also provided a significant impetus for specific studies of solar–weather and pressure–rainfall relationships involving North Africa and the Nile (Lyons, 1905), South Africa (Rawson, 1908) and Australia (Lockyer, 1909). Developing regression equations from the pattern of correlations found between Indian rainfall and remote surface observations, Walker issued the first forecast of monsoonal rainfall by this method in 1909. The initial regression equation for forecasting Indian monsoonal rainfall took the form:

{All India Monsoon Rainfall} = – 0.2 {Himalayan Snowfall Accumulation} – 0.29 {Mauritius Pressure} + 0.28 {Mean of South American [Buenos Aires, Cordoba and Santiago] pressure} – 0.12 {Zanzibar rainfall}

At the same time, Walker produced similar types of regression equations for forecasting Nile flood and Australian rainfall (Walker, 1910b). As with the research of the Lockyers and Bigelow, Walker had a strong interest in efforts to uncover solar–terrestrial atmosphere links and thus was drawn further towards a global view of climatic phenomena (Walker, 1915a, b). Walker's approach also drew on discussions of global weather links in a series of notes on seasonal relationships (graphically inferred) between Southern Hemisphere climate data by Mossman (1913). As the First World War began, Walker had begun to develop regression formulae for distinct geographical regions in India and parts of Burma (Walker, 1922). In an intense effort to improve and extend his forecast formulae, Walker accumulated and correlated as many meteorological and hydrological fields as he could obtain, and looked for statistically significant relationships amongst the variables. This involved a mass of computations that would now be undertaken using an electronic computer. Walker was fortunate to be able to call on a human resource that was available to him during the middle of his term in India. As noted by Sir Charles Normand (1953, p. 468):

> ...the first World War damped down the normal activities in the Indian Meteorological Office, most of the scientific staff being away on military duties and the few remaining senior members being over-burdened with routine. So, to find full occupation for members of his junior staff, Walker arranged this long-term programme of computation.

It was the synthesis of distinct statistical correlations linking widespread climatic relationships (Walker, 1923, 1924, 1928a, b, c; Walker and Bliss, 1930, 1932, 1937) that first led Walker to define and name, three coherent 'oscillations' in atmospheric variables between large regions of the earth's surface (Leon Teisserenc de Bort and Hildebrandsson's 'centres of action') (Figure 14). In 1924, Walker first used and defined the term Southern Oscillation (SO) as a 'see-saw' in atmospheric pressure and rainfall at stations across the Indo–Pacific region, where increased (decreased) pressure in locations surrounding the Indian region (Cairo, north-west India, Darwin, Mauritius, south-eastern Australia and the Cape Colony [South Africa]) tended to be matched by decreased (increased) pressure over the Pacific region (San Francisco, Tokyo, Honolulu, Samoa and South America) and decreased (increased) rainfall over India and Java (including Australia and Abyssinia [Ethiopia]). The other two 'oscillations' involved out-of-phase atmospheric pressure between the regions of the Azores and Iceland, named the North Atlantic Oscillation (NAO), and between Alaska and the Hawaiian Islands, termed the North Pacific Oscillation (NPO). These pressure 'oscillations' emerged from Walker's use of all available global records of long-term climatological and hydrological variables, which revealed that these phenomena also involved coherent seasonal fluctuations in air temperature and rainfall. As with his many regression formulae for forecasting, that had expanded to include equations for boreal winter temperatures in south-western and north-western Canada, and austral summer rainfall in South Africa and Australia (Walker and Bliss, 1930), Walker developed similar regression relationships to define the Southern Oscillation in terms of an index deduced on a seasonal basis (Walker and Bliss, 1932) (Figures 15 and 16). For the boreal summer (austral winter) months of June to August (JJA), and the boreal winter

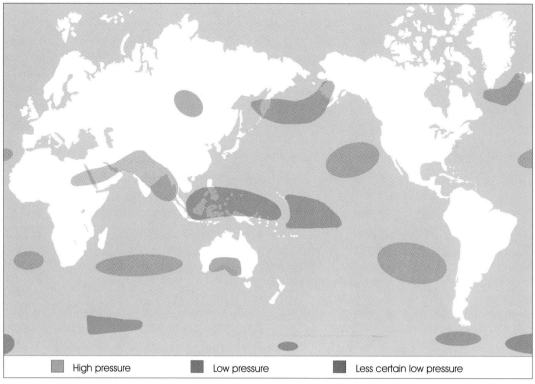

Figure 14: Centres of action of the atmosphere (after Mossman, 1913).

(austral summer) months of December to February (DJF) the Southern Oscillation Index (SOI) values were calculated as follows:

SOI (JJA) = {Santiago pressure} + {Honolulu pressure} + {India rainfall (Peninsula and northwest India)} + {Nile flood} + 0.7 {Manila pressure} − {Batavia [Jakarta] pressure} − {Cairo pressure} − {Madras temperature} − 0.7 {Darwin pressure} − 0.7 {Chile rainfall (mean of 9 stations between 30°S and 42°S)}

SOI (DJF) = {Samoa pressure} + {North-east Australia rainfall (Derby and Halls Creek in Western Australia, 7 stations in north Australia, 20 throughout Queensland)} + 0.7 {Charleston pressure} + 0.7 {New Zealand temperature (Wellington, Dunedin)} + 0.7 {Java rainfall} + 0.7 {Hawaii rainfall (12 stations)} + 0.7 {South Africa rainfall (15 stations, Johannesburg the most northern)} − {Darwin pressure} − {Manila pressure} − {Batavia [Jakarta] pressure} − {South-west Canada temperature (Calgary, Edmonton, Prince Albert, Qu'Appelle, Winnipeg)} − {Samoa temperature} − 0.7 {North-west India pressure (Lahore, Karachi)} − 0.7 {Cape Town pressure} − 0.7 {Batavia [Jakarta] temperature} − 0.7 {Brisbane temperature} − 0.7 {Mauritius temperature} − 0.7 {South America rainfall (Rio de Janeiro and 2 stations south of it in Brazil; 3 in Paraguay, Montevideo; 15 in Argentina, of which Bahia Blanca is the southernmost)}

By the 1930s, statistically significant simultaneous (contemporary), leading and lagging correlation patterns had emerged from this work and become the basis for Walker's development of regression formulae (using, principally, pressure, rainfall and air temperature data) for statistical forecasting of seasonal climatic conditions in the regions affected by each of the 'oscillations'. It soon became apparent that the SO provided the most potential in terms of long-range forecasting, in that it displayed marked interannual variability in its lead and lag correlations with climatic conditions in each season over a large part of the earth's surface (Walker and Bliss, 1932, 1937). In extending their analysis to examine correlation patterns in the autumn and spring seasons, Walker and Bliss (1937) were the first to detail the significant drop in lag correlations between the boreal spring (austral autumn) and subsequent

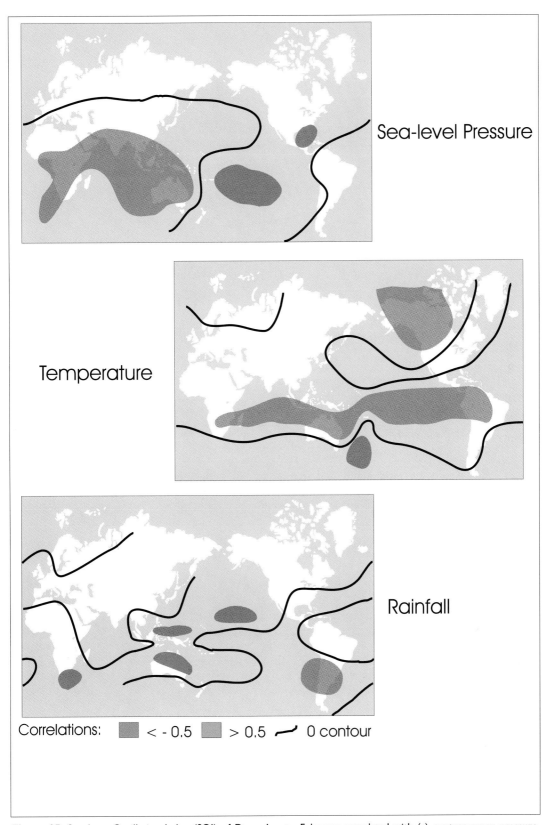

Figure 15: Southern Oscillation Index (SOI) of December to February correlated with (a) contemporary pressure, (b) contemporary temperature, and (c) contemporary rainfall (after Walker and Bliss, 1932).

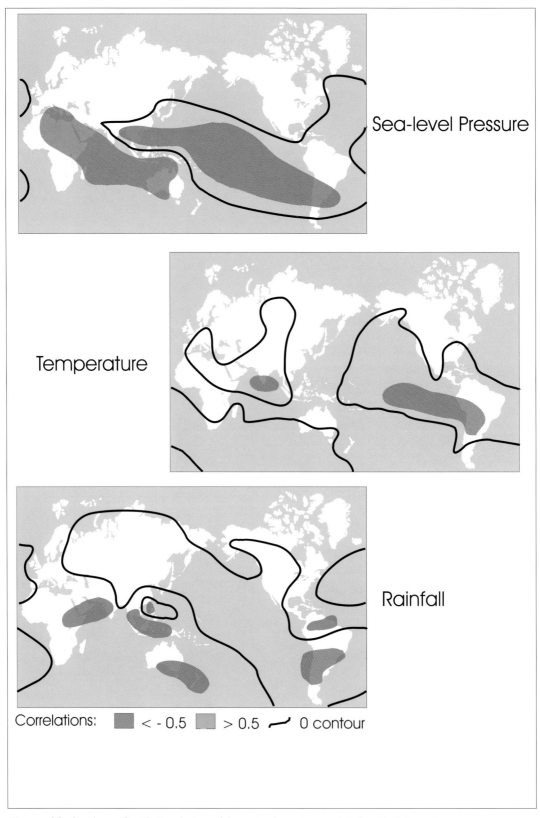

Figure 16: Southern Oscillation Index of June to August correlated with (a) contemporary pressure, (b) contemporary temperature, and (c) contemporary rainfall (after Walker and Bliss, 1932).

seasons. Reducing and/or overcoming this factor has become a major challenge to contemporary researchers. A comprehensive review of the bulk of Walker's early academic career and full publications list can be found in Taylor (1962), his contributions to statistical forecasting and world weather research are given by Montgomery (1940), while his later research into links between conditions in the Arctic and global weather patterns is detailed in Walker (1947). Further information on the contributions of Blanford, the Lockyers, Eliot and Walker to long range forecasting and monsoon meteorology is presented in Kutzbach (1987).

Independent research on precipitation anomalies (dry and wet spells) over the equatorial Pacific Ocean (Brooks and Braby, 1921; Leighly, 1933), and attempts to evaluate and refine links between the SO and weather patterns in South America (Mossman, 1923; Walker, 1928d), Canada (Walker and Bliss, 1930), China (Tu, 1936, 1937a, b), Burma (Maung Po, 1942), Australia (Kidson, 1925; Quayle, 1929; Walker and Bliss, 1930, Rimmer and Hossack, 1939), Indonesia (Berlage, 1927, 1934), the Caribbean (Bliss, 1930, 1936) and southern Africa (Walker, 1930; Walker and Bliss, 1930), appeared to provide a strong foundation for continued research on the SO and its potential in seasonal forecasting. In fact, these studies flourished during the meteorological community's initial response to Walker's techniques. Discussion and criticism of Walker's statistical methods had already received attention in the papers of Dines (1916, 1917), and further evaluations of his approach and statistical significance testing methods were given by Wishart (1928), Normand (1932) and Sellick (1932). However, during the 1940s–1960s period, there was a marked drop in SO research due to the combination of a number of factors. Of particular concern was the lack of any physical mechanisms that could explain pressure fluctuations such as the SO, NAO or NPO, let alone growing efforts to link numerous climatic patterns to lunar, solar and planetary influences. In addition, the correlations and algorithms described and used by Walker and others were often found to have diminished when the original data sets were extended as more data became available.

Despite occasional papers on the SO and forecasting by Schell (1947, 1956), Willett and Bodurtha (1952) and Normand (1953), it was not until the work of Hendrik Berlage in the then Dutch East Indies and the Netherlands (Figure 17), Sandy Troup in Australia (Figure 18) and Jacob Bjerknes in the United States (Figure 19), and the availability of data from the 1957–58 and 1965–66 El Niño events, that a new wave of concerted research on the SO occurred. Berlage (1957, 1961, 1966), Berlage and de Boer (1959, 1960), Schove (1961), Troup (1965), Schove and Berlage (1965) and Bjerknes (1966, 1969, 1972) produced comprehensive re-analyses of global climate and the SO, that showed details of annual and seasonal fluctuations in the spatial structure and dominant patterns of the phenomenon, and the various indices employed (Figure 20). Schove and Berlage (1965) focused on the Indo–Australasian 'centre of action' of the SO and a pressure fluctuation that Schove (1961) named the 'pressure parameter' (the out-of-phase atmospheric pressure see-saw between Greenland and the Indian Ocean region). Unlike the earlier work of Walker, which focused on the development of algorithms or regression

Figure 17: Hendrik Petrus Berlage (KNMI Library, 1995 pers. comm.)

Figure 18: Alexander James 'Sandy' Troup (CSIRO DAR, 1995 pers. comm).

Figure 19: Jacob Aall Bonnevie Bjerknes (Bull. Amer. Meteor. Soc., 56, 1975).

formulae using several atmospheric variables to make seasonal forecasts, Berlage (1957, 1961, 1966) used simple two-station monthly pressure differences from stations representative of the Indo-Australasia and south-eastern Pacific–South American 'centres of action' to construct indices of SO behaviour (i.e. Santiago–Jakarta; Juan Fernandez–Jakarta; Easter Island–Jakarta). The Troup (1965) re-analysis of the SO strongly emphasised the seasonal structure of global MSLP correlations with Walker's SOI, and the nature of persistence in lag correlation patterns (Figures 21 and 22). Although both Troup and Berlage included spatial and seasonal aspects of the phenomenon, Berlage also used the individual station MSLP record from Jakarta to define aspects of the SO, while Troup investigated relationships with Darwin MSLP. Subsequent research has investigated the use of Darwin, Tahiti, Easter Island, Rapa Island and Samoan MSLP alone and in combination (Quinn and Burt, 1970; Trenberth, 1976; Chen, 1982; McBride and Nicholls, 1983), and two-station difference or summation indices such as Easter Island–Darwin, Santiago–Bombay, Santiago–Darwin, Totegegie–Darwin, Rapa Island–Darwin, Juan Fernandez–Darwin, and Rapa Island + Easter Island (Quinn and Burt, 1972; Quinn and Zopf, 1975; Quinn *et al.,* 1978; Chen, 1982; Quinn and Neal, 1983). A return to the Walker-like concept of weighted formulae or algorithms to define an SOI was made by Wright (1977) using principal component analyses of MSLP from five and eight stations situated in the Indo–Pacific basin. However, it was Troup who first suggested the use of a simple normalised two-station difference in MSLP between Tahiti and Darwin as an 'abbreviated' SOI (Pittock, 1974). Given the strong relationships between Darwin and global MSLP, and with Jakarta pressure being difficult to obtain because of the political situation in Indonesia during the 1960s–1970s, the 'Troup' index (or slight variants of it) has tended to become the standard measure of the SO. A systematic time-series analysis of various single-station and two-station MSLP indices of the SO, and a recommendation to use the normalised Tahiti–Darwin index for diagnostic studies, is presented in Chen (1982). Wright (1989) used a mixture of his earlier multi-station index and the Troup-type approach to reconstruct an SOI back to 1851.

In the last 20–25 years, SO research has increased substantially. During this time, there has been a growing emphasis on and understanding of the physical linkages between SO and El Niño phases, an expansion of monitoring and diagnostic studies to encompass surface and upper air observations, and the development and increasing refinement of numerical models of the phenomenon.

The Ocean–Atmosphere System: ENSO

The term ENSO was first used by Rasmusson and Carpenter (1982) to describe the interaction of the oceanic El Niño with the negative phase of the atmospheric Southern Oscillation phenomenon. This step in climatic research stemmed from the strong focus on ENSO events that followed the major global impacts of the events in 1972–73 and 1982–83. However, as noted above, much of the basis for our current physical understanding of the climate system that has led to the acronym ENSO must be traced back to the efforts of Berlage (1957, 1961, 1966), Troup (1965) and Bjerknes (1966, 1969, 1972). The early physical explanations for ocean–atmosphere interactions at the heart of the phenomenon deduced by Bjerknes, and the subsequent oceanographic contributions by Klaus Wyrtki, University of Hawaii, have formed the basis for much of what is known about ENSO. However, it should be noted that there is considerable concern in some quarters about the nomenclature used to describe the whole or parts of the phenomenon. Disquiet ranges from scientific to religious grounds (when terms such as anti-El Niño are used). Chavez (1986) has questioned whether the local, regular El Niño warm current intensifies during what are now seen as wider ENSO events, and suggests that there is a distinction between the currents described by Schott (1931) and Schweigger (1945). He proposes that there are two southward flowing warm currents. In a wider context, some researchers use the terms El Niño and La Niña (or anti-El Niño) for the two phases of the ENSO phenomenon, some talk of ENSO and anti-ENSO, others of low and high

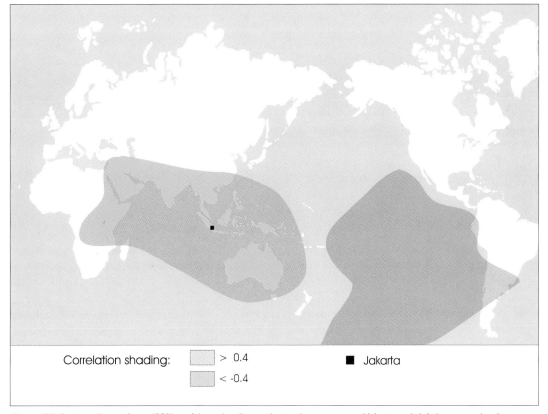

Correlation shading: > 0.4   ■ Jakarta

< -0.4

Figure 20: Statistically significant (95% confidence level) correlations between annual Jakarta and global mean sea level pressure (MSLP) (after Berlage, 1966).

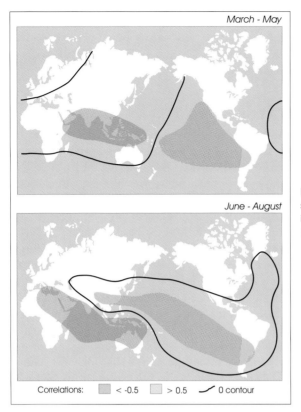

Figure 21: Statistically significant (95% confidence level) seasonal correlations of global MSLP with the Troup SOI in (a) March to May and (b) June to August (after Troup, 1965).

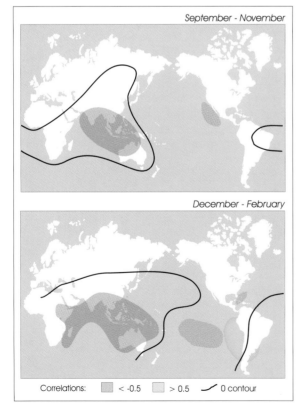

Figure 22: Statistically significant (95% confidence level) seasonal correlations of global MSLP with the Troup SOI in (a) September to November and (b) December to February (after Troup, 1965).

phases in the Southern Oscillation, while other scientists speak of warm and cold events. It seems that none of these terms are entirely satisfactory, and some may even cause distress in Spanish speaking countries (Aceituno, 1992). In this book we follow the trend of more recent work, which tends to refer to ENSO as the complete climatic entity (large-scale, ocean–atmosphere interaction) that embraces two distinct phases, defined for the oceanic component of the phenomenon as El Niño and La Niña.

In a descriptive sense, the ENSO phenomenon can be defined in the following manner (see schematic in Figure 23). During the phase associated with El Niño, anomalous cloudiness and convection are generated by a band of anomalously warm water that develops across the central to eastern equatorial Pacific, with lower than normal atmospheric pressure occurring in the eastern Pacific, and higher than normal atmospheric pressure in the Australasian region. As a consequence, widespread rainfall

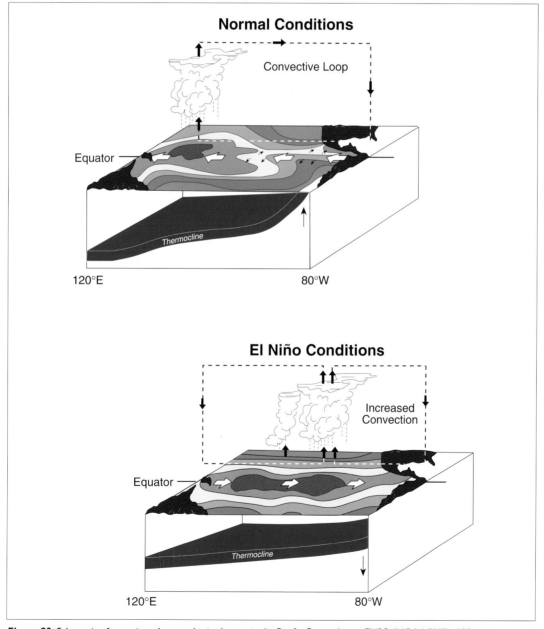

Figure 23: Schematic of oceanic and atmospheric changes in the Pacific Ocean due to ENSO (NOAA PMEL, 1996 pers. comm.).

develops over much of the equatorial Pacific and tropical cyclone tracks and genesis regions tend to be displaced to the north-east of their average locations in the south-western Pacific. Although usually not as extreme or extensive as in the Pacific Ocean, warmer oceanic temperatures and anomalous convection patterns also occur in the central tropical Indian Ocean. These patterns cause widespread drought across many parts of Australia, southern Africa, northern India, Sahelian Africa, Indonesia and South-East Asia (Ropelewski and Halpert, 1987, 1989). During the phase associated with La Niña (an amplification of the 'normal' conditions in Figure 23), impacts are generally the opposite to those described above, with very wet conditions occurring in Australasia, southern Africa, northern India, Sahelian Africa and South-East Asia, and a low rainfall region covering the central to eastern equatorial Pacific (Ropelewski and Halpert, 1987, 1989). Of these ENSO-sensitive regions, the Sahelian is perhaps one of the most complicated as it is also influenced markedly by multidecadal fluctuations in the climate system (Hastenrath and Wolter, 1992), and thus ENSO impacts wax and wane over time.

Other aspects of ENSO have been revealed by recent research. ENSO and its manifestations display a tendency to be locked to the seasonal cycle. Persistence of climatic patterns caused by the phenomenon is found to be strongest (weakest) in the boreal summer–autumn (spring) or austral winter–spring (autumn). This has profound implications for statistical forecasting during ENSO events. The season of weakest persistence of climatic anomalies has been seen as something of a 'predictability barrier'. No two El Niño or La Niña events are exactly the same, and there are usually differences in the timing, extent and magnitude of impacts during individual ENSO episodes. Interestingly, relationships between either of these phases and the planetary monsoon system appear to be more 'selectively interactive', varying seasonally and between various monsoon regions (Yasunari, 1990; Webster, 1995). Although several theories exist that explain aspects of ENSO, and are detailed later, the exact dynamics of its genesis and cessation remain unresolved.

The linking of oceanic and atmospheric aspects of the phenomenon under the banner of ENSO has not gone unquestioned. Although they find that many ENSO phases show closely timed oceanic and atmospheric responses, Deser and Wallace (1987) suggest that El Niño and the SO are not always as well linked and have occurred separately, often leading and lagging one another. Such fluctuations in ENSO appear to be related to regional conditions in at least the eastern equatorial Pacific Ocean. Allan and Pariwono (1990) have reported evidence for El Niño-like conditions persisting in the Australasian region, with no response in the Pacific basin, following the strong, global-impacting El Niño event in 1965–66. We still have only pieces of the puzzle.

ENSO Structure, Nature and Physical Characteristics

Teleconnections

In the course of a series of studies that first demonstrated plausible dynamical ocean–atmosphere processes underlying the phenomenon (Bjerknes, 1966, 1969, 1972), Bjerknes (1969) linked the vertical equatorial zonal circulation over the Pacific Ocean with SO pressure fluctuations and named the circulation cell the 'Walker Circulation'. Although first used by Angstrom (1935) when discussing the NAO, it was Bjerknes (1969) who popularised the word 'teleconnections' and used it to describe the complex of overturning cells consisting of enhanced Hadley (north–south) and east–west circulations that direct excess mass and energy (in the vertical plane) from the equatorial Pacific to other regions of the globe. For the first time, the correlations between climatic events in widely separated locations that had been investigated by Walker and others could be attributed to distinct physical mechanisms, providing a dynamical explanation for phenomena such as ENSO, the NAO and the NPO. Bjerknes (1966) also provided the basis for future work on one particularly important teleconnection modulating North American rainfall and air temperature. Later named the Pacific–North American (PNA) pattern (Horel and Wallace, 1981; Wallace and Gutzler, 1981), it was found to be an atmospheric 'wave train' linking North American climate patterns to both ENSO-linked and other fluctuations in the equatorial Pacific. Subsequent work by Karoly (1989) has indicated the presence of a comparable 'wave train' extending into the Southern Hemisphere high latitudes (Figure 24). An extension of the Bjerknes concept to encompass the global structure of mass and energy exchange from the regions of tropical convection in the vicinity of South America, Africa and the 'maritime continent' of Indonesia (Ramage, 1968) was developed by Flohn and Fleer (1975). The above research has provided the basis for continuing investigations of the spatial pattern of ENSO teleconnection structure, and its relationships to other features of the global climate system. In fact, research on the dynamical mechanisms driving extratropical teleconnections, such as the PNA pattern, has become a significant climatic theme. Comprehensive reviews of tropical and higher latitude teleconnections are found in Yarnal (1985) and Yarnal and Kiladis (1985), while the full global extent of teleconnection patterns and their physical nature and environmental impacts is given in Glantz *et al.* (1991).

Ocean–Atmosphere Interaction

The development of our understanding of ENSO has been made possible by consideration of the physical mechanisms underlying the oceanic and atmospheric interactions that drive the phenomenon. In the late 1950s–early 1960s, Berlage (1957, 1966), Berlage and de Boer (1960) and Schove (1963) attempted to describe a possible physical basis for the SO. The research of Berlage (1957, 1961, 1966) and Troup (1965) indicated that a close link between oceanic and atmospheric variables was a prime factor in a physically based explanation for the SO and relationships with the climate system. Schell (1968, p. 747) remarked that:

> Berlage was the first to show the relation of the Southern Oscillation with the Peru Current and to call attention to the Oscillation's perturbations, which cause it to accelerate and decelerate with a periodicity which he tried to establish, succeeding only partially.

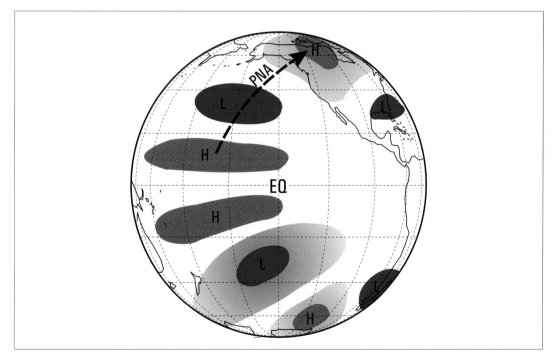

Figure 24: Schematic of upper tropospheric pressure anomalies over the Pacific basin in the boreal summer (austral winter) during the El Niño phase. Wave trains extending into both hemispheres are shown, including the Pacific–North American (PNA) pattern (after Horel and Wallace, 1981, Karoly, 1989; Rutllant and Fuenzalida, 1991).

However, as noted earlier, it was the research by Jacob Bjerknes (Bjerknes, 1966, 1969, 1972) that provided the first physically based synthesis of SST fluctuations over the wider equatorial Pacific, and their relationship to the SO. This work built on the earlier findings of a link between the episodic nature of the Pacific trade winds, central equatorial Pacific (CEP) island rainfall and shifts in equatorial low pressure towards either side of the dateline (Brooks and Braby, 1921), as well as indications of a correlation between precipitation at Malden Island in the Line Island chain and the pressure gradient between Darwin and Apia in Samoa (Leighly, 1933). Bjerknes (1969) proposed the existence of a large-scale, zonal (east–west) circulation cell in the vertical plane, that was driven by the SST gradient between the relatively colder eastern equatorial and warmer western equatorial waters of the Pacific Ocean. He named this zonal cell the 'Walker Circulation', and suggested that it provided a link between the 'maritime continent' in the west and South America to the east. The circulation structure consisted of lower level trade winds blowing to the west over progressively warmer SSTs and being heated and moistened, so that on reaching the western equatorial Pacific (WEP) region of high SST (the 'warm pool' located north of Papua New Guinea and east of the Philippines), deep rising motion (convection), condensation and rainfall were favoured. The Walker Circulation was completed by a return flow of this, now drier, air towards the east near the top of the atmospheric troposphere that eventually reached the eastern equatorial Pacific (EEP), where it subsided over the cold waters off South America. Bjerknes saw this dynamical circulation as the physical link between the atmospheric SO and the oceanic El Niño. Using the wider oceanic and atmospheric observations of the 1957–58 and 1965–66 El Niño events, he postulated that during such episodes there was a weakening of the SST gradient across the Pacific, brought about by warming of the eastern equatorial region, and that the trade-wind field across the entire equatorial Pacific slackened in response to the oceanic changes. Weakened trade winds implied a reduction in the MSLP gradient and thus the SO. In this dynamical framework, a distinct change to El Niño conditions could be initiated by a disturbance in one part of the large-scale, ocean–atmosphere system. If the disturbance became amplified and grew, it would lead to strong positive feedbacks between the ocean and the atmosphere. As the atmosphere is a gas and the ocean a fluid, the latter responds much more slowly to any forcings (in months as opposed to days or weeks with the atmosphere) and has a higher heat capacity, giving the system a 'memory' and a means whereby, once established, anomalous conditions could persist. As noted by Rasmusson (1984, p. 8):

> ...Bjerknes organised and intertwined elements of the SO, global-scale teleconnections, and the large-scale sea/air interactions associated with the Pacific warmings into a new conceptual framework, supported by plausible dynamic and thermodynamic reasoning.

However, there were aspects of the Bjerknes mechanism that were questionable. In his model, slackening of the trade winds along the South American coast and across the Pacific, and the coastal and Pacific warming of the ocean, occurred simultaneously. In addition, El Niño events occurred and ceased as the result of a transition from one near-equilibrium state to another. Furthermore, there was no explanation for the dynamics that brought an event to an end.

It was Klaus Wyrtki, researching sea level and wind observations at the University of Hawaii, and Stuart Godfrey, using numerical modelling simulations, at the CSIRO Division of Oceanography, who provided the basis for modifications to the Bjerknes model (Wyrtki, 1973, 1974, 1975a, b, 1976, 1977; Godfrey, 1975). Through sea level patterns and observations of ocean currents, Wyrtki showed that there was a Pacific-wide response to El Niño events, and that initial fluctuations in the trade-wind field occurred in the central to western equatorial portions of the basin. Sea level responses to the trade-wind changes in the WEP were found to occur a few months prior to, and be negatively correlated with, those along the west coast of South America. What emerged was that El Niño events were characterised by high (low) western Pacific sea levels prior to the onset of (during) events. In the year or so before an episode occurred, the trade winds strengthened, spinning up the South Pacific oceanic gyre, intensifying the South Equatorial Current and increasing the zonal (east–west) gradient of sea level across the Pacific by building up the waters of the western Pacific. Any prolonged relaxation of central Pacific trade winds would then lead to a dynamical sequence in which the accumulated western Pacific water could not be maintained under a reduced zonal sea level gradient, and thus would travel eastwards as a wavelike body of upper-layer oceanic water mass. Such features were shown in the model simulations of Godfrey (1975), and had the effect of 'draining' the warm waters from the western Pacific, raising the thermocline (the region around 50–150 metres depth where the sharpest near-surface temperature gradient in the vertical plane occurs) and reducing the sea level in that region. On entering the waters off the west coast of South America, Wyrtki suggested that the water mass overrode the local cold waters brought from southern latitudes by the Humboldt Current, deepening the thermocline and raising the sea level of the EEP. During strong El Niño episodes, the trade-wind field across the WEP-CEP was even found to reverse. Contrary to the Bjerknes hypothesis, Wyrtki found little change in the local wind field along the Pacific coast of South America, and thus no local wind-driven mechanism to suppress coastal upwelling and raise SSTs. The findings of Godfrey (1975) and Wyrtki (1975a, b) suggested that the wavelike oceanic water mass response was as an internal Kelvin wave that was confined to the near-equatorial belt by the earth's rotation. Such waves were found to take 2–3 months to traverse the equatorial Pacific and, upon reaching the South American continent, were shown to generate branches that travelled both north and south into higher latitudes as coastally trapped features. Later numerical modelling studies (Busalacchi and O'Brien, 1981; O'Brien et al., 1981; Busalacchi et al., 1983) provided further theoretical support for the Kelvin wave theory, and also revealed and confirmed the Godfrey (1975) proposition that reflected Rossby waves, generated by the impact of Kelvin waves on the South American coast were directed back westward along the equatorial belt. Other westward propagating Rossby wave responses were found in the WEP-CEP, on the eastern margins of the region of trade wind relaxation, and seen to enhance the shallowing of the thermocline in the western Pacific. It was from this modified Bjerknes model, and the compositing of observations during the 1957–58, 1965–66 and 1972–73 El Niño events in overview papers such as Rasmusson and Carpenter (1982) and Gill and Rasmusson (1983), that the picture of the 'canonical ENSO' event was established.

In the canonical model, an El Niño phase was seen as being synchronous with the seasonal cycle, and it required a period in which there was a build-up of sea level and warm SSTs in the WEP before an event could occur. Each event was then an anomaly from the 'normal' ocean–atmosphere state that was triggered by distinct forcings that destabilised the climate system over the Pacific basin. One of the prime candidates for an El Niño trigger was the presence of westerly wind bursts over the WEP. Many of these wind bursts were linked to the occurrence of a pair of low latitude cyclones positioned on either side of the equator during the major El Niño events of the 1950s–1970s (Keen, 1982). Under the canonical framework, the period of some 10–15 months before the ocean–atmosphere system 'collapsed' into an El Niño event came to be seen as the opposite phase of ENSO. The concept of La Niña was developing. However, no sooner had this model of ENSO become established, than a massive El Niño event occurred in 1982–83 that provided the grounds for some serious reassessment of the canonical concept of the nature and structure of ENSO.

The 1982–83 El Niño event became known not only as the strongest event this century, but also one in which there was little or no 'build up' of the WEP sea level and SST in the preceding year. It was also an event that developed later in the year than the canonical perspective dictated. Thus, ENSO was seen to possess more variable characteristics than had previously been perceived. Although the 1982–83 El Niño revealed that there was still much to be understood about the phenomenon, it also served to direct considerable research attention towards the problem. Improvements in numerical models of ENSO, better observations of the climate system, and stronger theoretical applications during the latter part of the 1980s, led to some important modifications to the understanding of ENSO.

Observationally, perhaps the single most important development at this time was the establishment of the Tropical Ocean and Global Atmosphere (TOGA) international program. This program has provided the most detailed data available on the ocean–atmosphere system during the period from 1985 to 1994, and established or reaffirmed several important aspects of ENSO. Firstly, the phenomenon is an irregular fluctuation of the global climate system that varies in a continuum between extremes of its two phases. Once set up, an event tends to show a quasi-biennial nature, although that is not always readily apparent. Secondly, an understanding of oceanic dynamics is vital, as they dictate the 'memory' of ENSO. Thirdly, it is evident that although there are often characteristics common to events, no two

ENSO events are the same in terms of genesis, life cycle and cessation. Fourthly, a physically consistent mechanism to explain the cessation of ENSO events is needed. Finally, further reassessments of the modified Bjerknes model are required.

In recent years, the strongest developments have come from numerical modelling and theoretical perspectives (Philander 1992b, Neelin *et al.*, 1994; Battisti and Sarachik, 1995; Schneider *et al.*, 1995).

One of the central challenges in developing a fully coupled ocean–atmosphere model that can simulate ENSO phases is to determine the physical processes or mechanisms that maintain ENSO and dictate its periodicity, magnitude and duration. During the early to mid 1980s, investigations using simple, linearised coupled ocean–atmosphere models of the near-equatorial Pacific basin resolved slowly-propagating disturbances or modes that were distinct from Kelvin or Rossby waves trapped in the near-equatorial region. It was found that these propagating modes were a product of the equations in the model that calculated SST, and hence were often referred to as 'SST modes' (Neelin *et al.*, 1994). The periodicity and direction of propagation of these 'SST modes' varied with the nature of the oceanic perturbations generating them. Thus, dominance of the model thermocline depth term promoted eastward moving disturbances, while the zonal wind and vertical velocity terms favoured westward propagation. Although some observed ENSO events show features similar to the simulated disturbance modes in the simple models, and also in more recent and elaborate intermediate and general circulation models (GCMs), the lack in the simple models of important non-linear processes and natural fluctuations in seasonal and zonal structure found in the real ocean–atmosphere system have cast doubt on these modes as the key ENSO mechanism (Battisti and Sarachik, 1995).

By the late 1980s, Schopf and Suarez (1988), Suarez and Schopf (1988) and Battisti and Hirst (1989) had invoked ocean wave dynamics to construct what has become known as the 'delayed action oscillator' model of ENSO (Figure 25). Essentially, in a large, bounded ocean where the timescale of responses to forcings in the ocean basin is similar to that of the ocean–atmosphere interactions occurring in it, the interplay between various types of subsurface oceanic Rossby and Kelvin waves propagating and reflecting back and forth between the eastern and western boundaries can be shown to dictate an intrinsic temporal mode for interannual variability (Graham and White, 1991; Wakata and Sarachik, 1991; Mantua and Battisti, 1994). Modelling refinements, detailed by Neelin *et al.* (1994) and Battisti and Sarachik (1995), have provided further support for the application of the 'delayed action oscillator' model to ENSO dynamics. These models suggest that the El Niño phase follows a sequence in which a disturbance affecting the low latitudes in the WEP-CEP (such as a westerly wind burst) (time step 0 on the left hand panel in Figure 25) causes the immediate off-equatorial thermocline in both hemispheres to become shallower, as it sets up an Ekman transport of water mass in the upper layers of the ocean towards the equator. This water mass is then acted on by the convergence near the equator, which dictates that westerly winds also set up subsurface, equatorially trapped Kelvin waves in the form of wave 'packets' of a downward propagating type, that deepen the thermocline as they travel eastwards towards South America in the near-equatorial plane (time step 4 on the right hand panel in Figure 25). In deepening the thermocline in the EEP, cold upwelled waters that would otherwise cool the near-surface wind-mixed oceanic layer of the region as they are brought in from depth, become relatively less effective on the overall water temperature, and the SST increases relative to that in the western margins of the equatorial Pacific (time steps 3–6 on the left hand panel in Figure 25). Sea level also rises in the EEP as the influence of the downward propagating Kelvin wave becomes established. Within about 5° of the equator, in the western margin of the region of reduced trade winds and enhanced SST, a dynamical response to equatorial Kelvin waves is the generation of near-equatorial, westward moving and upward propagating Rossby waves that travel into the WEP, tending to reduce the depth of the regional thermocline (time steps 4–15 on the right hand panel in Figure 25). These Rossby waves can take 6–12 months to reach the western boundary of the equatorial Pacific, where they reflect again, back towards the east in the equatorial wave guide, as upward propagating Kelvin waves that progressively decrease the thermocline depth as they travel towards South America (time steps 7–17 on the right hand panel in Figure 25) (Kessler, 1991; White and Tai, 1992). Such wave dynamics provide the negative feedback required to end the El Niño event (time steps 16 on the left hand and 17 on the right hand

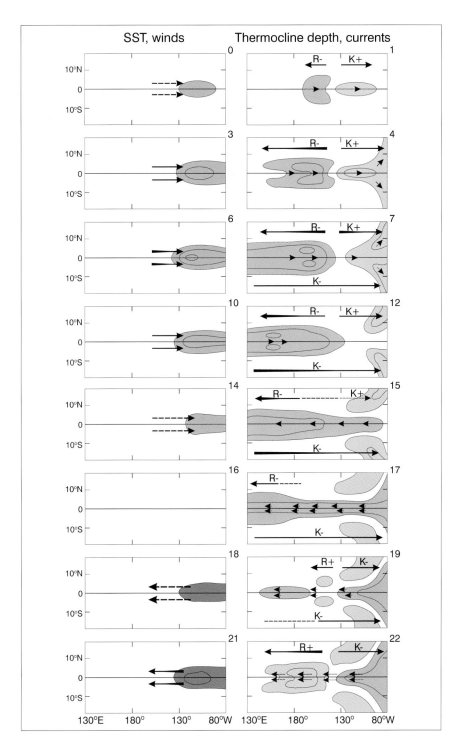

Figure 25: Schematic 'delayed action oscillator' behaviour of a coupled ocean–atmosphere model over the Pacific basin in response to the introduction of a westerly wind anomaly across the CEP. In the left-hand panels, arrows indicate wind anomalies (thickness indicates relative strength) and shadings indicate SST anomalies (positive in red and negative in blue). In the right-hand panels, thin dashed and solid arrows are ocean current anomalies and thicker arrows labelled K+ and K–, R+ and R– indicate the longitudinal progress of positive and negative Rossby and Kelvin Waves respectively (thickness of arrows indicates relative strength). Orange shaded regions are positive (deeper than average) and green shaded regions are negative (shallower than average) thermocline depth anomalies. Numbers outside each of the individual panels indicate the elapsed time in months since the wind anomaly forcing in the top right-hand panel was applied to the coupled model simulation (Hirst, 1989).

panel in Figure 25), and lead to relatively cooler SSTs, reduced sea level and a shallower thermocline in the EEP, with the reverse situation in the WEP as La Niña conditions develop (time steps 18–21 on the left hand and 19–22 on the right hand panels in Figure 25).

Overall, the delayed oscillator hypothesis was built on the earlier work of Bjerknes, Godfrey and Wyrtki, and is seen as the best explanation of the physical mechanisms responsible for interannual ENSO behaviour to date (Neelin *et al.*, 1994; Battisti and Sarachik, 1995; Schneider *et al.*, 1995). This modelling research suggests that the successive regeneration of Rossby and Kelvin waves only occurs for ocean basins with east–west dimensions of greater than 13 000 km. It has been argued that this is one reason why such dynamical interactions are restricted to the Pacific Ocean basin. Other studies have investigated the possible role of off-equatorial Rossby waves up to 12° from the equator in ENSO dynamics (Pazan and White, 1986; Pazan *et al.*, 1986; White *et al.*, 1989; Graham and White, 1991; Kessler, 1991; White and Tai, 1992; White, 1994), and links between ENSO and the westward propagating, near-equatorial tropical cloud and convective features associated with the 30–60 day oscillation (Lau and Chan, 1985; Singh *et al.*, 1992; Yi and Longxun, 1992; Ronghui, 1994). Some suggestions have also been made that both the 'SST modes' and 'delayed action oscillator' responses can occur under different states of the climate system, and that a mixture of both may be required to explain ENSO dynamics (Neelin *et al.*, 1994).

The importance of observational studies in providing the data needed to confirm and interpret numerical modelling and theoretical research continues to be stressed in specifically focused projects such as the World Ocean Circulation Experiment (WOCE) and the TOGA The Coupled Ocean–Atmosphere Response Experiment (COARE) (1991–93) — which includes the Tropical Atmosphere–Ocean (TAO) buoy array across the near-equatorial Pacific (Figure 26), and the Joint Global Ocean Flux Study (JGOFS) of processes in the equatorial Pacific. Both of the TOGA projects have provided high resolution oceanic and atmospheric data over the WEP, where important ENSO mechanisms occur (McPhaden, 1993; Murray *et al.*, 1992, 1995; Lukas *et al.*, 1995). These projects have been particularly timely, in that they have been operating during the most recent phase of attempts to understand and model the dynamics of ENSO. As has occurred many times in the course of the evolution of research on the phenomenon, an ENSO event or a climatic fluctuation has produced physical characteristics that are difficult to explain by existing models or theory. What might be termed a 'persistent' El Niño sequence affected the global climate system from August 1990 to June 1995. To date, its nature has been construed in different ways: as several events following one another without a distinct return to the opposite phase; as a period in which El Niño or El Niño-like conditions persisted

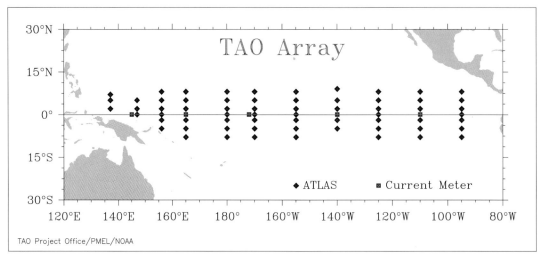

Figure 26: Observational network making up the TOGA TAO array across the equatorial Pacific Ocean (NOAA PMEL, 1996 pers. comm.).

for some 4–5 years; or as a long-term fluctuation in the climate system (Latif *et al.,* 1995; Allan and D'Arrigo, 1996). Efforts to understand this latest climatic phenomenon continue (McPhaden, 1993; Kerr, 1994; Bigg, 1995; Kessler and McPhaden, 1995; Liu *et al.,* 1995). With the completion of TOGA, a more unified approach across the various disciplines is being sought through the new CLIVAR program studying climatic variability and prediction (WCRP, 1995).

More detailed reviews of the above evolution in understanding of concepts associated with ENSO are given by Ramage and Hori (1981), Rasmusson and Wallace (1983), Philander (1983, 1985, 1989, 1990, 1992a, b), Lockwood (1984), Rasmusson (1984, 1985), Philander and Rasmusson (1985), Rasmusson and Arkin (1985), Cane (1986), Ramage (1986), Enfield (1987, 1989, 1992), Graham and White (1988), Chao and Philander (1993), Latif and Neelin (1994), Neelin *et al.* (1994), Battisti and Sarachik (1995), Efimov *et al.* (1995), Schneider *et al.* (1995) and Webster (1995).

Recent Statistical Forecasting

The resurgence since the 1950s of scientific interest in the ENSO phenomenon has included an increasing emphasis on empirically based forecasting of events (Rasmusson, 1987; Philander, 1992b). Although the development of numerical models has been seen as the approach with the greatest potential to make significant advances in ENSO prediction, statistical forecasting techniques continue to be used extensively around the world. Most statistically based approaches draw on the findings of correlation studies that can be traced back to the pioneering work of Walker and colleagues. One of the most important findings of Walker's research (Walker and Bliss, 1937) that has major ramifications for ENSO forecasting is that climatic conditions during the boreal spring (austral autumn) period display poor correlation with those in the following seasons. This was also noted by Troup (1965) and Wright (1977) and seen as a time of lowest persistence in the ENSO signal. It is variously referred to now as the 'predictability barrier' or 'spring frailty' (Webster, 1995), and is currently seen as a challenge to researchers attempting to improve forecasting techniques through the use of numerical modelling approaches.

The mid to latter part of the 1980s saw the first equatorial Pacific SST forecasts from what is still the most widely used intermediate model (see 'Modelling ENSO' p. 39), that of Cane and Zebiak (Cane and Zebiak, 1985; Cane *et al.,* 1986; Zebiak and Cane, 1987; Barnett *et al.,* 1988). Further developments in both modelling and statistical forecasting techniques have relied on continual observational and theoretical studies of ENSO and the climate system. The majority of detailed research on spatial distributions of simultaneous and lagged seasonal climatic correlations with ENSO variables (especially the SOI or Pacific SSTs) in Indo–Pacific countries has occurred since the 1982–83 El Niño event (Streten, 1981, 1983, 1987; McBride and Nicholls, 1983; Nicholls, 1984a, b, 1989; Cadet, 1985; Gordon, 1985, 1986; van Loon and Shea, 1985; Hackert and Hastenrath, 1986; Holland, 1986; Ropelewski and Halpert, 1986; Tyson, 1986; van Loon and Henry, 1986; Behrend, 1987; Williams, 1987; Aceituno, 1988, 1989; Drosdowsky, 1988; Lindesay, 1988; National Climate Centre, 1988; van Heerden *et al.,* 1988; Allan and Pariwono, 1990; Bigg, 1990, 1995; Jose, 1990, 1992; Drosdowsky and Williams, 1991; Ward and Folland, 1991; Seleshi *et al.,* 1992; Beltrando and Camberlin, 1993; Drosdowsky, 1993a, b; Hastenrath *et al.,* 1993; Marengo and Hastenrath, 1993; Nicholls and Kariko, 1993; Barros and Scasso, 1994; Cayan and Redmond, 1994; Pisciottano *et al.,* 1994; Sittel, 1994a, b; Yin, 1994; Dettinger and Cayan, 1995; Mullan, 1995; Drosdowsky, 1996). Several other papers have shown lagged relationships between Indian and the following Australian monsoonal rainfall characteristics and ENSO phases (Joseph *et al.,* 1991, 1994; Gregory, 1991; Nicholls, 1995). Syntheses of the global extent of ENSO-induced climatic anomalies were produced in the studies of Ropelewski and Halpert (1987, 1989), Halpert and Ropelewski (1992), Kiladis and Diaz (1989) and Bradley *et al.* (1987). The persistence of many ENSO correlations with variables such as rainfall and air temperature in the boreal summer–autumn or austral winter–spring seasons has become a strong element in statistical forecasting approaches. More recently, statistical studies have examined climatic correlations with the 'phase of regular cycles' or tendencies in the values of the SOI (Stone and Auliciems, 1992; Stone and McKeon, 1992; Zhang and Casey, 1992).

Other climatic features impacted by ENSO phases, which have also begun to receive increasing attention, include simultaneous and lagged correlations between ENSO phases and tropical cyclones/hurricanes. Most diagnostic and statistical studies have concentrated on ENSO-related SST and atmospheric circulation influences on Atlantic hurricane frequency (Gray and Sheaffer, 1991; Landsea *et al.*, 1992) and on tropical cyclone activity (tracks, genesis regions and numbers) in both the Australian and Pacific Ocean basin regions (Nicholls, 1979, 1984b, 1992; Chan, 1985, 1994, 1995; Revell and Goulter, 1986a, b; Dong, 1988; Hastings, 1990; Solow and Nicholls, 1990; Evans and Allan, 1992; Lander, 1994). Of these regions, the largest amount of work using long data records has been undertaken in the Atlantic basin, although that basin appears to have the most indirect and variable links with ENSO (Delecluse *et al.*, 1994; Polonsky, 1994; Moron *et al.*, 1995). The relationship in the Atlantic Ocean is one of suppression of hurricane frequency during El Niño events. For the northwestern Pacific Ocean, El Niño events tend to be associated with changes in ocean–atmosphere interactions that lead to a westward displacement in seasonal tropical cyclone distributions. In the southwestern Pacific, tropical cyclone genesis regions and tracks have been found to be displaced eastward (westward) of their climatological locations during El Niño (La Niña) phases. The extension of such findings into the Australian region has shown that tropical cyclone genesis regions and tracks shift further eastwards towards the northwestern coast (and away from the northeastern coast) of Australia during El Niño events. During La Niña events, westward displacements occur relative to both Australian coasts. Studies by Evans and Allan (1992) and Vermeulen and Jury (1992) have indicated modulations in eastern and southwestern Indian Ocean tropical cyclone activity and tracks during ENSO events. Global examinations of tropical cyclone frequencies and genesis influences are detailed in Ryan *et al.* (1992) and Watterson *et al.* (1995). The potential application of these relationships to seasonal statistical forecasting has concentrated on using either antecedent upper atmospheric conditions or links to the persistence inherent in broadscale patterns of oceanic and atmospheric variables such as SST and indices such as the SOI. Details of the current tropical cyclone/hurricane forecasting approaches for the Atlantic Ocean and northern Australian regions are given in Gray and Sheaffer (1991), Nicholls (1992) and Climate Prediction Center (1994–1996).

Research on ENSO influences on streamflow/river discharge, lake levels, frost and snow has begun to provide varying potential for forecasting relationships. The impetus for such research has its roots in early studies of SO relationships to the Nile River in Egypt (Lyons, 1905; Bliss, 1926; Brooks, 1928a). In particular, attempts to detail streamflow responses to ENSO phases and/or develop forecasting strategies, as has occurred with rainfall and air temperature, have been reported for the United States (Diaz and Markgraf, 1992; Kahya and Dracup, 1993, 1994), Australia (Kuhnel *et al.*, 1990; Whetton *et al.*, 1990; Simpson *et al.*, 1993a, b), South America (Marengo, 1995), Argentina (Mechoso and Iribarren, 1992), New Zealand (Moss *et al.*, 1994) and Brazil (Richey *et al.*, 1989; Marengo and Hastenrath, 1993). Lake level relationships to ENSO have been less evident in the literature; the most prominent are those involving the Great Salt Lake and the Laurentian Great Lakes of North America via the PNA teleconnection (Rodionov, 1994; Lall and Mann, 1995; Mann *et al.*, 1995a) and Lake Eyre in central Australia (Allan, 1985, 1988, 1991). Links between ENSO phases and frost are best represented by the studies of Allen (1989), Allen *et al.* (1989), and Brookfield (1989) in the Indonesian and New Guinea region. A study by Stone *et al.* (1996) which examines relationships between frost in northeastern Australia and tendencies in the values of the SOI, and the forecasting potential of such links is one of the most advanced in the literature. Research on snow accumulation in the Andes (Thompson *et al.*, 1984; Michaelsen and Thompson, 1992; Thompson *et al.*, 1992; Thompson, 1993) has shown negative correlations with the SOI, while similar studies for the Tibetan Plateau in China (Thompson, 1992; Thompson *et al.*, 1992), across southern Australia (Budin, 1985) and in New Zealand (West and Healy, 1993), have indicated significant positive correlations with the SOI. The spatially confined nature of ENSO influences on snowfall should not be seen to downgrade their importance, but rather makes them more crucial to regional or local forecasting situations. At present, the potential of using sea-level correlations with various stages of ENSO phases in direct forecasting of rainfall patterns has received little attention outside studies in the Australasian region (Pariwono *et al.*, 1986; Allan *et al.*, 1990; Mitchell, 1994).

A number of meteorological services now regularly produce regional, real-time empirically based seasonal bulletins, general outlooks and distinct forecasts of rainfall and other climatic impacts. The evolution of these approaches in specific countries is detailed in Glantz *et al.* (1991), Hastenrath (1991, 1995) and Battisti and Sarachik (1995). Various research groups are also making specific 'experimental long-lead' forecasts for features such as tropical Pacific and Indian Ocean SSTs, United States air temperature, and the amount and/or onset of rainfall for various seasons over northern, eastern, Sahelian and southern Africa, northeastern Brazil, India, Australia and tropical Pacific islands (Climate Prediction Center, 1994–1996). As noted earlier, simple indices have evolved that are used widely to measure the atmospheric and oceanic components of the ENSO phenomenon. The state of the SOI and SST in the Niño 1, 2, 3 and 4 regions are now combined with indices measuring anomalies in variables such as zonal wind in lower and upper layers of the atmosphere, outgoing longwave radiation (OLR) (indicative of the presence or absence of deep convective activity), upper oceanic water volume over the near-equatorial Pacific Ocean basin, oceanic temperature sections with depth and spatial fields of sea level, SST, winds, OLR, MSLP etc. These data are regularly available in the monthly issues of the CPC *Climate Diagnostics Bulletin* and in products from many other meteorological services. Publications of this type also contain some ENSO forecasts and outlooks that detail various sophisticated analyses that use trends, analogues, and life-cycle stages in ENSO variables to construct probability distributions and skill scores for seasonal rainfall. This has been extended to include regular compilations of the latest statistical and numerical model forecasts from groups in the USA, Germany, Australia, United Kingdom, Chile and South Africa (Climate Prediction Center, 1994–1996). Detailed synopses of the nature and current status of simple SOIs and Niño 3 and 4 region SSTs can be found in Allan *et al.* (1996) and Simpson *et al.* (1996).

Even in the first half of this century, broad syntheses and examinations of long-range forecasting techniques have been produced many times (Abbe, 1901; Brooks, 1928b, 1938; Normand, 1932; Schell, 1947). With the more recent resurgence in ENSO research, concerted attempts to verify all types of forecasting approaches have become an integral part of the ENSO research and applications focus of all groups. Statistically based forecasts are assessed in various ways, ranging from the examinations of various El Niño events in the 1980s by Barnett (1984a), Nicholls (1985a), Philander (1986) and Rasmusson (1987), to more recent appraisals by Barnston and Ropelewski (1992), Smith (1994a), Casey (1995), Hastenrath and Greischar (1993) and Hastenrath *et al.* (1995). However, with the rapid expansion of model-based forecasts, verifications of various model projections have been reported extensively in the literature (Latif and Graham, 1992; Miller *et al.*, 1993b; Nigam and Shen, 1993; Penland and Magorian, 1993; Balmaseda *et al.*, 1994a, b, c; Brankovic *et al.*, 1994; Davey *et al.*, 1994, 1996; Hunt *et al.*, 1994; Ji *et al.*, 1994a, b; Kleeman and Power, 1994; Latif *et al.*, 1994a; Stockdale *et al.*, 1994; Wu *et al.*, 1994; Xue *et al.*, 1994; Annamalai, 1995; Graham and Barnett, 1995; Mantua and Battisti, 1995; Mo and Wang, 1995; Hoerling *et al.*, 1996; Kumar *et al.*, 1996; Perigaud and Dewitte, 1996). Overviews of current forecasting approaches, schemes and prospects are detailed in Nicholls (1991c), Barnett *et al.* (1994), Barnston *et al.* (1994), Latif *et al.* (1994c) and Palmer and Anderson (1994).

Although the ENSO phenomenon has the capacity to impact globally on the environment, it is important to realise that current forecasting techniques have a strong Pacific Ocean emphasis in ENSO research and forecasting signatures. In addition, such forecasting approaches are evolving as research reveals more about the highly variable nature of the climate system and, in particularly, the ENSO phenomenon. The patterns of global impacts caused by ENSO phases, and their longer term stability, are discussed in later sections.

Indo–Pacific Regional Focus

Although the historical evolution of ENSO research has important roots in the Indian Ocean region, the growing realisation of the dominant role of the Pacific basin in the physical ocean–atmosphere processes and mechanisms explaining ENSO behaviour has tended to direct most research towards the Pacific Ocean. This can even be seen in studies of particular features of the climate system in the Pacific. Although similar cloud bands linking the 'centres of action' in the tropics with extratropical regions

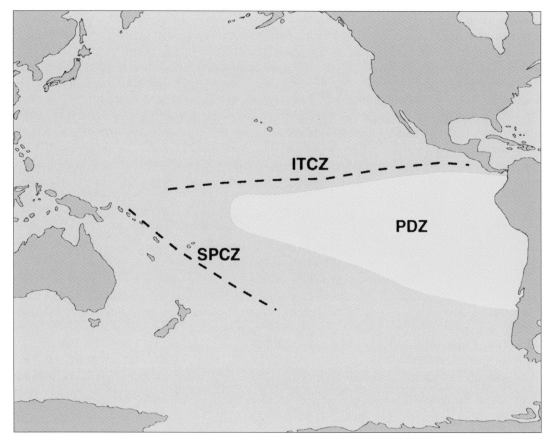

Figure 27: Annual mean positions of Pacific Dry Zone (PDZ), Intertropical Convergence Zone (ITCZ) and the South Pacific Convergence Zone (SPCZ).

occur over both the Indian and Atlantic Ocean basins, the greatest focus has been on the strongest global feature, the South Pacific Convergence Zone (SPCZ), which connects the convective region that vacillates between the 'maritime continent' and the CEP during ENSO phases, with the higher latitudes of the south–southwestern Pacific (Streten, 1973, 1975; Trenberth, 1976; Vincent, 1994). This is also the case for the Intertropical Convergence Zone (ITCZ) in the Pacific (Reiter, 1978, 1979), and features such as the 'Pacific Dry Zone' (PDZ) (Quinn and Burt, 1970, 1972; Quinn, 1974), which covers the region from the CEP to the EEP, from where it extends southward to the mid latitudes of the South American west coast (Figure 27). However, some of the most recent ENSO research (Yasunari, 1990; Webster and Yang, 1992; Penland and Matrosova, 1994; Webster, 1995) has indicated that there may be more direct dynamical influences from the Indian Ocean region than are observed in its SST patterns and climatic impacts, which are usually seen to be driven solely by ocean–atmosphere interactions in the Pacific basin during ENSO phases.

As noted previously, links between what we now know as the ENSO phenomenon and Indian monsoon rainfall have been at the heart of the earliest attempts to understand global climatic patterns. Expansion of the original Blanford (1884) study on Indian monsoon relationships with Himalayan snow accumulation to include snow cover over the whole of Eurasia, through both statistical and modelling approaches, are detailed in Khandekar (1991), Parthasarathy and Yang (1995), Peiji (1995) and Vernekar *et al.* (1995). Similar research on Chinese rainfall and Eurasian snow cover is found in Yang and Xu (1994). This emphasis has been part of a concerted examination of Indian and Asian monsoon interactions with ENSO phases (Chongyin, 1990; Meehl, 1993, 1994; Latif *et al.*, 1994b; Nigam, 1994; Verma, 1994; Shen and Lau, 1995). Recent syntheses of such theoretical, numerical modelling and observational studies aimed at improving ENSO forecasting potential by Webster and Yang (1992) and

Webster (1995) have shown the need to include the planetary monsoon system, particularly the major component given by the Indian Monsoon, in efforts to improve and extend prediction of the climate system. This research suggests that the Walker Circulation and the Indo–Asian Monsoon system are 'selectively interactive' and in 'quadrature' (Figure 28). This means that when the Walker Circulation is at its maximum strength during the boreal autumn to winter (austral spring to summer), the monsoon is close to the equator and at its weakest. By the boreal spring to summer (austral autumn to winter), the roles are reversed and the monsoon system dominates large-scale Indo–Pacific circulation. Thus, there appears to be a need to incorporate the Indian Ocean and the planetary monsoon system in both statistical and numerical models in order to better understand and predict ENSO. In fact, Webster and Yang (1992) and Webster (1995) propose that the so called 'predictive barrier' during the boreal spring (austral autumn), when statistically based ENSO correlation structure is weakest and predictive ability diminishes, is a time when the coupled ocean–atmosphere system exhibits a low signal-to-noise ratio. As current coupled ocean–atmosphere forecasting models have been constructed around the dynamics of the Pacific Ocean basin, it is not surprising that they experience the largest random error growth in their ENSO simulations during the boreal spring (austral autumn). In addition, Webster (1996, pers. comm.) has suggested that the 'predictive barrier' has fluctuated in both temporal extent and intensity over the historical record. Thus, forecasting models need to be developed that take account of variations in the longer term ENSO structure.

The other argument for taking an Indo–Pacific view of ENSO dynamics is emphasised in the following sections and in this book as a whole. Fluctuations in the global climate system at decadal–multidecadal and lower frequencies provide a larger temporal and spatial 'envelope' in which interannual phenomena such as ENSO are embedded. Even simple low-frequency modulations of features of the large-scale ocean–atmosphere system, such as the monsoonal regimes, are likely to produce changes to the physical conditions in which ENSO dynamics operate, to the extent that important characteristics of this interannual phenomenon may be affected.

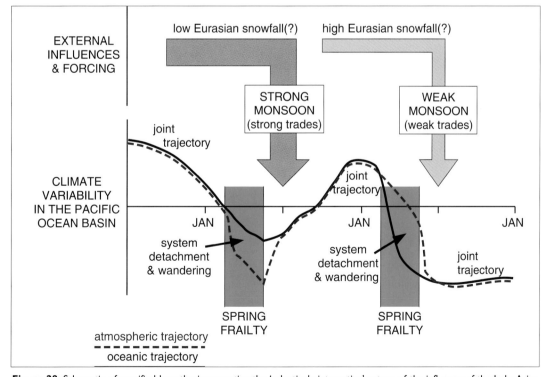

Figure 28: Schematic of a unified hypothesis suggesting the 'selectively interactive' nature of the influence of the Indo-Asian Monsoon system on ENSO, involving the combined effect of the 'predictive barrier' or 'spring frailty' and external influence on the climate of the Pacific Ocean basin (Webster, 1995).

Interestingly, research on ENSO as an Indo–Pacific phenomenon has also led to the resolution of other interannual climate signals in the Indian Ocean region. The study of Nicholls (1989) suggested the presence of a north-east–south-west oriented SST dipole pattern in the eastern Indian Ocean off Australia that was independent of the Indian Ocean ENSO signal. When this SST dipole was warm to the north and cool to the south, it was found to be important as a moisture source for tropical–temperate cloud bands across central to eastern Australia (Drosdowsky, 1993b, c). However, as a forecasting tool, this pattern is restricted to the austral summer to autumn/early winter period (Drosdowsky, 1993b, c; Smith, 1994b) and the dipole structure itself is very similar to patterns found in ENSO phases and on longer interdecadal time scales. Several modelling studies have, to varying degrees, simulated processes underlying this SST influence (Simmonds, 1990; Simmonds and Rocha, 1991). Much of the oceanic focus on this dipole has centred on the role of the 'throughflow' of water mass between the Pacific and Indian Oceans via the Indonesian straits (Wyrtki, 1987; Hirst and Godfrey, 1993, 1994; Clarke and Liu, 1994; Qu et al., 1994; Wajsowicz, 1994, 1995). Other independent studies have resolved a similar, but more north–south SST, dipole pattern operating on interannual scales in the western Indian Ocean, again in a configuration similar to the major ENSO-phase responses in that region (Walker and Lindesay, 1989; Mason, 1990; Walker, 1990; Lindesay and Allan, 1993). Both the ENSO and the western Indian Ocean SST dipole patterns modulate southern African summer rainfall (Mason, 1995; Rocha and Simmonds, 1996a, b). Work on resolving the nature of all of these interannual signals in the Indian Ocean and their influence on Australian and southern African rainfall is continuing.

Climatic Variability

The studies of Schell (1956), Berlage (1957, 1961, 1966) and Troup (1965), were amongst the first to detail evidence for periods in which the SO influence was stronger or weaker (Figures 29 and 30). Since then, numerous studies have identified changes in correlation patterns between rainfall and SOIs, El Niño activity, near-global teleconnections of the ENSO phenomenon, the coherence of El Niño and SO events, and in SOIs over the periods of historical and instrumental record (Berlage, 1957, 1966; Schove and Berlage, 1965; Kidson, 1975; Wright, 1977; Quinn et al., 1978; Reiter, 1983; Quinn and Neal, 1983, 1992; Kousky et al., 1984; Lough and Fritts, 1985, 1990; Flohn, 1986; Deser and Wallace, 1987, 1990; Quinn et al., 1987; Trenberth and Shea, 1987; Yasunari, 1987a, b; Allan, 1988, 1989, 1991; Elliot and Angell, 1988; Enfield, 1988, 1989, 1992; Kiladis and van Loon, 1988; Philander, 1990; Whetton et al., 1990; Enfield and Cid, 1991; Parthasarathy et al., 1991, 1992; Lindesay, 1992; Quinn, 1992, 1993; Zhang and Casey, 1992; Diaz and Pulwarty, 1994; Whetton and Rutherfurd, 1994; Solow, 1995).

A number of contemporary and historical studies have shown that, on interannual time scales, climatic variability over much of the globe is influenced, to varying degrees, by the ENSO phenomenon. Some of this research has also detected apparent changes in both mean climatic conditions and historical ENSO characteristics, and related teleconnection patterns. Most of the latter findings are documented in regionally specific studies, with evidence from Australasia (Allan, 1985, 1988, 1989, 1991; Isdale and Kotwicki, 1987; Nicholls, 1988a, 1992; Allen, 1989; Allen et al., 1989; Brookfield, 1989; Allan et al., 1991; Kotwicki and Isdale, 1991; Lough, 1991, 1992, 1993, 1994; Fitzharris et al., 1992; Cordery and Opoku-Ankomah, 1994; Opoku-Ankomah and Cordery, 1993), Asia (Yoshino and Yasunari, 1986; Fu and Fletcher, 1988; Pant et al., 1988; Xiangong et al., 1989; Zhang et al., 1989; Wang and Li, 1990; Parthasarathy et al., 1991, 1992; Thapliyal and Kulshrestha, 1991; Liu and Ding, 1992; Shaowu, 1992; Yongqiang and Yihui, 1992; Liang et al., 1995; Vijayakumar and Kulkarni, 1995), Africa (Nicholson and Entekhabi, 1986; Hulme, 1992; Nicholson, 1993; Oladipo, 1995), Europe (Fraedrich and Muller, 1992; Fraedrich et al., 1992; Wilby, 1993; Fraedrich, 1994), the Americas (Quinn et al., 1978, 1987; Hamilton and Garcia, 1986; Deser and Wallace, 1987; Quinn, 1992, 1993; Quinn and Neal, 1992; Martin et al., 1993; Ortlieb and Machare, 1993), the Indian Ocean (Blazejewski et al., 1986; Allan et al., 1995; Reason et al., 1996; Shinoda and Kawamura, 1996), the Pacific Ocean (Graham, 1994; Latif and Barnett, 1994a, b; Miller et al., 1994a, b; Nitta and Kachi, 1994; Trenberth and Hurrell, 1994; Gu and Philander, 1995; Wang and Ropelewski, 1995) and Antarctica and the Southern Ocean (Fletcher et al., 1982; Enomoto,

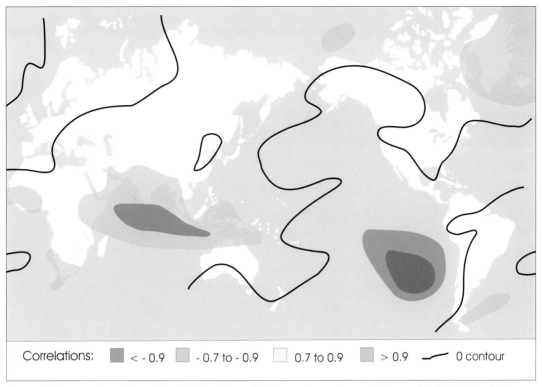

Figure 29: Statistically significant (95 and 99% confidence levels) correlations between Jakarta and global MSLP in the January 1931 to December 1939 epoch (after Berlage, 1961).

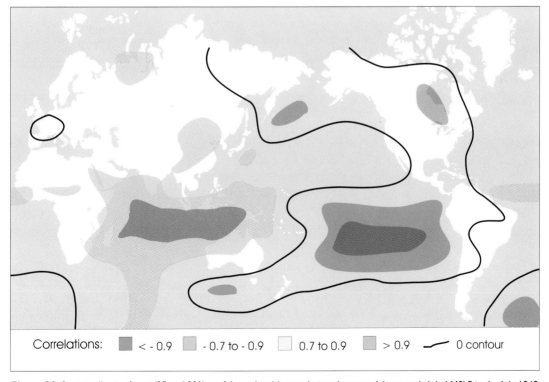

Figure 30: Statistically significant (95 and 99% confidence levels) correlations between Jakarta and global MSLP in the July 1949 to July 1957 epoch (after Berlage, 1961).

1991). Broader and more globally oriented assessments are detailed in Fu (1986), Fu *et al.* (1986), Kiladis and Diaz (1986), Elliott and Angell (1987, 1988), Trenberth and Shea (1987), Yasunari (1987a, b), Enfield (1988, 1989, 1992), Sarker and Thapliyal (1988), Folland *et al.* (1990), Lough and Fritts (1990), Stockton (1990), Whetton *et al.* (1990), Enfield and Cid (1991), Nitta (1992), Nitta and Yoshimura (1993), Rasmusson and Arkin (1993), Diaz and Pulwarty (1994), Kawamura (1994), Klaben *et al.* (1994), Nitta and Kachi (1994), Whetton and Rutherfurd (1994), Diaz and Kiladis (1995), Solow (1995) and Wang (1995a, b). Once assembled, these papers indicate that, although the major relationships with ENSO occur across the Indo–Pacific basin, the ENSO influence is near-global in extent but appears to wax and wane on decadal–multidecadal time scales (Ramage, 1983; Pittock, 1984; Allan, 1985, 1993).

More detailed investigations of particular historical oceanic and atmospheric variables have tended to confirm suggested variations in ENSO, despite some concerns over data quality and interpretation (Chang and Yasunari, 1982; Yasunari, 1985; Krishnamurti *et al.*, 1986; Bigg *et al.*, 1987; Whysall *et al.*, 1987; Wright, 1988; Cooper and Whysall, 1989; Diaz *et al.*, 1989; Posmentier *et al.*, 1989; Richey *et al.*, 1989; Cardone *et al.*, 1990; Pan and Oort, 1990; Bigg, 1992, 1993; Bigg and Inoue, 1992; Zhang and Casey, 1992; Allan, 1993; Gu and Philander, 1995; Inoue and Bigg, 1995; Wang, 1995a, b; Wang and Ropelewski, 1995; Karoly *et al.*, 1996; Kestin *et al.*, 1996). Fluctuations in ENSO teleconnection patterns are probably caused by changes in the frequency, magnitude and spatial characteristics of ENSO events between different climatic epochs. Low frequency, decadal–multidecadal fluctuations of the climate system are often linked to large-scale ocean–atmosphere interactions involving changes in water mass transport, the oceanic thermohaline circulation, and deep water formation, particularly in the North Atlantic and Southern Oceans (Street-Perrott and Perrott, 1990; Broecker, 1991; Greatbatch *et al.*, 1991; Marotzke and Willebrand, 1991; Mehta, 1991, 1992; Rind and Chandler, 1991; Barnett *et al.*, 1992; Gordon *et al.*, 1992; Hughes, 1992; Meyers and Weaver, 1992; Wunsch, 1992; Greatbatch and Xu, 1993; Lenderink and Haarsma, 1994; Levitus *et al.*, 1994; von Storch, 1994; Ezer *et al.*, 1995; Pierce *et al.*, 1995). These studies suggest that there may be a propensity for the climate system to exist in one of several stable climatic states. However, there is still the question of cause and effect. Are ENSO phases able to modulate the basic climate state and thus influence the transition between multidecadal periods, or is the reverse the case? Or in such a closely coupled system, do both situations occur? Answers to these questions would have profound implications for any assessment of possible changes in the ENSO phenomenon.

Broad synopses of decadal–multidecadal fluctuations in various features of the climate system are covered in Oort *et al.* (1987), Wunsch (1992), Crowley and Kim (1993), Mann and Park (1993, 1994a, b), Allen and Smith (1994), Schlesinger and Ramankutty (1994), Mann *et al.* (1995b), Saravanan and McWilliams (1995), Schimel and Sulzman (1995), Unal and Ghil (1995) and von Storch and Hasselmann (1995). An overview of oceanic interdecadal climate-variability research in the North Pacific and Atlantic oceans up to 1992 can be found in Ad Hoc Study Group on Oceanic Interdecadal Climate Variability (1992). A detailed synopsis of global thermohaline circulation is given in Schmitz (1995), and recent GCM simulations of interdecadal variability have been performed by Luksch and von Storch (1992), Schneider and Kinter (1994), von Storch (1994) and Kawamura *et al.* (1995a, b). Attempts to force GCMs with observed global SST fields in order to assess model simulations of interannual variability and ENSO in the longer term are presented in Smith (1995) and Stephenson and Royer (1995a).

Theories involving volcanic dust as a factor in contemporary and historical ENSO phases and variability have been put forward in the literature over the last decade. Most recent proponents of such links are exemplified by the studies of Handler (1984, 1986), Strong (1986), Hirono (1988) and Handler and Andsager (1990). However, such claims cannot be substantiated when the data are placed under close scrutiny (Nicholls, 1988c, 1990; Robock *et al.*, 1995). With the recent persistent El Niño sequence, any links to Mount Pinatubo, which erupted in mid-June 1991, and Mt Hudson, with eruptions in August 1991, are not apparent. More incisive research in this area is in studies of the influence of separate or combined volcanic and ENSO signals in United States surface air temperature and rainfall patterns by Portman and Gutzler (1996), on Northern Hemisphere winter tropospheric temperature, winds and geopotential height by Kirchner and Graf (1995), and on zonal mean atmospheric circulation by Ulbrich *et al.* (1995).

The influence of climatic variability on global carbon fluxes, stratospheric ozone and other trace gases has increasingly become a feature of research priorities, with concerns about the enhanced greenhouse effect and polar ozone depletion. Some of the earliest evidence for ENSO modulation of the carbon cycle was found by Bacastow (1976) and Bacastow *et al.* (1980). Across the CEP–EEP region, the largest natural global oceanic source of CO_2, oceanic dynamics tend to favour a reduction in CO_2 content in the upper layers of the ocean and reduced flux into the atmosphere during the early stages of El Niño phases. In the cessation stages of an El Niño phase, upper ocean CO_2 concentration increases and a higher flux to the atmosphere is re-established. With improved and expanded sampling networks, a concerted campaign has been mounted to resolve the full extent of ENSO influence on CO_2 fluxes (Conway *et al.*, 1994; Feely *et al.*, 1994; Meyers and O'Brien, 1995; Francey *et al.*, 1995; Murata and Fushimi, 1996). An overview of this work and its links to biological processes in the equatorial Pacific is given in Murray *et al.* (1994).

Relationships between ENSO phases and stratospheric ozone show tropical and higher latitude responses (Stephenson and Royer, 1995b). The tropical pattern in stratospheric ozone mirrors the 'centres of action' seen in the Southern Oscillation phenomenon, with reduced ozone levels in the EEP and increased ozone levels in the WEP. The higher-latitude ozone response is most evident in the Southern Hemisphere south of 25°S. In general, the ozone response at the higher southern latitudes is the opposite to that in the tropics. This research refines the findings of earlier studies by Shiotani (1992) and Zerefos *et al.* (1992). An overview of both global trace gas and aerosol responses on interannual time scales is provided by Kane (1994).

Other studies of relationships with ENSO have focused on associations with intraseasonal oscillations such as the Madden-Julian or 30–60 day oscillation (Lau, 1985; Singh *et al.*, 1992; Wang, 1992; Yi and Longxun, 1992; Meehl, 1993, 1994; Ronghui, 1994), its influence on mean global and hemispherical air temperature time series (Jones, 1988; Privalsky and Jensen, 1995), links to Northern Hemisphere stratospheric circulation (Baldwin and O'Sullivan, 1995), with the stratospheric Quasi-Biennial Oscillation (QBO) (Gray *et al.*, 1992; Mason and Tyson, 1992; Xu, 1992; Mason and Lindesay, 1993; Mason, 1995) and with atmospheric angular momentum (Rosen *et al.*, 1984; Ponte *et al.*, 1994).

ENSO Components (Quasi-Biennial and Low Frequency)

Spectral analyses of atmospheric or oceanic variables influenced by ENSO phases show significant energy in tropospheric quasi-biennial (QB) (around 18–35 months) and lower frequency (LF) (around 32–88 months) bands (Yasunari, 1985; Jinghua *et al.*, 1988; Rasmusson *et al.*, 1990; Jingxi *et al.*, 1990; Barnett, 1991; Ropelewski *et al.*, 1992; Tomita and Yasunari, 1993). This can be seen when a long gap-filled SOI, recently constructed using interpolated MSLP values for missing data at Tahiti and Darwin (Young, 1993), and Niño 3 and 4 SST anomalies (Raynor, 1996, pers. comm.) are filtered using a recursive Butterworth (bandpass) filter in both the QB and LF bands over the periods 1876–1996 and 1871–1994, respectively (see 'Calculation and filtering of Niño 3, Niño 4 and SOI series' p. 57) (Figure 31). Values of 50%, 66% and 56% of the raw SOI signal series, and Niño 3 and 4 index variance, respectively, are captured by combining their QB and LF bands. Individual ENSO phases in both SOI and SST variables are dominated by either or both of these bands. The LF band represents a relatively broad frequency envelope, and can dominate for longer than the more usual interannual ENSO time scales (Allan *et al.*, 1995; Reason *et al.*, 1996). These data suggest that there is a 'family' of ENSO types in either phase, and that the phenomenon can vary in frequency, magnitude and duration. Most recently, Allan and D'Arrigo (1996) have emphasised that this includes three other 'persistent' El Niño and six La Niña sequences in the historical record prior to the 1990–1995 situation.

Recent oceanic and coupled GCM simulations by Barnett *et al.* (1995) indicate that any spectral peaks found in ENSO variables are very dependent on factors such as the length of data record, data sample location, mean climatic state and strength of ocean–atmosphere interactions. As a consequence,

Barnett *et al.* (1995) suggest that the characteristic frequency band in their study of 26–40 months is highly analysis- and data-dependent. An application of three time-frequency spectra methods to the historical SOI and the Wright (1989) EEP-CEP SST and CEP precipitation time series in Kestin *et al.* (1996) indicates that although the combined QB and LF bands carry the bulk of the ENSO signal, the phenomenon does not appear to possess fixed oscillation modes. These findings are acknowledged here, and although the QB band used in this book covers much of the characteristic frequency band found by the analyses of Barnett *et al.* (1995), the ENSO phenomenon is better represented by the interaction between both the QB and LF bands.

At decadal–multidecadal frequencies, there are also periods when both the QB and LF components of the SOI are particularly weak or strong. The most notable weak epoch is the period 1921–41, whereas

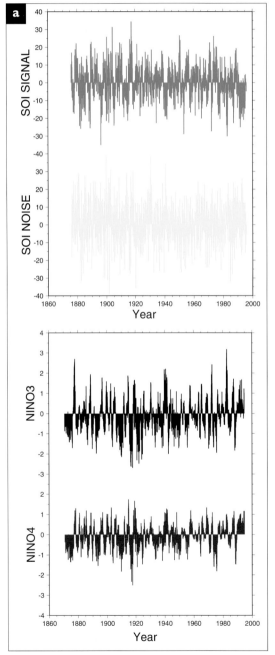

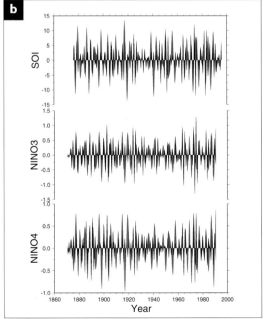

Figure 31: (a) Raw time series of SOI signal (normalised Tahiti minus Darwin MSLP anomalies), SOI noise (normalised Tahiti plus Darwin MSLP anomalies) (following Trenberth, 1984; Trenberth and Hoar, 1996), and Niño 3 and Niño 4 SST anomalies (reconstructed from GISST2.3), since 1871; and

(b) bandpass filtered time series of the SOI signal, SOI noise, and Niño 3 and Niño 4 SST anomalies in the 18–35 month (blue) and 32–88 month (red) bands since 1871. All anomalies and normalisations are with respect to the 1961–1990 period.

the decades up to c.1920, and since the mid 1960s, show more robust signals. Varying ENSO behaviour has also been reported by other studies using individual time series and station differences (Trenberth and Shea, 1987; Elliott and Angell, 1987, 1988; Trenberth and Hurrell, 1994; Kestin *et al.*, 1996). One important remaining concern is whether simple SOI and/or SST indices are a viable measure of ENSO over time. Such questions are addressed in this book.

Stationary and Propagational Aspects

Recent research suggests that, although often displaying the characteristics of a standing wave, ENSO phases also exhibit propagational characteristics and evolve in response to an interaction between the annual cycle and the quasi-biennial and lower frequency components discussed above. Other studies indicate that the core region of ENSO in the Indo–Pacific basin is influenced by decadally to multidecadally varying and progressive atmospheric pressure signals, which appear to fluctuate between Indo–Asian and southern African–Antarctic source regions (Barnett, 1983, 1984b, c, 1985, 1988; Yasunari, 1985; Krishnamurti *et al.*, 1986; Yasunari, 1987a, b; Jinghua *et al.*, 1988; Kiladis and van Loon, 1988; Jingxi *et al.*, 1990; Barnett *et al.*, 1991; Mehta, 1992). This research is schematicised in Figure 32, which suggests that in the year before an El Niño phase, a pressure signal, appearing to originate out of Asia (DJF −1), propagates across the Indian Ocean to Australasia where it becomes intensified during the boreal summer-autumn (austral winter-spring) period (JJA −1 and SON −1) and is matched by a pressure signal of the opposite sign that becomes intensified over the southeastern Pacific Ocean to form the 'classic' Southern Oscillation pressure dipole pattern. By the boreal autumn (austral spring) (SON −1), there is a suggestion that part of the Pacific pressure signal 'folds back' towards Asia across higher latitudes of the North Pacific Ocean, and although the dipole structure collapses, provides an impetus for the generation of a pressure signal of the opposite sign in the beginning of the year of the El Niño phase (DJF 0). In the El Niño year (DJF 0 to SON 0), a similar propagation and evolution structure is proposed, but with pressure anomalies now of the opposite configuration across the Indian and Pacific regions.

However, the above structure has recently come under scrutiny. After examining uncoupled oceanic and atmospheric GCM simulations and hybrid-coupled model experiments, Latif and Barnett (1995) have questioned the extension of the above by Barnett (1983, 1985), Barnett *et al.* (1991) and Tourre and White (1995) to include the concept of a global zonally propagating wave. Latif and Barnett (1995) propose that the global wave concept is an 'illusion' caused by the different responses of each ocean basin to low-frequency fluctuations in ocean–atmosphere characteristics. Nevertheless, both zonal and meridional features with apparent propagational properties in each of the global ocean basins have been reported widely, and such structures require further examination.

Even lower frequency modes in the climate system may provide a modulation of the boundary conditions in which the ENSO phenomenon operates and thus influence teleconnection patterns (Mehta, 1991, 1992). Examinations of the historical instrumental record suggest that ENSO characteristics wax and wane, with epochs when little more than the core regions of ENSO influence remain intact and others when ENSO influences are extensive and robust (Allan, 1993). Such changes appear to be most pronounced when contrasting recent climatic regimes with those in the 1920s–1930s epoch. Thus the ENSO phenomenon and the climate system are highly interactive on a number of time and space scales.

Modelling ENSO

Three types of models have been used in attempts to model the ENSO phenomenon. Terminology for these models may vary, but they are classified in this book as of intermediate, limited area General Circulation Model (GCM) and global GCM type. The development of intermediate and limited area models has been aimed at capturing the essence of the ENSO phenomenon in the Pacific, and thus producing better forecasts of El Niño and La Niña phases (Latif *et al.*, 1994b). Attempts have also

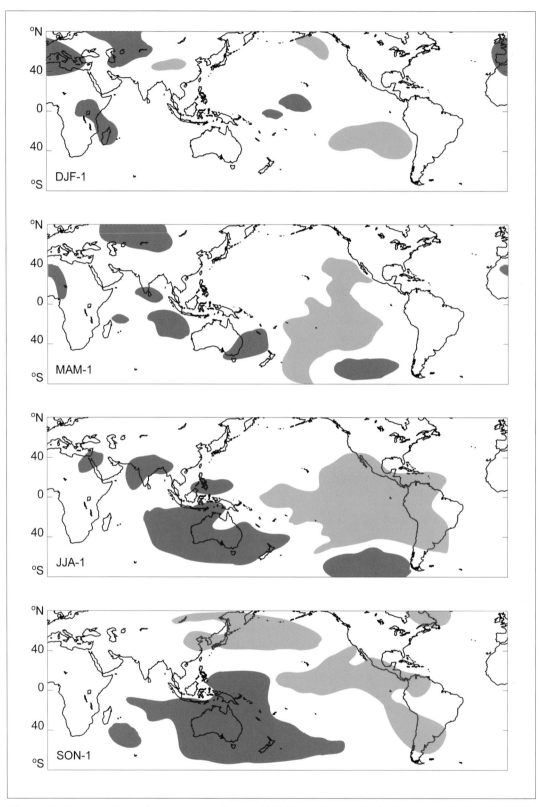

Figure 32: Schematic of the global propagation of seasonal MSLP anomalies associated with a composite of major El Niño phases (covering the year before [marked –1] and during [marked 0] the El Niño events) since the 1950s. Blue areas are MSLP anomalies less than –0.5 hPa and red areas are MSLP anomalies greater than +0.5 hPa (after Barnett, 1988).

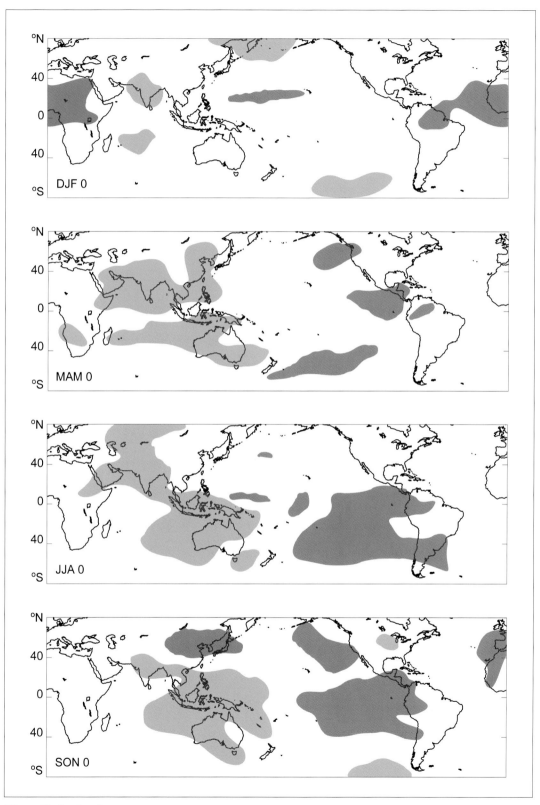

Figure 32: Continued

been made to examine the ENSO phenomenon in GCMs. However, the known aspects of ENSO dynamics are not specifically contained within GCMs, and they are yet to produce the full range of observed ENSO characteristics. Ongoing model development promises to remedy this situation (Mechoso *et al.*, 1995).

Intermediate models employ simplified physics and tight dynamical constraints. They are best represented by the models discussed in Chen *et al.* (1995) and Kleeman (1991, 1993). These models have minimal physical representation of the atmosphere (often only two layers in the vertical) and a construction designed to resolve the basic dynamics of the ENSO phenomenon alone. The building of models, in this context, has required substantial resources and particular recourse to observational studies and dynamical theories that provide a conceptual framework for the physical processes likely to underlie the phenomenon. As described earlier, there is no firm theory describing all aspects of ENSO events, but rather a dynamically consistent understanding of some of the important processes governing its behaviour. More recent developments in these simple coupled ocean–atmosphere models have seen improvements in the ability of these tools to predict ENSO. At present, a number of groups around the world are using such coupled models to predict changes in variables indicative of ENSO events, such as Pacific SSTs. Forecasts from these groups are detailed regularly in the *Experimental Long-Lead Forecast Bulletin* (Climate Prediction Center, 1994–1996).

In general, limited area GCM simulations of ENSO are higher resolution models, with explicit physics that embed a fine meshed grid of the Pacific Ocean region within a coarser global representation of the remaining tropical and subtropical regions in both hemispheres. Under forcing by observed wind and heat flux fields, they produce realistic representations of both the mean and perturbed states of the Pacific represented by ENSO phases. This type of model is best represented by the studies of Philander *et al.* (1992) and Chao and Philander (1993).

An important advance in the development of global GCMs is the coupling of atmospheric and full oceanic models. This has produced numerical tools for simulating and understanding the essential elements responsible for observed climatic patterns, particularly the mean state and seasonal cycle of the climate system. Both of these climatic components have been shown to influence the temporal structure and spatial characteristics of ENSO. Over the last few years, a number of attempts have been made to capture the known physical characteristics of the ENSO phenomenon in fully coupled ocean–atmosphere GCMs. Although ENSO-like and lower frequency features are resolved in some of these models, and show some similarities to observed fields, more research and model development is required to capture the full essence of these vacillations in the climate system.

Some of the best initial model results from limited area and global GCMs have been identified in simulations from the Geophysical Fluid Dynamics Laboratory (GFDL) (Lau *et al.*, 1992; Philander *et al.*, 1992). Several other high resolution GCMs from the University of Hamburg–Max Planck Institute (Latif and Villwock, 1990; Latif *et al.*, 1990; Oberhuber *et al.*, 1990) also showed significant differences from observations, in that equatorial SSTs were too cool, or only weak ENSO-like events were resolved. The best simulations from this group were found when their coupled GCMs were flux-corrected, so that the model was artificially kept close to the observed climatology. Other higher resolution models from the UKMO (Gordon, 1989; Ineson and Gordon, 1989), the National Center for Atmospheric Research (NCAR) (Meehl, 1990b, 1991) the University of California (Mechoso *et al.*, 1990), the Meteorological Research Institute, Japan (Tokioka *et al.*, 1984, 1988; Nagai *et al.*, 1992, 1995) and the Max Planck Institute (Cubasch *et al.*, 1990), have had problems with climate drift (Moore, 1995) and display major difficulties such as poor zonal SST gradients in the Pacific Ocean, too cool equatorial SSTs, or unreasonable oceanic upwelling fields. Similar problems appear to be inherent in the lower resolution Oregon State University/Lawrence Livermore National Laboratory (OSU/LLNL) simulations (Sperber *et al.*, 1987. 1992; Hameed *et al.*, 1989; Sperber and Hameed 1991).

Another coupled GCM focus that has aided the development of models of ENSO is the concern about possible changes in the phenomenon under enhanced greenhouse conditions. Meehl *et al.* (1993), and more recently Smith *et al.* (1996), have analysed the effects of warm EEP SST forcings in GCM experiments with global and mixed-layer ocean models. A number of the responses in their simulations were suggested to be ENSO-like events. Transient CO_2-coupled model experiments by Knutson and Manabe (1994, 1995) and Tett (1995) have simulated Pacific SST features exhibiting ENSO-like nature. A coupled GCM with enhanced near-tropical resolution run by Tokioka *et al.* (1995) has also detected features indicative of ENSO operation. Results from recent coupled-GCM simulations by Gordon and O'Farrell (1996) show ENSO-like phases that are somewhat weaker in magnitude when compared with observed features.

Modelling studies in the last three years have resulted in improvements to the simulations documented above. Mechoso *et al.* (1995), in a detailed review of 11 coupled ocean–atmosphere GCMs that did not use flux correction, report that although they generally resolve the equatorial Pacific Ocean 'cold tongue' of SSTs, the model feature is too strong and narrow and extends too far to the west. The 'warm pool' in the WEP is also reproduced in a number of the models, although in some it is too narrow. However, several particular concerns exist. SST patterns are poorly simulated in parts of the EEP, as is the spatio–temporal structure of the ITCZ and the seasonal cycle of EEP SSTs. Recent simulations by Roeckner *et al.* (1995) are an example of efforts using minimal flux correction to overcome the problem of climate drift (Moore, 1995) and unrealistic climate states becoming established in models. They claim to have produced realistic annual and seasonal cycles in their simulations, and acceptable ENSO-like features that have the frequency, magnitude and life-cycle characteristics and phase locking of Pacific SSTs to the annual cycle that are seen in observations. Specific studies of fully coupled GCM model simulations of oceanic and atmospheric components of ENSO are beginning to be undertaken. Schneider *et al.* (1996) discuss the results for the 'warm pool' in the WEP from two such models of varying resolution with no flux correction. The higher resolution model is reported to provide the best results, in that it resolves an SST structure with seasonal location and linked atmospheric cloudiness, moisture and heat fluxes similar to observations. Other efforts have focused on various components of the ocean–atmosphere system in GCMs (von Storch *et al.*, 1994; Wang and Weisberg, 1994a, b; Chen *et al.*, 1995; Lunkeit *et al.*, 1996). All of these recent results are encouraging for the ongoing development of ENSO modelling.

The most comprehensive reviews of the evolving status of ENSO modelling are given in Meehl (1990a), McCreary and Anderson (1991), Neelin *et al.* (1992, 1994), Stockdale *et al.* (1993), Latif *et al.* (1994b), Battisti and Sarachik (1995) and Mechoso *et al.* (1995). Further progress can be expected as the underlying dynamics of ENSO episodes become better understood, and the capabilities of models to represent such complex processes develop further.

2. DATA AND METHODS

This section provides a detailed technical overview of the construction of the various data sets used in this book. The synopsis of the development of the gridded monthly MSLP data draws heavily on the step by step procedure given in Basnett and Parker (1996). However, technical knowledge of such procedures and data is not necessary to appreciate the information presented in the analysed Niño 3, Niño 4 and SOI series or the global MSLP and SST fields presented here.

Global Monthly Gridded SST Data from 1871 to 1994

Global monthly gridded SST data from 1871 to 1994 are from the UKMO Global sea-Ice and Sea Surface Temperature version 1.1 (GISST1.1) compilation. These data are part of the UKMO's ongoing historical SST reconstructions that have been documented extensively in the literature (Reynolds *et al.*, 1989; Bottomley *et al.*, 1990; Parker *et al.*, 1995a, b). The GISST specific data on a 1° x 1° grid (Parker *et al.*, 1994, 1995a) are the best currently available historical SST compilation back to 1871 in terms of completeness. The GISST1.2, 2.1 and 2.2 versions have been released in the last few years, but, although they have improved data interpolation using empirical orthogonal functions, they cover at best only the period from 1903 to 1994 (Rayner *et al.*, 1995). The latest version being completed as this book was published (GISST2.3) will provide coverage back to 1871. For this book, both actual and anomalies of monthly SST fields were required. Monthly mean GISST data were formed into gridded monthly normalised SST anomalies relative to 1961–1990 monthly means in the standard manner. Thorough appraisals of the reconstruction of long-term data compilations, including SST, are given in Oort *et al.* (1987), Parker and Folland (1994) and Folland and Parker (1995). Other ongoing efforts to improve global historical SST fields can be found in Reynolds and Smith (1994) and Smith *et al.* (1994, 1996).

Global Monthly Gridded MSLP Data From 1871 to 1994

The global monthly MSLP fields on a 5° x 5° grid are a new product, technically referred to as the Global Mean Sea Level Pressure (GMSLP) version 2.1 (GMSLP2.1) data set, developed in a collaborative venture by the authors, in conjunction with Dr M.J. Salinger at the National Institute of Water and Atmospheric Research (NIWA) in New Zealand. The global monthly MSLP data were constructed by blending together existing gridded data sets from the UKMO, NCAR, Scripps Institute of Oceanography (SIO), CSIRO DAR, and the Climatic Research Unit (CRU) at the University of East Anglia (UEA), with land station and ship observations from the UKMO Marine Data Bank, UKMO CLIMAT Archives, the Comprehensive Ocean–Atmosphere Data Set (COADS) (Woodruff *et al.*, 1987), and from the CSIRO DAR and NIWA MSLP archives (New Zealand and south-west Pacific station details in Collen 1992 and Fouhy *et al.* 1992) (Figure 33). As with the GISST compilation, both monthly and anomaly MSLP fields were used in this book. Monthly MSLP data were formed into gridded monthly normalised MSLP anomalies relative to 1961–1990 monthly means in the standard manner. However, considerable data collation, manipulation and quality control were required before the creation of the mean and anomaly fields.

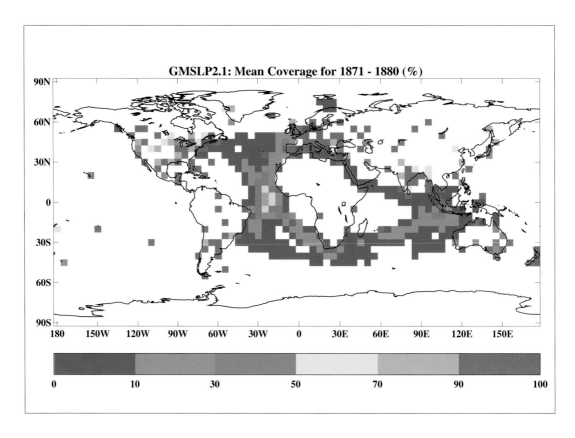

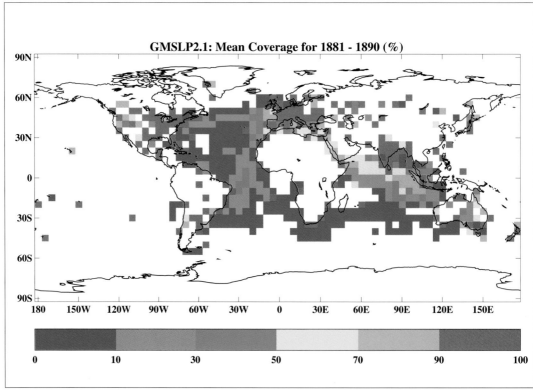

Figure 33: Distribution of land, island and ship observations of MSLP used in construction of the GMSLP2.1 gridded data fields during various time slices between 1871 and 1994.

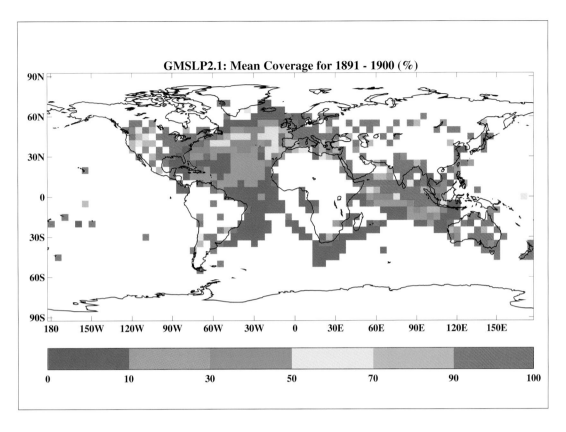

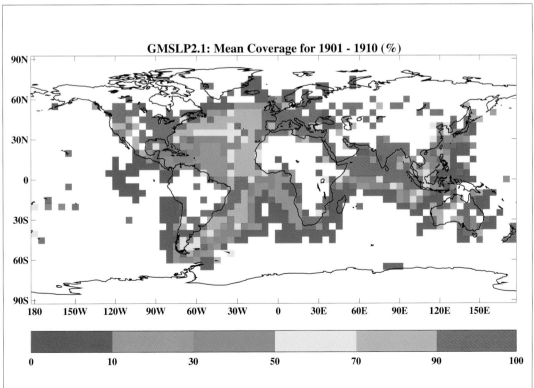

Figure 33: Continued

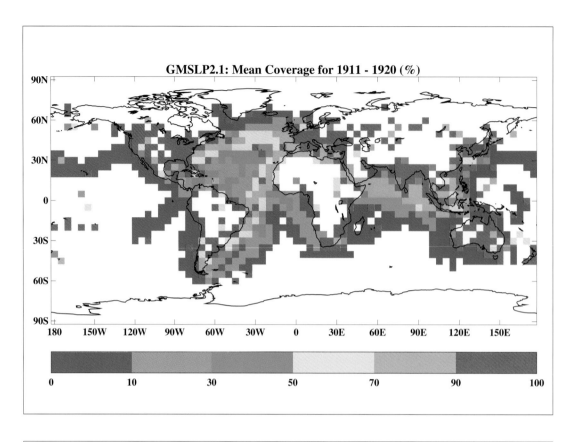

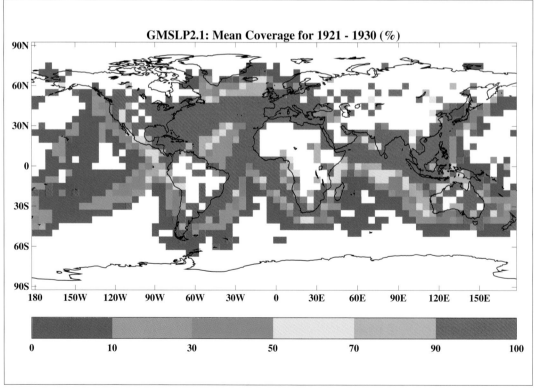

Figure 33: Continued

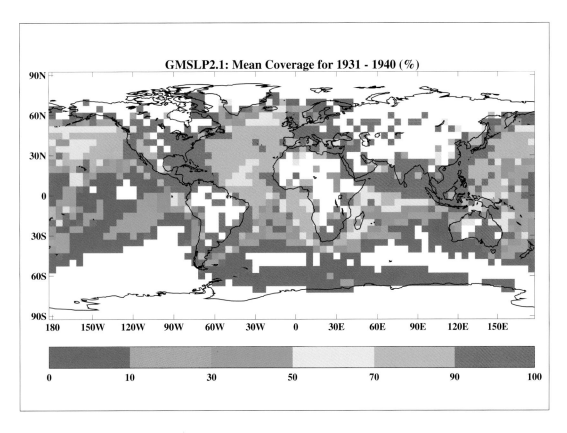

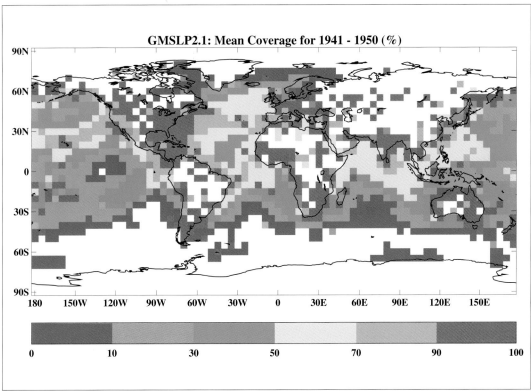

Figure 33: Continued

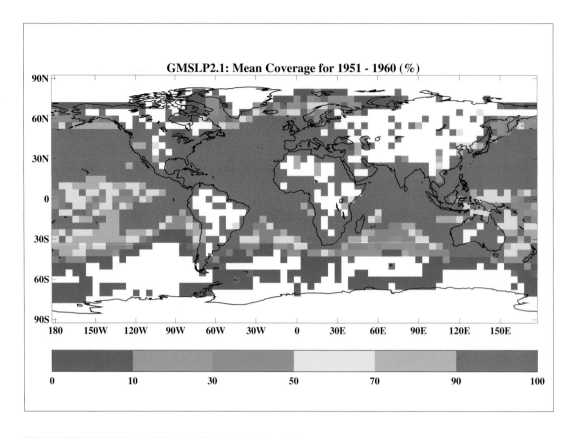

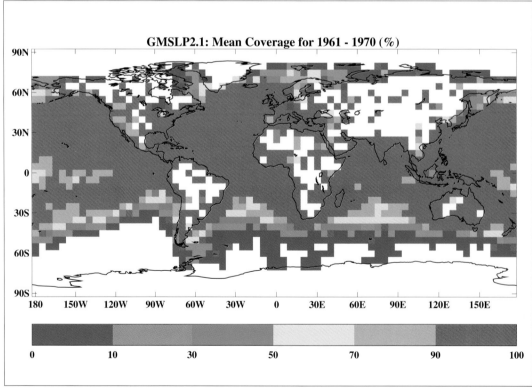

Figure 33: Continued

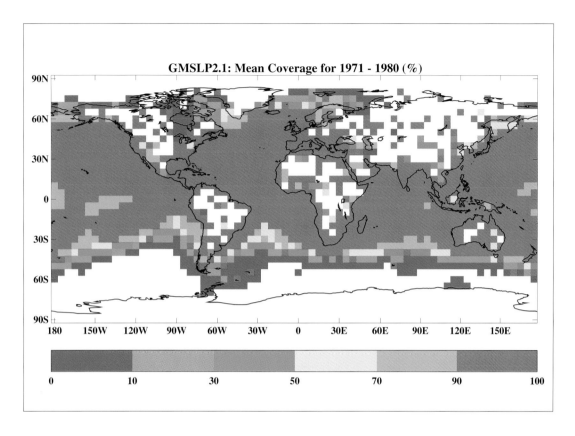

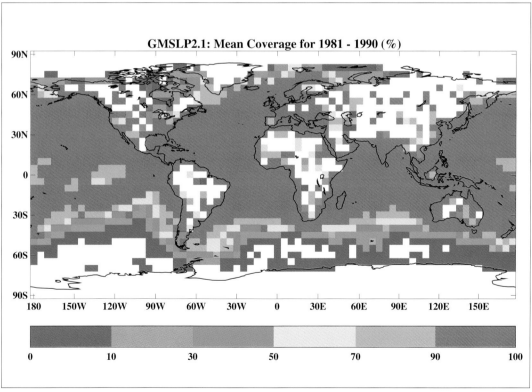

Figure 33: Continued

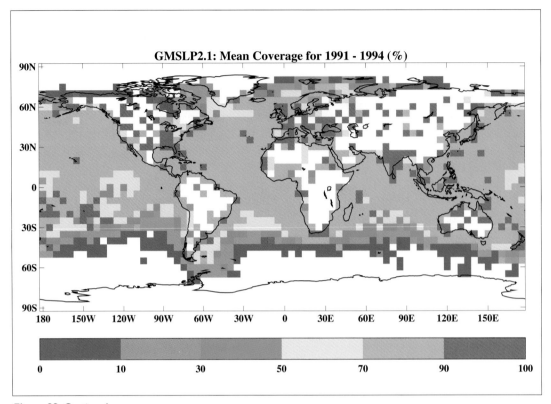

Figure 33: Continued

Land and Island MSLP Data from 1871 to 1994

A data set of land and island stations compiled at CSIRO DAR using the World Weather Records (WWR), Monthly Climatic Data for the World, Reseau Mondial, and numerous data compilations and records held by meteorological services around the globe, together with MSLP observations from the NIWA MSLP archive, provided the bulk of long historical land and island MSLP observations used in the gridded MSLP reconstruction. Table 1 provides a detailed listing of the sources of MSLP data in the CSIRO DAR archive. One of the most important initial tasks after collection of the individual land and island station and ship MSLP data series was the development of procedures for checking records for inhomogeneities and correcting them.

Checking and Correcting Land and Island Station MSLP Data Inhomogeneities

Processing and checking of the raw land and island station MSLP data from the CSIRO DAR MSLP archive was done in a number of steps. Although much routine 'number crunching' can be programmed once various testing and checking approaches have been established, there was still a very important human element involved in checking station records, files and histories, and in checking each station individually before the complete data set was available for a computer inhomogeneities checking routine.

The following approach was undertaken with all CSIRO DAR MSLP archive land and island pressure stations used to create the global MSLP fields.

A reduction of all station pressure time series with sufficient reduction information to MSLP was made. Stations with no reduction information were retained as station level pressure and processed with MSLP series. The reduction involved checking station height above mean sea level, making corrections for attached thermometer, index error and standard gravity (especially pre-1900 data), and ensuring the series was corrected to 24-hour means. As gravity varies slightly with latitude, it is necessary to make a gravity correction to take account of the gravitational influence on barometric pressure readings. The departure of local gravity from the standard gravity acceleration value of 9.8 m s^{-2} is calculated with respect to a standard latitude of 45°. This leads to negative adjustments to pressures for stations at latitudes equatorward of 45° north or south, and positive adjustments for stations poleward of 45° north or south.

Each individual MSLP time series was then checked for jumps and trends and corrected, where possible. When they were available, a check of details of instrument/measurement changes in individual station records was made. A 'findjumps' programme (listing maximum differences in means for two adjacent periods in a time series for lengths 12 to 120 months) was then used to check for jumps and trends. Corrections for these problems were then initiated with regard to the long-term mean MSLP data statistics of individual stations.

Next, checks were made against nearest neighbours within 1000 to 3000 km of each individual MSLP time series for discontinuities, trends, outliers etc. A simple 'two station differences' routine was used to check for discontinuities, and an 'outlie' programme (listing mean and standard deviation of a time series and any points more than 3 standard deviations from the mean) was used on each time series and its nearest neighbours to check for outliers. Corrections were made to series where discontinuities were detected, and nearest neighbours had homogeneous long-term mean MSLP data statistics.

Table 1: CSIRO DAR MSLP Archive Source List

GLOBAL:

- Smithsonian Institute (1927, 1934, 1947), US Weather Bureau (1959, 1967), WeatherDisc Associates (1994): World Weather Records.
- Oak Ridge National Laboratory (1992): The Global Historical Climatology Network: Long-Term Monthly Temperature, Precipitation, Sea Level Pressure, and Station Pressure Data.
- National Environmental Satellite Data and Information Service (1951–1995): Monthly Climatic Data for the World.
- Young (1993): Kenneth C. Young, Institute of Atmospheric Physics, University of Arizona, Tucson, Arizona, USA.
- Reseau Mondial (1910–1934).
- Meteorologische Zeitschrift (early volumes).
- Deutschen Seewarte, Deutscher Wetterdienst Seewetteramt (Heft I-Heft XII). Deutsche Ueberseeische Meteorologische Beobachtungen.
- Central Meteorological Observatory (1954): Data for Japan and the Far East Area.
- Jones (1992–1996, pers. comm.): Climatic Research Unit, University of East Anglia.
- Climatological Table for the British Empire (1923–1939): The Meteorological Magazine.
- Lockyer (1908): Barometric Pressure for 73 Selected Stations over the Earth's Surface.
- Meteorological Council (1890): Meteorological Observations at the Foreign and Colonial Stations of the Royal Engineers and the Army Medical Department, 1852–1886.

REGIONAL:

- Jose (1994, pers.comm.): Philippine Atmospheric, Geophysical and Astronomical Services Administration (PAGASA) — Climate Data Section.
- van Bemmelen (1913), Boerema (1939), Berlage (1939, 1940, 1941): Observations made at Secondary Stations in Netherlands East-Asia .
- Doraiswamy Iyer and Francis (1941): Data for the Seychelles.
- Department of Meteorological Services (1952), Rhodesia and Nyasaland Meteorological Service (1951): Data for Southern and Northern Rhodesia.
- Walter (1948): Observations of atmospheric pressure in East Africa.
- Weather Bureau (1941–1952): Union of South Africa, Weather Bureau, Reports .
- Russell (1871–1886, 1904, 1905, 1906), Ellery (1873–1891), Post and Telegraph Department (1889–1896), Cooke (1901), Griffiths (1910), Hunt (1910–1913, 1911, 1914, 1916, 1918, 1929), Watt (1936, 1940), Warren (1940, 1948), Bureau of Meteorology (1945–1953, 1954–1956), Bureau of Meteorology National Climate Centre Data Services Section (1990–1996, pers. comm.), Bureau of Meteorology, Queensland Regional Office (1993, pers.comm.), Mullan (1994, pers. comm): Data for Australia, Papua New Guinea and some South Pacific Islands.
- Campbell (1879), Robb (1880), Brooks (1918, 1919), Henry (1925): Miscellaneous data records.

Once the above had been done to all stations, the data were processed by computer to remove outliers of more than 3 standard deviations. A version of the 'Potter' method (Potter, 1981) was then employed to objectively check for discontinuities between each individual time series and its nearest neighbours. In the 'Potter' method, differences between each time series and its nearest neighbours were examined in terms of means for all time portions in the difference series, excluding the end points, and a student t-statistic was then calculated for the temporal difference in the means; a plot of the t-statistic as a function of time reveals a discontinuity as a maximum in the t-statistic. A programme was then employed to make corrections to remove the detected inhomogeneities. It was at this stage that any remaining station level pressure series were adjusted to sea level with regard to the surrounding nearest MSLP neighbour stations.

Not all of the CSIRO DAR and NIWA MSLP archive data were incorporated into GMSLP2.1. For each station, annually averaged deviations from the GMSLP2.0 background field (see 'Collation of ship MSLP data, and checking and correcting their inhomogeneities', p. 54) were plotted as time series, using both the CSIRO DAR and NIWA MSLP archive data and the UKMO CLIMAT data. Stations whose deviations showed biases, jumps or extrema inconsistent with those of neighbouring stations were, in general, passed over for further checking and held back for use in future versions of GMSLP. The existence of biases, jumps or extreme deviations in themselves did not disqualify stations' data, as they sometimes occurred at whole groups of stations and revealed errors in the background field. For example, GMSLP2.0 showed a discontinuity around 1899 over much of southeastern Asia when the NCAR gridded fields began. In some cases of doubt, simultaneous plots of actual station values and actual collocated GMSLP2.0 values confirmed or disproved the veracity of the station data. For individual extreme annual deviations, the monthly deviations were examined also and a few unlikely values were deleted. For a few stations, the earliest parts of the record appeared suspect and were not used. A few other stations with broken, erratic records were also excluded.

Most of the retained station MSLP data for stations north of 60°N were from the UKMO CLIMAT archive as these areas were not, in general, included in the CSIRO DAR and NIWA MSLP archives. These data were checked against the GMSLP2.0 background field and neighbours in the same manner as described above. Over the rest of the globe, the CSIRO DAR and NIWA MSLP archive data were used, but updated into the 1990s with UKMO CLIMAT data when these were consistent with the earlier records in terms of deviations from GMSLP2.0. The NIWA MSLP archive data were mainly for New Zealand and the southwestern Pacific. In order to give adequate coverage in the earlier years, a few long period stations which had been discarded in the checks against GMSLP2.0 were re-examined, adjusted where necessary, and used in GMSLP2.1.

The accepted station data were assigned to the appropriate 1° x 1° area and month, combined with the ship data in this form (detailed in the following section) and then blended with GMSLP2.0 as described in 'Construction of the gridded monthly MSLP fields' (p. 55).

Collation of Ship MSLP Data, and Checking and Correcting their Inhomogeneities

UKMO Marine Data Bank and COADS MSLP observations from ships were used to create a 1° x 1° global gridded MSLP data set. This task required considerable data processing, including a number of important steps to check and correct the data observations. Such quality control steps were integrated into the data operations.

Data compatibility in time and space was a major consideration. The UKMO Marine Data Bank contains individual ship MSLP observations, while the COADS MSLP data exist as a 2° x 2° grid of global monthly MSLP over the oceans. As a result, the UKMO Marine Data Bank observations were processed first. For a particular or 'target' month, all UKMO Marine Data Bank observations made in that month were extracted and subjected to corrections for standard gravity. Data for the period 1871–1920 required 100% standard gravity correction, those in the 1921–1935 period needed a partial correction for standard gravity (ranging linearly from 93.75% in 1921 to 6.25% in 1935), while from 1936 onwards there was no correction necessary (Basnett and Parker, 1996). In the next step, all of the corrected MSLP observations in the target month that differed from the relevant 1° x 1° area monthly GMSLP2.0 background field (detailed in the following section) value by greater than, or equal to, 10 hPa were rejected.

The corrected target month MSLP observations were then processed in terms of a selected 1° x 1° target box. All corrected target month observations lying within a concentric 7° box around the target box were converted into differences relative to the nearest GMSLP2.0 field value. If less than 5 pentads (5 day periods) contained data in the 7° box, the 1° x 1° target box was rejected. Pentads were strictly pseudo-pentads, in that there were always 6 per target month (the final pentad in a month was the period from the 26th until the end of the month). Where data were available for more than 5 pentads, the mean daily difference from GMSLP2.0 values using all acceptable observations in the 7° area for each day in the month was calculated (up to a maximum of 31 days). The sample error of the mean daily differences was also calculated as prescribed by Parker (1984). A 1° x 1° target box was then accepted if the sample error was less than, or equal to, 20% of the standard deviation of mean daily differences, or the standard error was less than 1 hPa. The median of the mean daily differences was then deduced for the accepted 1° x 1° target box. Finally, this median value was added to the relevant 1° x 1° monthly GMSLP2.0 field value. The procedure outlined in this paragraph was then repeated for all 1° x 1° target boxes in the target month, and the whole process in this and the previous paragraph was applied to all months in the 1871–1994 period.

Having quality controlled and effectively put the UKMO Marine Data Bank MSLP observations on a 1° x 1° grid, it was then necessary to merge these data with the COADS MSLP data which were on a 2° x 2° grid. In a target month during the 1871–1994 period, all 1° x 1° UKMO Marine Data Bank MSLP and 2° x 2° COADS MSLP differences relative to GMSLP2.0 were calculated. A target 2° x 2° COADS box was then examined for any coincidence with 1° x 1° UKMO Marine Data Bank MSLP observations. If no coincidence was found, then the mean difference for all 1° x 1° UKMO Marine Data Bank MSLP boxes lying within 10° latitude and longitude of the centre of the 2° x 2° COADS target box was calculated. In cases where the mean 1° x 1° difference was not missing, then the 2° x 2° COADS target box value was accepted provided the corresponding difference deviated by less than 1 hPa from the mean 1° x 1° UKMO Marine Data Bank MSLP difference. If the target 2° x 2° COADS box value was accepted, then the actual value was placed into all coinciding 1° x 1° UKMO Marine Data Bank MSLP boxes. The COADS data were not used if the mean 1° x 1° UKMO Marine Data Bank MSLP difference was missing. The procedure outlined with the target 2° x 2° COADS box was repeated for all 2° x 2° COADS boxes, and then the whole process in this paragraph was repeated for all target months.

Construction of the Gridded Monthly MSLP Fields

As no long-term global gridded monthly MSLP data set existed to use in this book, a new product needed to be constructed. This task involved the combination of individual land and island station and ship MSLP observations with a blend of existing gridded MSLP fields for the Northern, and parts of the Southern, Hemisphere. The details of the steps involved in creating the latest version of the compilation, GMSLP2.1, are given below, and the steps from GMSLP1.2 to the final GMSLP2.1 version are summarised in Figure 34.

The initial GMSLP data set (GMSLP1.0) was developed by blending together, back to 1951 and north of 15°N, the UKMO historical, NCAR and SIO data sets (detailed in Table 2) and interpolating them onto a 5° latitude x 5° longitude grid. In this process, the median values at each grid point for each month were calculated. However, to maximise both spatial and temporal coverage, the availability of only one grid value was accepted, while in the presence of only two grid values the mean of their values was taken. In areas of serious mismatching of various gridded data values, observed station data were used to resolve such problems where available. This technique was drawn from Parker *et al.* (1994). Further developments used the above approach with all available historical gridded data sets (Table 2) over the globe, and led to the creation of GMSLP1.1. Although encouraging, GMSLP1.1 had problems at the boundaries of blended sets over the Central Asian region, and with a lack of extreme MSLP values at extratropical latitudes in both hemispheres. Comprehensive discussion of these problems can be found in Basnett and Parker (1996). A revised blend, GMSLP2.1, with the area of the CSIRO DAR gridded imput restricted, showed some improvement, especially in the Northern Hemisphere.

Development of the GMSLP2.1 version involved blending the existing gridded MSLP sets with the land, island and oceanic data described in the previous three sections. To achieve best data coverage, quality and quantity, reliable MSLP 'background' fields (GMSLP2.0) were first constructed. To do this, GMSLP1.2 north of 20°N, and all of the individual data sets employed in it further south, were converted to monthly MSLP anomalies relative to each of their 1984–1989 monthly MSLP means (but 1951–1985 and 1957–1985 for the UEA Southern Hemisphere and Antarctic which had no later data). The period 1984–1989 was chosen because it is covered by the best global monthly MSLP climatology available. This climatology was constructed by using the UKMO operational analysis data covering 20°N to 90°S, together with the GMSLP1.2 data climatology from 90°N to 25°N. Blending together each of the individual anomaly fields added to the 1984–1989 monthly MSLP climatology yielded first-guess, incomplete (Antarctica was often missing as shown by Table 2) background fields. To enhance these blended data even further, they were interpolated onto a 1° x 1° global grid, spatially smoothed and then subjected to a Laplacian interpolation with respect to the 1984–1989 climatology. The result was the GMSLP2.0 background fields.

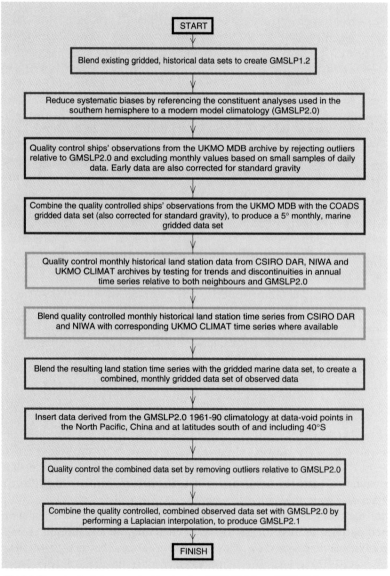

Fig 34: Flow chart of steps required to develop GMSLP2.1 from GMSLP1.2

Table 2: Gridded analyses contributing to GMSLP

DATA SOURCE	AREA COVERED	PERIOD	RESOLUTION
UKMO Historical (Basnett and Parker, 1996)	15°N–90°N	1873–1994	Regular 5°lat. x 10°long
UKMO Operational (Basnett and Parker, 1996)	Globe	1985–1994	Regular 2.25°lat. x 3.75°long
NCAR (Basnett and Parker, 1996)	15°N–90°N	1899–1994	Regular 5° lat. x long
SIO (Barnett and Jones, 1992)	42.5°S–72.5°N	1951–1993	Regular 5° lat. x 10° long
CSIRO DAR (Allan, 1993)	60°S–60°N	1871–1989	Regular 10° lat. x long
CRU UEA (Jones,1991)	15°S–60°S	1951–1985	Staggered 5° lat. x 10° long
CRU UEA (Jones and Wigley, 1988)	60°S–75°S	1957–1985	Staggered 5° lat. x 10° long

However, additional modifications were necessary as the earlier GMSLP2.0 background fields were found to have persistent biases in the high latitudes of the Southern Ocean with respect to the 1961–1990 climatological average. As a consequence, the 1961–1990 GMSLP2.0 climatological average values were inserted at grid points south of and including 40°S, with no data within a distance equivalent to 10° latitude. It was also necessary to insert climatology data prior to 1950 in the North Pacific (the region 60°N–20°N, 130°E–130°W), to remove a strong positive bias, and China (the regions 50°N–30°N, 70°E–100°E and 30°N–10°N, 90°E–105°E;), to remove negative and positive biases, respectively. As with the higher latitude Southern Hemisphere, climatological data were inserted only at grid points at more than 10° latitude equivalent distance from marine or land station data. Some anomalous Russian station data (to the north of Japan) that displayed a strong positive bias in their anomalies (relative to neighbours) were removed from the data analysis. These 'bogus' data insertions act like stations and work to prevent the above biases from entering GMSLP2.1.

At this stage, an interactive checking scheme was applied to further ensure the quality of GMSLP2.1. This consisted of a routine in which all the monthly data from both land and ocean were mapped on a screen as 5° x 5° gridbox anomalies from the GMSLP2.0 background field in a colour code ranging through the spectrum according to the sign and magnitude of each anomaly. Gridded values could then be printed out for investigation, or deleted, by clicking on them and on a menu with a mouse. Each month's data were then saved with any deleted values omitted. Unlikely gridded anomalies were readily detected and checked or deleted. Finally, the GMSLP2.0 background field was used in a Laplacian interpolation with the accepted gridded anomalies to create GMSLP2.1 on a 5° x 5° degree global grid.

Calculation and Filtering of Niño 3, Niño 4 and SOI Series

The 1871–1994 SST anomalies for the Niño 3 and Niño 4 regions used to construct Figure 31, are with respect to the 1961–1990 climatology of Parker *et al.* (1995b) and were provided by Rayner (1996, pers. comm.). Up to 1948, the anomalies are based on a new experimental version of the GISST data set. This version was created by using the available SST data for each month, with centred 3-month averaging in data-sparse years, to estimate the coefficients of predetermined seasonally invariant ocean-basin eigenvector patterns defined for 1950–1994 plus one predetermined seasonally invariant global low-pass filtered pattern defined for 1901–1990. Therefore, an assumption is implied that the characteristic patterns have remained stable throughout the record. From 1949 to 1994, the GISST2.2 data set (Rayner *et al.,* 1995) is used. This is also an eigenvector-based reconstruction up to 1981, but the ocean-basin eigenvectors were defined for 1951–1990. For 1982–1994, a blend of *in situ* and satellite-based Advanced Very High Resolution Radiometer (AVHRR) SST data is used in GISST2.2. Further details are found in Rayner *et al.* (1995).

The 1876–1996 SOI signal and noise series used to construct Figure 31 are also calculated with respect to the 1961–1990 mean statistics. The SOI signal series is formulated in the manner attributed to Troup (Pittock, 1974), with monthly index values calculated by taking the difference between each of the monthly MSLP anomalies at Tahiti and Darwin and normalising each of the differences by dividing them by the standard deviation of the monthly differences in MSLP anomalies at Tahiti and Darwin. The value obtained in this manner is then multiplied by 10 to create the monthly SOI. This approach differs from other Tahiti minus Darwin MSLP SOI signal series calculated by Trenberth (1984) and by the then Climate Analysis Center in Washington (*Climate Analysis Center,* 1986). Trenberth (1984) removes the annual cycle from the MSLP data and then normalises each of the monthly Tahiti and Darwin MSLP anomalies, by dividing them by the annual mean standard deviations at each location. The difference of the standardised monthly Tahiti minus Darwin MSLP anomalies is used as the SOI signal series. A measure of the noise in the SOI is also suggested by Trenberth (1984), and takes the form of the normalised Tahiti plus Darwin MSLP anomalies. It is from a calculation of the noise in the SOI back to the 1930s that Trenberth (1984) rejects the early part of that record. This is reiterated in Trenberth and Hoar (1996) with regard to the full historical SOI record, with the Tahiti data seen as most suspect prior to the 1930s. As a consequence, both the SOI signal and noise series are calculated in Figure 31. This shows that the SOI is somewhat noisier prior to the 1940s, but much of this noise is due to a few outliers in the series beyond values of + or – 30. It is felt that this does not warrant the rejection of the SOI signal prior to the 1940s, but rather that such data must be used with care. Further work on the Tahiti MSLP record may act to remove some of this problem.

The now named Climate Prediction Centre (CPC) calculate an SOI in the manner of Trenberth (1984), but they go a step further and do a final normalisation by dividing by the standard deviation of the difference between the standardised monthly Tahiti minus Darwin MSLP anomalies. Comparisons of each of the above SOIs for the period 1935–1986 can be found in the *Climate Diagnostics Bulletin* of March 1986. Early assessments of SOIs are detailed in Chen (1982), and adjustments for changing observing times at Darwin are addressed in Parker (1983). Within the last ten years, Ropelewski and Jones (1987) and Rebert and Donguy (1988) have recovered Tahitian MSLP data from the French Archives, and Allan *et al.* (1991) discovered usable Darwin MSLP records back to the 1870s in the Northern Territory Archives in Darwin. These efforts have permitted the extention of the Tahiti minus Darwin SOI back to 1876. Searches continue for even earlier Tahitian MSLP data, especially given the existence of British Board of Trade (1861) records of monthly Tahiti MSLP from 1855 to 1860.

However, gaps in monthly MSLP data for Tahiti and Darwin result in all of the above SOIs having periods of missing values. Jones (1988) made the first attempt to construct gap-filled MSLP series at Tahiti and Darwin and to calculate a continuous monthly SOI series using the CPC method. This approach infilled missing Tahiti MSLP data with values obtained from regression relationships with Apia in Samoa, and did similarly for Darwin using regression relationships with Jakarta in Indonesia. As this still left early gaps in the Tahitian record, infills were made using relationships with Santiago, Chile. More recently, Banglin (1994) published a gap-filled SOI constructed by performing a singular spectrum analysis (SSA) on the then Climate Analysis Centre's (CAC) monthly SOI. The results of the SSA were then used to produce a 'fully data-adaptive filter', which was applied iteratively to the CAC SOI series to fill the missing monthly values. For this book, a new gap-filled Tahiti-Darwin SOI which shows extremely high correlation with the Jones (1988) index is used (Allan *et al.*, 1996). The major difference from the Jones (1988) SOI is that the Allan *et al.* (1996) gap-filled SOI uses Tahiti and Darwin monthly MSLP series with missing values estimated by Young (1993) using several regression techniques involving a global set of station time series. In addition, the Tahitian MSLP series has been improved through analyses of data from the CPC, Jones (1988) and WWR sources, so that it has a better signal to noise ratio in the early part of the record.

The raw monthly Niño 3, Niño 4 and SOI time series are all filtered in the time frames where most of the ENSO signal is observed (e.g. 18–35 and 32–88 months). This involved each of the raw series being subjected to two passes of a recursive Butterworth (bandpass) filter (Stearns and Hush, 1990) with high and low cutoff frequencies at 18 and 35 months in the first pass of the filter through the raw data, and at 32 and 88 months in the second pass.

Calculation of Correlation and Filtered Fields of MSLP and SST

The GMSLP2.1 and GISST data fields were examined first in terms of seasonal correlation patterns over the period of record, and in various epochs when major climatic fluctuations occurred, using monthly anomalies from the 1961–1990 monthly means. Only field significant areas of positive and negative point correlations at the 95% confidence level are shown in this part of the analysis. For the bulk of the figures shown in the book, normalised anomaly fields of GMSLP2.1 and GISST were also filtered in the time frames where most of the ENSO signal is observed (e.g. 18–35 and 32–88 months). This involved each of the raw monthly normalised anomaly fields of MSLP and SST being subjected to two passes of a recursive Butterworth (bandpass) filter (Stearns and Hush, 1990) with high and low cutoff frequencies at 18 and 35 months in the first pass of the filter through the raw data, and at 32 and 88 months in the second pass. These filtered results are all shown in the book CD-ROM, but, due to constraints of space, only the major results of the 32–88 month filtering of the MSLP and SST fields are presented in the hard copy form of the book.

Data Sources for ENSO Phase Composites and Impact Maps

The ENSO phase composites of physical responses of the ocean–atmosphere system, hydro-climatological responses and air temperature patterns, and the November 1994 and 1995 sea level distributions were constructed using data from the following sources. Figure 35 was developed by the authors from existing schematics of ENSO and published in an early form in O'Neill (1995). In Figure 36, anomalous rainfall (dry and wet) distributions are adapted from Ropelewski and Halpert (1987, 1989), with additional data added to the compilation from Aceituno (1988), Rutllant and Fuenzalida (1991), Martin et al. (1993) and Whetton and Rutherfurd (1994). River discharge data are from Whetton et al. (1990) with additional information on the Amazon (Brazil) from Richey et al. (1989), the Orange River (South Africa) from the South African Journal of Science (1987), the Parana River (Argentina) from Aceituno (1988) and Mechoso and Iribarren (1992), and the River Murray (Australia) from Close (1990). Information on frost impacts is from Allen (1989), Allen et al. (1989), and Brookfield (1989) and for fire and flooded lake from Allan (1985, 1991). The timing of the rainfall anomalies during composited El Niño and La Niña phases is shown by the month indicated, where the 0 and + symbols in brackets following indicate that the printed month is in the year of the event or that the printed month is in the year after the event, respectively. Anomalous temperature patterns in Figure 37 are reproduced from the distributions in Halpert and Ropelewski (1992). Figure 38 is taken from regular sea level products produced by the Center for Space Research, University of Texas, Austin.

A wide range of material was used to construct the series of individual El Niño and La Niña phase impact maps in Section 7. Global precipitation and air temperature percentile series from ENSO-sensitive regions were taken from Ropelewski and Halpert (1986, 1987, 1989) and Halpert and Ropelewski (1992), while major river discharge records for the Nile, Krishna and Senegal rivers were drawn from Whetton et al. (1990) and Quinn (1992, 1993). Specific regional information came from the following sources: South American river discharge for the Amazon and Parana rivers (Aceituno, 1988; Richey et al., 1989; Mechoso and Iribarren, 1992), northeastern Brazil, Peru and Surinam precipitation (Berlage, 1966; Jones, 1995, pers. comm.), Atlantic hurricane frequency (Gray and Sheaffer, 1991; Landsea et al., 1992), CEP rainfall (Wright, 1989), Indian and Sri Lankan monsoon rainfall (Pant et al., 1988; Suppiah, 1989; Parthasarathy et al., 1991, 1992; Vijayakumar and Kulkarni,

1995), Chinese rainfall (Tu, 1936, 1937a, b; Wang and Li, 1990; Whetton and Rutherfurd, 1994), Indonesian drought (Quinn *et al.,* 1978), Lake Eyre flooding (Allan, 1985), Australian Murray-Darling basin discharge (Close, 1990; Allan *et al.,* 1996), southern African rainfall (Nicholson and Entekhabi, 1986; South African Journal of Science, 1987; Vogel, 1989; Lindesay and Vogel, 1990; Hulme, 1992; Nicholson and Palao, 1993; Nicholson, 1993) and East African rainfall (Nicholson and Entekhabi, 1986; Hulme, 1992; Hutchinson, 1992; Nicholson, 1993).

3. Oceanic, Atmospheric and Hydrological Variable Responses to ENSO

ENSO and its teleconnections generate impacts in both marine and terrestrial environments over a large part of the globe. The following section deals with the evidence accumulated and published to date for physical impacts and manifestations resulting from ENSO and teleconnection activity with near-global coverage. As a consequence, relationships with features with incomplete coverage, such as tropical cyclones/hurricanes, ocean currents, and features in the high latitudes of the Arctic and Antarctic, are not detailed. However, Atlantic Ocean tropical cyclone numbers during individual ENSO phases are shown in the historical ENSO phase impact maps in Section 7. Overviews of tropical cyclone/hurricane relationships can be found in Gray and Sheaffer (1991), various ocean current responses in Godfrey *et al.* (1995) and WCRP (1995) and links to the Arctic and Antarctic regions in Carleton (1989), Smith and Stearns (1993), Gloersen (1995) and Simmonds and Jacka (1995).

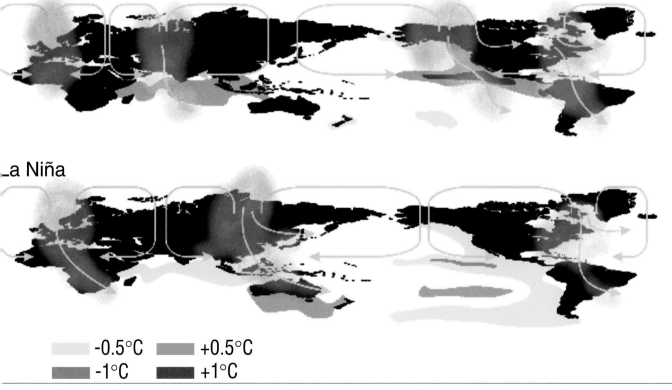

Figure 35: Schematics of global physical ocean-atmosphere interactions in the boreal winter (austral summer) seasons during strong El Niño and La Niña phases (after O'Neill, 1995).

Schematicised portrayals of the most documented impacts and responses are summarised in Figures 35–38, which show composites based on data gleaned from a number of separate and pronounced El Niño and La Niña events. Figure 35 illustrates the physical responses of the ocean–atmosphere system for composites of strong events at the average peak of ENSO phases during the boreal winter (austral summer). The global hydro-climatological and air temperature patterns in Figures 36 and 37 show climate responses during the peak impacts of composite El Niño and La Niña phases. Global sea level patterns during ENSO phases in Figure 38 are shown for two months of very recent extremes, as the satellite data archives of this variable providing global coverage extend back only until late 1992. Data from remotely sensed platforms with tight ground truthing from standard observational networks are already providing a new generation of detailed spatial and temporal information on ENSO and its impacts (Tsimplis and Woodworth, 1994).

SST, Convective Regions and Atmospheric Teleconnections

The near-global extent of the physical influences of the ENSO phenomenon in both phases is shown schematically for composites of strong events during the boreal winter (austral summer) in Figure 35 The core region for ocean–atmosphere interactions is located in the Indo–Pacific domain, which contains the

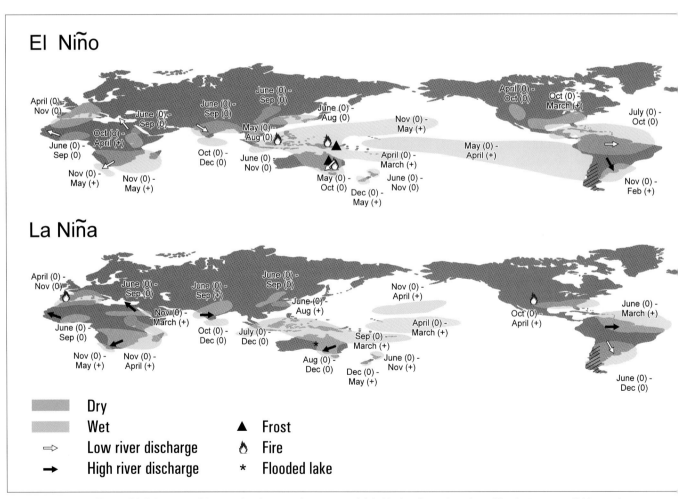

Figure 36: Schematics of the spatial and temporal responses of global hydro-climatological variables during strong El Niño and La Niña phases (From various sources detailed in the text).

most pronounced SST and convective regions across the Pacific Ocean and related responses in the Indian Ocean. This diagram synthesises much of the current knowledge about the physical structure of ENSO phases that was detailed in 'ENSO Structure, Nature and Physical Characteristics' p. 22.

During strong El Niño phases, broadscale convective regions over the 'maritime continent'/Australasia, equatorial Africa and northern South America are displaced from their climatological locations. During this phase, anomalous convective regions develop over the CEP, central-western equatorial Indian Ocean, off/near the Atlantic equatorial coast of Africa and northwestern South America. This redistribution in the major areas of tropical convection is a response to variations in atmospheric circulation and wind fields. There is a low level convergence of wind flow towards the convective regions that is particularly evident through anomalous winds in the trade wind regimes, and a compensatory divergence of upper atmospheric winds away from the displaced convective regions. As a consequence, there are significant changes in global teleconnection patterns, as the displaced tropical sources of the teleconnections now link to midlatitude locations that differ from climatology. The preferred paths for these teleconnections are given by regions of enhanced meridional circulations, some of which are associated with tropical-to-temperate latitude oriented cloud bands, emanating from the displaced convective regions around the tropics. Such variations in teleconnections carrying the outflow of mass and energy from the tropics to higher latitudes can lead to anomalous jet stream and long atmospheric wave patterns, and thus change weather patterns in either or

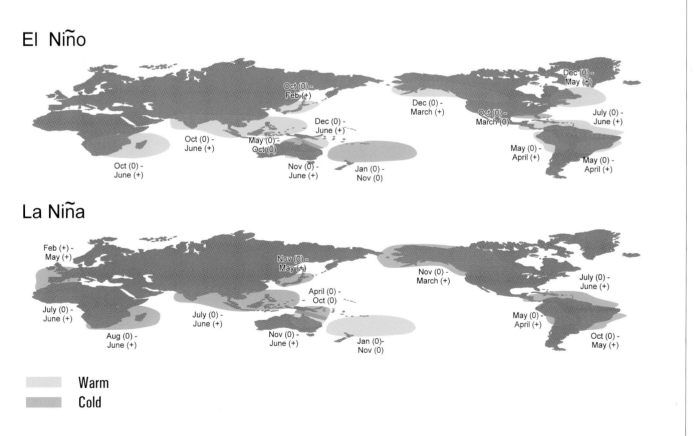

Figure 37: Schematics of the spatial and temporal responses of global air temperature during strong El Niño and La Niña phases (adapted from Halpert and Ropelewski, 1992).

both hemispheres. As ENSO phases reflect the interactions of a closely coupled ocean–atmosphere system, changes also occur in the ocean. The surface manifestation of SSTs is affected, and anomalously warm SST regions appear in the Pacific and Indian oceans. The warming is most pronounced over the equatorial Pacific Ocean, with the dominant SST maxima occurring over the CEP to EEP. The Pacific region of warmer than usual SST is linked to the major global area of enhanced tropical convection during this phase of ENSO. Warming is also evident over the equatorial Indian Ocean, and is closely allied with the formation of the convective region over the central-western portions of that basin. However, in the eastern equatorial Pacific convection is absent, and short wave forcing under reduced cloudiness raises the regional SST (Kiladis and Diaz, 1989). As noted earlier ('ENSO Structure, Nature and Physical Characteristics' p. 22), there are also changes in most physical aspects of the oceans, with displacements and modulations of ocean currents, thermocline patterns, upwelling regions and sea level.

During strong La Niña events almost the opposite pattern of oceanic and atmospheric variables occurs. Deep convective regions are concentrated over the 'maritime continent'/Australasia, equatorial Africa and northeastern South America with regions of enhanced meridional circulations, some of which are associated with tropical to temperate latitude oriented cloud bands, extending into higher latitudes. As with the El Niño phase, such regions of enhanced meridional circulations and cloud bands mark the teleconnections that carry mass and energy out of the tropics and provide a modulation to higher latitude long wave and jet stream patterns. However, during the La Niña phase, they are in markedly different locations and provide a source for changes in midlatitude weather. Atmospheric circulation and wind fields are then such that they tend to converge on the more continentally displaced convective regions in the lower levels of the atmosphere and diverge away from them in a compensatory manner at upper atmospheric levels. Over the oceans, the SST pattern across the Indo–Pacific region is also the reverse of that observed during the El Niño phase. Suppressed convection over the tropical regions of both ocean basins is allied with areas of anomalously cold SST, with again the most pronounced anomalies in the CEP to EEP. As with the El Niño phase, the La Niña situation results in significant fluctuations in ocean currents, thermocline patterns, upwelling regions and sea level.

Rainfall (Drought and Floods), River Discharge, Snow, Frost and Fire

The global pattern of hydro-climatological responses to ENSO phases from published studies is shown in Figure 36. These impacts are a direct result of the physical alterations in the ocean–atmosphere system shown in Figure 35. As with the physical variations in the ocean–atmosphere system during ENSO phases, hydroclimatological impacts display contrasting patterns in El Niño and La Niña events (Figure 36). In fact, many of the rainfall and river discharge anomaly patterns are almost mirror images of one another. In general, regions with strongest impacts and most stable temporal responses to ENSO tend to be found in countries bordering the Indo–Pacific basin. It should be noted that the major ENSO impacts are shown in Figure 36, and that in some cases the ENSO-sensitive regions shown display the opposite characteristics at other seasons during either ENSO phase.

The El Niño phase is dominated by rainfall deficiencies (dry or drought conditions) over southern and Sahelian Africa, the Indian subcontinent, northern China, Australasia, northern South America and the Caribbean. Dry or drought conditions also lead to low river discharge in the Nile (Egypt), Senegal (Senegal), Orange (South Africa), Krishna (India), Murray–Darling (Australia) and Amazon (Brazil) river systems. In general, rainfall deficiencies vary in their timing during the El Niño phase, with indications of a suppression of monsoonal activity or tropical rainfall-producing influences during the southern African and Indian summer seasons. Over Australasia, the major influence on rainfall patterns in this phase occurs during the austral winter to early summer period. Reductions in rainfall-producing systems promote dryness and clearer skies, which are important factors that favour conditions for wildfires and a tendency for more frosts in winter. During this phase, major wildfire problems occur in Indonesia, New Guinea and southeastern Australia, while increased influence of frosts is evident over New Guinea and the eastern half of the Australian continent.

The largest expanse of above-average rainfall in the El Niño phase is seen over the Pacific Dry Zone (PDZ), and is a result of the major redistribution of tropical convective regions shown in Figure 35. An indication of increased rainfall over the equatorial Indian Ocean is suggested by the precipitation maximum in the Sri Lankan area. Other significant regions of rainfall surplus (wet/flood conditions) are found over much of western Europe, parts of eastern Africa, the Western Cape of South Africa, Vietnam, central China to southern Japan, the South Island of New Zealand, the Great Basin and Gulf regions of the United States and northern Mexico, central Chile and southeastern Argentina. In these rainfall surplus areas, the principal enhancement of river flow occurs with the Parana river system in Argentina, while some flooding episodes occur in the Great Salt Lake in the United States during pronounced El Niño phases.

The La Niña phase shows almost the opposite configuration of responses to the El Niño. Enhanced rainfall patterns tend to dominate the major continental masses bordering the Indo–Pacific region, in particular southern and Sahelian Africa, the Indian subcontinent, northern China, Australasia, and northeastern South America. During this phase, the SPCZ is most active in the region extending from the 'maritime continent' into the southwestern Pacific. As with the El Niño phase, rainfall patterns in New Zealand, although complex, are often out-of-phase between the North and South islands. River discharge is high in the major river systems of the Nile, Senegal, Orange, Krishna, Murray–Darling and Amazon. Monsoonal/tropical-linked summer rainfall is enhanced over southern Africa and India, and prominent cloud band influences lead to rainfall surplus regions over southeastern Australia and southern Africa. During this phase, there is also a tendency for inland lake systems, even in arid to semi-arid regions, to experience major pluvial episodes. Extensive flooding occurs in the Lake Eyre basin in central Australia during pronounced La Niña events.

Suppression of rainfall is most evident over the equatorial Pacific and Indian oceans, with continental regions of southeastern Argentina, central Chile, central China, eastern Africa, the Western Cape of South Africa and western Europe now showing dry or drought conditions. The most notable river system where discharge is now low as a consequence of these rainfall changes is the Parana River in southeastern Argentina. The predominance of dry conditions in the south–southwestern United States and western Europe enhances the propensity for wildfires in both regions.

In general, the majority of the hydroclimatological responses shown in both phases result from zonal teleconnection changes in the tropical–subtropical belts of the globe. Rainfall changes over western Europe, southern Africa, China and Japan, the USA, southeastern Australia and Argentina tend to be influenced by the enhancement or suppression of meridional circulations and cloud bands linked to midlatitude teleconnections. The extent of snow in the southern Andes in South America is also a consequence of changes in mid to high latitude circulation patterns.

Air Temperature

Fluctuations in near-global air temperature patterns reported in the literature are depicted in Figure 37. As with the physical and hydroclimatological impacts (Figures 35 and 36), both phases of ENSO cause broadly contrasting air temperature anomalies.

During the El Niño phase, the distribution of air temperature anomalies in Figure 36 is dominated by warmer than normal conditions. Such increases in air temperature are found over southern Africa, India and south-east Asia, southeastern Australia, Japan, southern Alaska and northwestern Canada, eastern Canada, the Caribbean and parts of northern South America. Anomalously lower air temperatures are restricted to the Gulf states of the United States, northeastern Australia and New Guinea, and across the central southwestern Pacific Ocean. Lower air temperatures over northeastern Australia and New Guinea are also associated with a greater propensity for frosty conditions during this phase.

65

Air temperature anomalies during the La Niña phase are predominantly of colder than normal conditions. Such impacts are prevalent over western Europe, northwestern and Sahelian Africa, southern Africa, India and south-east Asia, Japan, Alaska and northwestern Canada, the Caribbean and parts of northern South America. Again, as with the El Niño phase, the opposite air temperature anomalies are restricted to the regions of northeastern Australia and the central southwestern Pacific, which experience warmer conditions. It should be noted that the major ENSO impacts are shown in

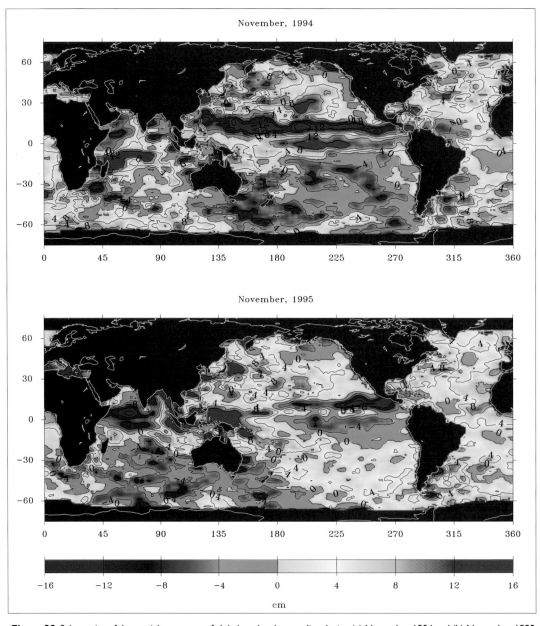

Figure 38: Schematics of the spatial responses of global sea level anomalies during (a) November 1994 and (b) November 1995 (University of Texas at Austin, Center for Space Research pers. comm, 1996).

Figure 37, and that in some cases the ENSO-sensitive regions shown display the opposite characteristics at other seasons during either ENSO phase.

Sea Level

Unlike the composites for other ENSO phase impacts (Figures 35–37), the data base available to construct global sea level responses to ENSO is limited to only the last three years or so. However, in conjunction with records from various station time series and Pacific patterns of sea level back over the last 20 years, it is possible to refer to the sea level anomalies in November 1994 and 1995 as illustrating El Niño and La Niña influences, respectively (Figure 38).

Current understanding of the physical interactions and processes underlying ENSO phases, detailed in 'Ocean–Atmosphere Interaction' (p. 23), emphasize the response of oceanic sea level to the phenomenon. In the El Niño phase, reflected in the November 1994 data, highest coherent positive sea-level anomalies occur in the CEP to EEP region, in the western equatorial Indian Ocean and over much of the mid to high latitude North Pacific Ocean. Lowest negative sea-level anomalies are found in the eastern equatorial Indian Ocean, in bands across the immediate off-equatorial zones in both hemispheres (especially the Northern Hemisphere), in the equatorial Atlantic Ocean and at higher latitudes in the southern Indian and Pacific and northern Atlantic oceans.

The La Niña phase presents basically the opposite sea-level pattern to that seen during the El Niño composite. Responses to equatorial wave dynamics generated by strong trade-wind fields across most of both ocean basins, take the form of a cohesive positive sea-level anomaly covering almost the whole eastern Indian Ocean and extending to the WEP. Coherent positive anomalies also extend from the 'maritime continent' in off equatorial bands in both hemispheres. This pattern is balanced by low negative sea-level anomalies in the western Indian Ocean, and in a thin tongue across the CEP to EEP.

Both sea-level distributions in Figure 38 are thus part of the broad pattern of physical relationships and manifestations detailed in Figure 35. With the increasing development and use of satellite-derived measurements of oceanic and atmospheric variables, the potential to provide a more complete picture of ENSO dynamics, teleconnections and impacts is heightened.

4. ENSO Teleconnection Patterns in Historical Records

Value of Simple ENSO Indices (i.e. SOI and Niño 3 and 4)

As has been documented previously, simple indices of ENSO have evolved from studies of atmospheric and/or oceanic aspects of the phenomenon over the last 100 years or more. They carry varying degrees of information about the state of ENSO and the climate system, and are particularly revealing when ENSO phases are strong and teleconnection patterns are extensive. However, like any simple measure they have limitations and provide long-range forecasters with somewhat of a 'blunt instrument'. By their very nature, simple indices encapsulate a dynamic and spatially extensive phenomenon in a single value. The bulk of statistical prediction research has focused on the SOI, with more emphasis on SST indices in the Niño 3 and 4 regions in recent years. As noted in 'Recent Statistical Forecasting' p. 29, the availability of other variables measuring the state of the ocean–atmosphere system holds the promise for the development of other ENSO indices. As was done with the regression formulae of Sir Gilbert Walker, attempts have been made to use EOFs to construct multivariable indices of the phenomenon (Kousky and Kayano, 1993). Thus, simple indices have their place, but care must be taken when using them and they must be seen in the light of the longer term fluctuations in ENSO and the climate system already noted (see pp.35–42).

Teleconnection Stability in Time and Space

Much of the current understanding and attempts to develop robust predictability of the ENSO phenomenon have focused on analyses of data collected over the last 30–40 years. Contemporary concerns about the longer term stability of the ENSO phenomenon and its teleconnection patterns were raised in the literature not long after the major 1982–83 El Niño event (McBride and Nicholls, 1983; Ramage, 1983; Pittock, 1984; Allan, 1985). The last 10 years has seen mounting evidence for decadal–multidecadal fluctuations in ENSO and the climate system (see 'Climatic Variability' p. 34 and overviews in Glantz et al., 1991; Diaz and Markgraf, 1992; Allan, 1993; Karoly et al., 1996). The problem has been made acute by evidence for a climatic regime shift over the Pacific Ocean since the mid 1970s (Graham, 1994; Latif and Barnett, 1994a, b; Miller et al., 1994a, b; Nitta and Kachi, 1994; Trenberth and Hurrell, 1994), over the middle to high latitudes of the southern hemisphere (van Loon et al., 1993; Hurrell and van Loon, 1994), in the middle to high latitude North Pacific (Chen et al., 1992; Jacobs et al., 1994; Tanimoto et al., 1993; Dettinger and Cayan, 1995; Lagerloef, 1995), with regard to the North Atlantic and the NAO (Deser and Blackmon, 1993; Hurrell, 1995), with regard to Northern Hemisphere circulation (Nitta and Yamada, 1989; Nitta, 1992; Burnett, 1993; Graham et al., 1994; Lejenas, 1995) and, more recently, following concerns about the nature of the long El Niño sequence or climatic situation from mid 1990 to mid 1995 (Latif et al., 1995). However, a thorough evaluation of the longer term stability of ENSO and teleconnection patterns requires a global examination of historical climatic patterns.

Historically, the above concerns were indirectly being broached as early as the turn of the century in a paper examining the extensive drought conditions prevailing in many parts of the British Empire between 1892 and 1902 (Eliot, 1904). Later studies such as Foley (1957), Allen (1989), Allen et al.

(1989), Brookfield (1989), Lindesay and Vogel (1990), Whetton *et al.* (1990), Parthasarathy *et al.* (1991, 1992), Vijayakumar and Kulkarni (1995) and Suppiah (1996) have provided further evidence for weaker ENSO phases and a period of protracted droughts in the Australasian, African and Indian regions around the turn of the century. The significance of these conditions in Australia is highlighted by the above period being known specifically as the 'Federation Drought'. Physical changes in climate at this time have centred on evidence for a shift from meridional to zonal monsoonal circulation over India (Fu and Fletcher, 1988; Parthasarathy *et al.*, 1991), and on evidence for a major cooling of SSTs over the globe (particularly the Northern Hemisphere) (Parker *et al.*, 1994). These changes have been seen to lead to significant spatial and temporal variations in statistically significant correlation patterns between ENSO and rainfall in India and Sri Lanka (Parthasarathy *et al.*, 1991; Vijayakumar and Kulkarni, 1995; Suppiah, 1996).

However, most attention to longer term fluctuations in ENSO teleconnections and the climate system this century has focused on the period centred around the 1920s to 1940s. Studies such as Chang and Yasunari (1982), Bigg *et al.* (1987), Elliott and Angell (1987, 1988), Trenberth and Shea (1987), Whysall *et al.* (1987), Cooper and Whysall (1989), Pan and Oort (1990), Enfield and Cid (1991), Parthasarathy *et al.* (1991, 1992), Bigg and Inoue (1992), Zhang and Casey (1992), Allan (1993), Ward *et al.* (1994), Gu and Philander (1995), Vijayakumar and Kulkarni (1995), Wang and Ropelewski (1995) and Karoly *et al.* (1996) have, to varying degrees, noted a significant change in climatic variables related to ENSO and its teleconnection patterns in the above period. Although they question the reliability of COADS SST data in some periods, Pan and Oort (1990) have performed decadal correlations between EEP and global SSTs that suggest that the earliest and latest data periods show more robust correlation structure than the intervening time. However, the most notable findings have been the marked weakening in the magnitude and spatial extent of correlation patterns between ENSO variables and rainfall (Parker *et al.*, 1994a, b), a reduction in the zonal SST gradient and trade-wind flow across the Pacific Ocean, and a marked trend towards warmer global SSTs since the 1920s–1940s. Parthasarathy *et al.* (1991) have also seen that this period marks the peak in more zonal circulation associated with the Indian monsoon, prior to its tendency to have been more meridional up until the 1960s. More general literature covering the period details the 'Dust Bowl' droughts in the United States and long runs of drought in southern Africa and Australia. Thus, decadal to multidecadal periods of drought appear to have occurred during epochs of reduced ENSO activity.

The periods around the turn of the century and in the 1920s–1940s can be put into perspective by analyses of filtered SOI and SST indices in the QB and LF bands over the historical instrumental period (see 'ENSO Components' p. 37). This approach suggests that ENSO and the climate system have undergone decadal–multidecadal 'shifts', with climatic relationships and ENSO patterns fluctuating between robust and weak conditions. When analyses of the 'phase of regular cycles' or tendencies in the SOI (Stone and Auliciems, 1992; Stone and McKeon, 1992; Zhang and Casey, 1992; Stone *et al.*, 1996) are extended over the historical instrumental period, they show variations between periods dominated by strong ENSO-phase activity and those when conditions are close to climatology. However, some indication is needed of the extent to which such fluctuations in indices of ENSO are a stable measure of actual decadal–multidecadal variations in the phenomenon. It is observed that variations in Tahiti and Darwin are not always out of phase, and there are many instances when the majority of the SOI signal is being carried by one station over the other. On historical timeframes, the pattern of ENSO correlations appears to have varied spatially (Berlage, 1961; Allan, 1991, 1993), so that either or both nodes of the SO may not necessarily be reflected in measurements at the fixed stations of Tahiti and Darwin. A simple running correlation between raw Darwin and Tahiti MSLP over the period of historical record can be seen to reveal important changes in the out-of-phase relationship between them (the corner-stone of their use as an SOI). Using a 21-year running correlation, to provide a data sample long enough for statistical analysis, the pattern of correlations over time reveals that the austral (boreal) winter–spring (summer–autumn) period of maximum statistical correlations and strong autocorrelation in the 'Troup' SOI in the last 20–30 years, the basis for statistical seasonal forecasting, is not sustained over the remainder of the record (Figure 39). Using the normalised Tahiti plus Darwin MSLP as an index of the noise in the SO, Trenberth (1984) and Trenberth and Hoar (1996) have also questioned the representativeness of the early part of this SOI record as a measure of ENSO phases (see 'Calculation and filtering of Niño 3, Niño 4 and SOI series', p. 57). Other features of the SO remain relatively intact over time, especially the boreal (austral) spring (autumn) season when ENSO correlations are known to weaken dramatically to the extent that the period has been referred to as the 'predictability barrier' (Webster, 1995). It would seem that a wider spatial analysis of ENSO and its teleconnections is required to resolve the above indications of historical fluctuations in the phenomenon.

Spatial Patterns of Correlations of Darwin MSLP with Global MSLP in 21-year Epochs: 1879–1899, 1900–1920, 1921–1941, 1942–1962, 1963–1983

The analysis of filtered QB and LF components of ENSO in 'ENSO Components' p. 37 resolves periods or epochs of varying strength in the phenomenon. The broadest period of suppressed ENSO activity in the various indices in Figure 31, the 21-year epoch 1921–1941, provides a reference period long enough for statistical analysis. Epochs of the same length have been used to cover most of the MSLP compilation; these are 1879–1899, 1900–1920, 1921–1941, 1942–1962 and 1963–1983 periods. Epoch correlations have been calculated between seasonal MSLP at Darwin and all other points in the data set (seasons are DJF, MAM, JJA and SON). Only field significant areas of positive and negative point correlations at the 95% confidence level are shown. Other studies have used shorter sets of annual MSLP data correlated with Darwin or Jakarta (Trenberth and Shea, 1987; Berlage, 1966). Darwin is selected, as it has the most robust correlation patterns with near-global MSLP of all stations with long MSLP records in the Indo–Australasian node of the SOI. The southeastern Pacific node, usually represented by Tahiti, is not used in the same way because a study of two epoch correlation fields with the annual Jakarta MSLP (see Figures 29 and 30), shows a distinct shift in this node between the Tahitian region and the west coast of South America (Berlage, 1961).

Areas of statistically significant positive and negative correlations between Darwin and global MSLP in the MSLP epoch fields have fluctuated considerably over the period of historical record (Figure 40).

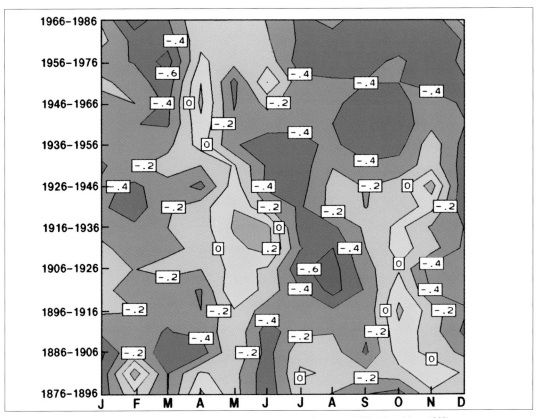

Figure 39: A 21-year running correlation between Darwin and Tahiti MSLP since 1876 (after Allan, 1993).

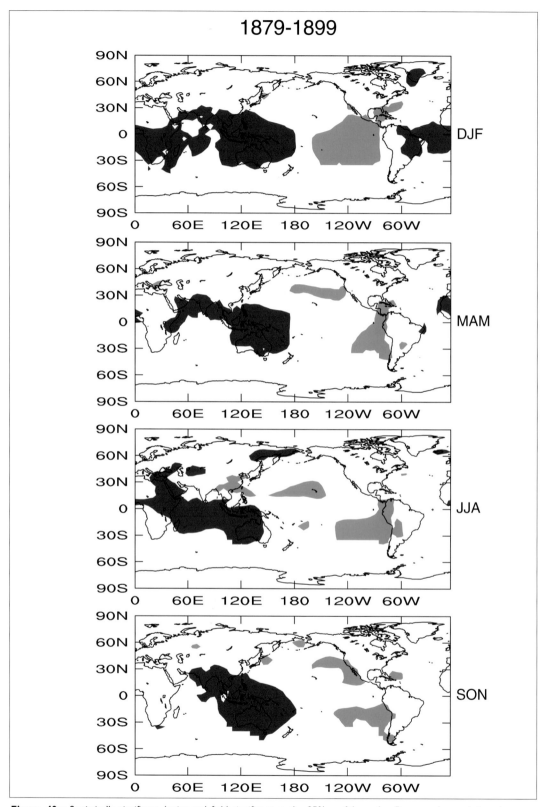

Figure 40a: Statistically significant (point and field significant at the 95% confidence level) seasonal correlations between Darwin and global MSLP during the 1879–1899 epoch. Red areas are significant positive correlations and green areas significant negative correlations.

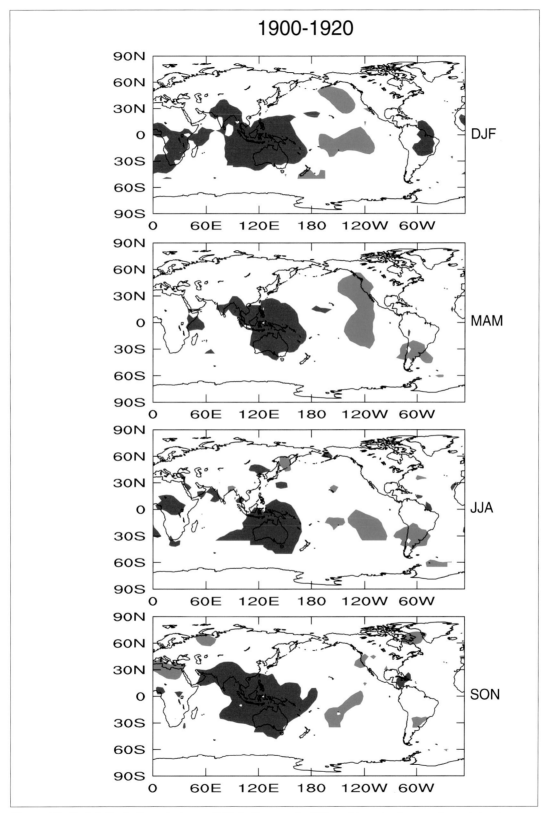

Figure 40b: Statistically significant (point and field significant at the 95% confidence level) seasonal correlations between Darwin and global MSLP during the 1900–1920 epoch. Red areas are significant positive correlations and green areas significant negative correlations.

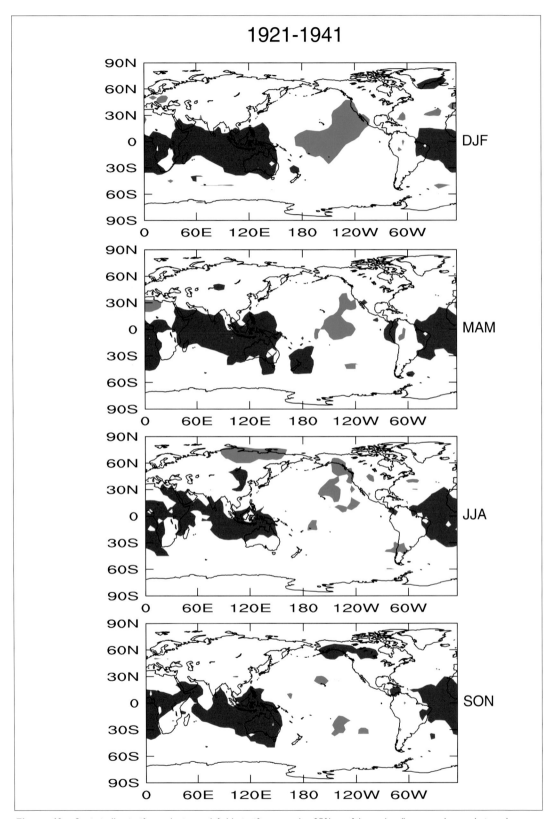

Figure 40c: Statistically significant (point and field significant at the 95% confidence level) seasonal correlations between Darwin and global MSLP during the 1921–1941 epoch. Red areas are significant positive correlations and green areas significant negative correlations.

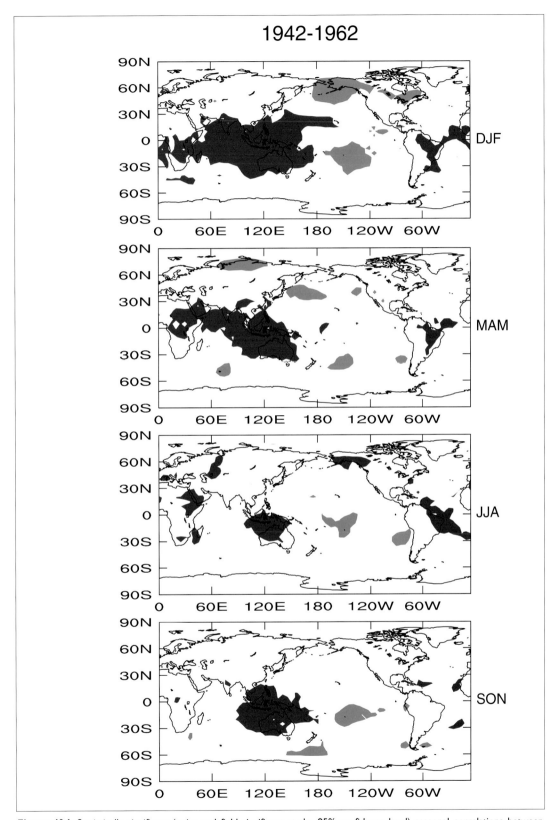

Figure 40d: Statistically significant (point and field significant at the 95% confidence level) seasonal correlations between Darwin and global MSLP during the 1942–1962 epoch. Red areas are significant positive correlations and green areas significant negative correlations.

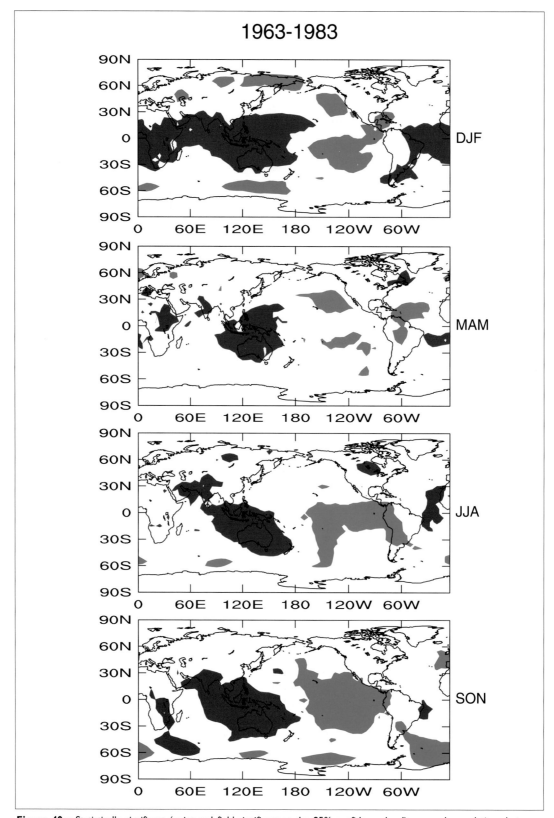

Figure 40e: Statistically significant (point and field significant at the 95% confidence level) seasonal correlations between Darwin and global MSLP during the 1963–1983 epoch. Red areas are significant positive correlations and green areas significant negative correlations.

Fluctuations in ENSO links across the Indian Ocean are seen in all seasons when contrasting epochs, but are particularly marked in MAM, JJA and SON. In general, the most robust patterns of positive correlations with regard to ENSO-sensitive regions (Figures 36 and 37) in the MSLP epoch fields occur during 1879–1899 and 1963–1983. There is a marked fluctuation in such relationships during several periods this century, particularly during JJA and SON in 1900–1920 and 1942–1962 when the area of positive MSLP correlations across the Indian Ocean contracts towards Australasia. The 1921–1941 epoch is notable for an enhanced zonal circulation associated with the Indian monsoon (Parthasarathy *et al.*, 1991), and is also the period with most weakness in the pattern of out-of-phase MSLP across the Indo–Pacific region that is characteristic of the operation of the SO. Weakest SOI correlations with southern African (Whetton *et al.*, 1990) and eastern Australian (McBride and Nicholls, 1983; Lough, 1991, 1992, 1993, 1994; Cordery and Opoku-Ankomah, 1994) rainfall occur throughout much of this time, and into the 1942–1962 epoch. Links to ENSO-sensitive regions over southern and Sahelian Africa, the Indian subcontinent and northeastern Brazil wax and wane when contrasting the MSLP epoch correlations.

The Pacific focus loses much of its integrity prior to the 1963–1983 epoch, although this may reflect a data problem in the earlier epochs. Nevertheless, the broad negative correlation pattern in the Pacific during SON is most coherent in 1879–1899 and 1963–1983, and the poorest in 1921–1941. The net effect of these changes, is that global teleconnection fields are weakened, fragmented and their spatial scales tend to be most contracted during the 1900–1920, 1921–1941 and 1942–1962 epochs. In contrast, teleconnection patterns are generally much stronger, more coherent and spatially extensive in both the 1879–1899 and 1963–1983 periods. Despite the short historical instrumental record, the findings from Figure 40 could be extended to suggest that the ENSO phenomenon is modulated by low frequency natural variability in the climate system.

Spatial Patterns of Correlations of EEP SST with Global SST in 21-year Epochs: 1879–1899, 1900–1920, 1921–1941, 1942–1962, 1963–1983

Following the work of Pan and Oort (1990) with decadal COADS SSTs, the GISST data are analysed seasonally in epochs in a similar way to MSLP in the above subsection. Seasonal mean SSTs in the eastern equatorial Pacific region (5°N–5°S, 150°W–120°W), which are most representative of SST fluctuations during ENSO phases, are correlated with global values during each epoch. As with the MSLP analysis, only the areas of positive and negative correlations that are both point and field significant at the 95% confidence level are shown.

Relationships with the eastern Pacific node of ENSO are best observed in the SST correlations, particularly with regard to the important zonal SST gradient across the Pacific basin noted in 'Ocean–Atmosphere Interaction' (p. 23). Figure 41 indicates a weaker gradient of correlation between the positive values in the CEP-EEP SST and the negative values in the WEP correlations during the JJA and SON seasons in the 1921–1941 epoch than is seen in the 1879–1899 and 1963–1983 periods. This also applies to the structure of off-equatorial negative SST correlations that are seen as integral to our current understanding of ENSO dynamics. This fits with the changes in ENSO persistence discussed above, and in 'Climatic Variability' (p. 34), and 'ENSO Components' (p. 37). In the Indian Ocean, fluctuations in the region of positive correlation with EEP SST during DJF and SON are most evident. In general, the 1963–1983 epoch shows the strongest correlation patterns. ENSO links to southern Africa in the MSLP analysis are not clear in the SST correlations. However, this may reflect the fact that precipitation in the region is linked closely to the location of SST anomalies in the central western off-equatorial Indian Ocean (Walker and Lindesay, 1989; Walker, 1990; Jury and Pathack, 1991, 1993; Lindesay and Jury, 1991; Jury *et al.*, 1994a, b, 1995; Rocha and Simmonds, 1996a, b).

Overall, there is considerable support for coherent fluctuations in major elements of the ocean–atmosphere system over the period of historical instrumental record. In particular, the 1921–1941 epoch stands out as one in which ENSO and the climate system appear to have been functioning in a less robust manner than in recent or earlier periods over the last 120 or so years. Such longer term variations have significant implications for attempts to forecast ENSO phases.

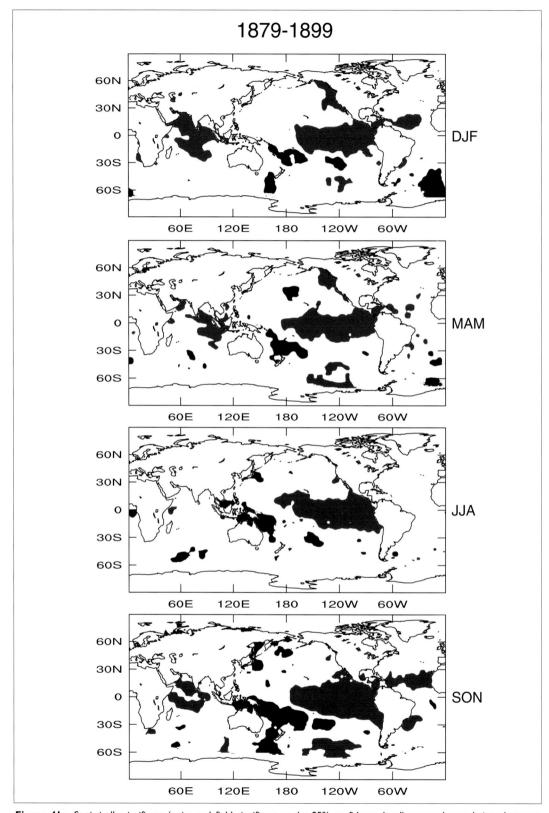

Figure 41a: Statistically significant (point and field significant at the 95% confidence level) seasonal correlations between eastern equatorial Pacific (EEP) (5°N–5°S; 150°W–120°W) and global SST during the 1879–1899 epoch. Red areas are significant positive correlations and blue areas significant negative correlations.

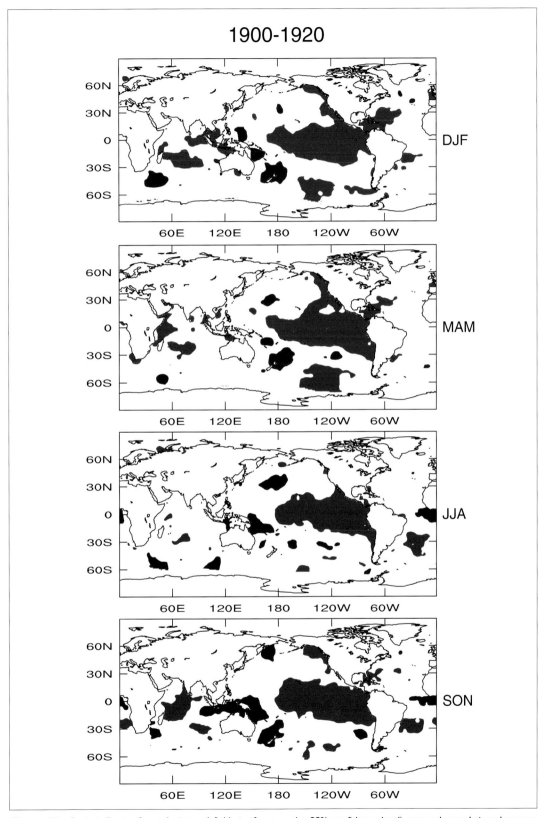

Figure 41b: Statistically significant (point and field significant at the 95% confidence level) seasonal correlations between eastern equatorial Pacific (EEP) (5°N– 5°S; 150°W–120°W) and global SST during the 1900–1920 epoch. Red areas are significant positive correlations and blue areas significant negative correlations.

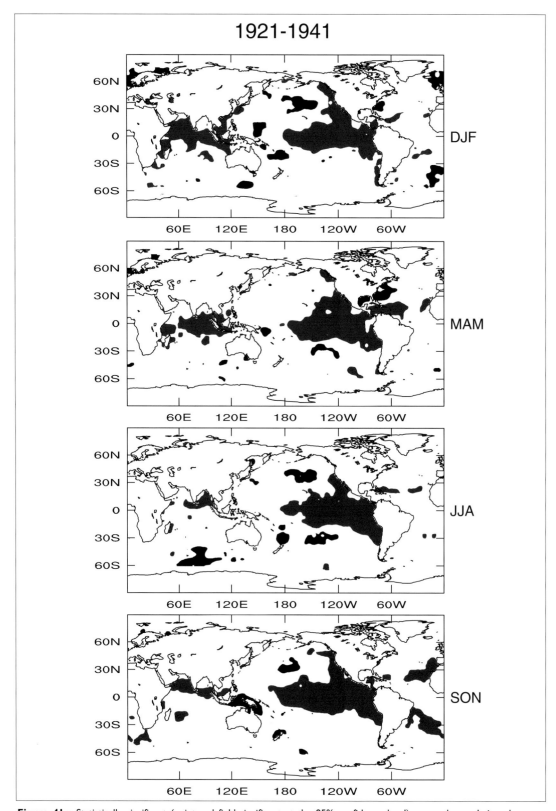

Figure 41c: Statistically significant (point and field significant at the 95% confidence level) seasonal correlations between eastern equatorial Pacific (EEP) (5°N– 5°S; 150°W–120°W) and global SST during the 1921–1941 epoch. Red areas are significant positive correlations and blue areas significant negative correlations.

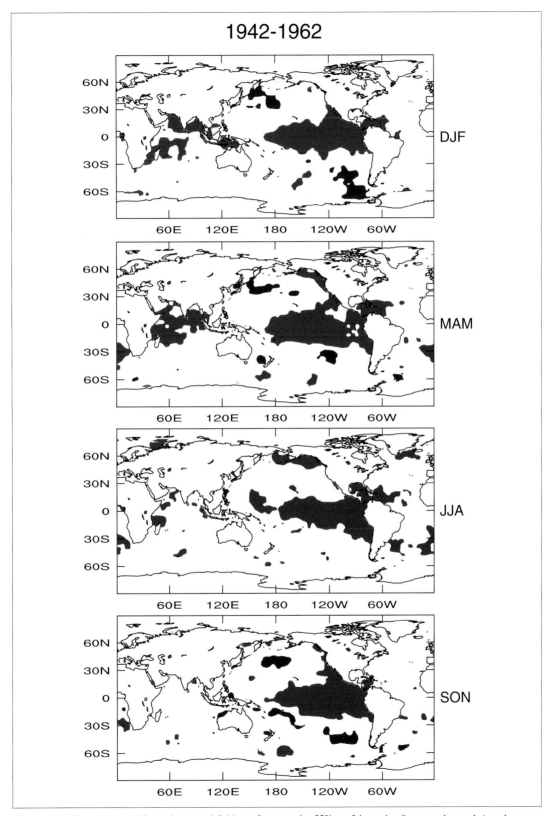

Figure 41d: Statistically significant (point and field significant at the 95% confidence level) seasonal correlations between eastern equatorial Pacific (EEP) (5°N– 5°S; 150°W–120°W) and global SST during the 1942–1962 epoch. Red areas are significant positive correlations and blue areas significant negative correlations.

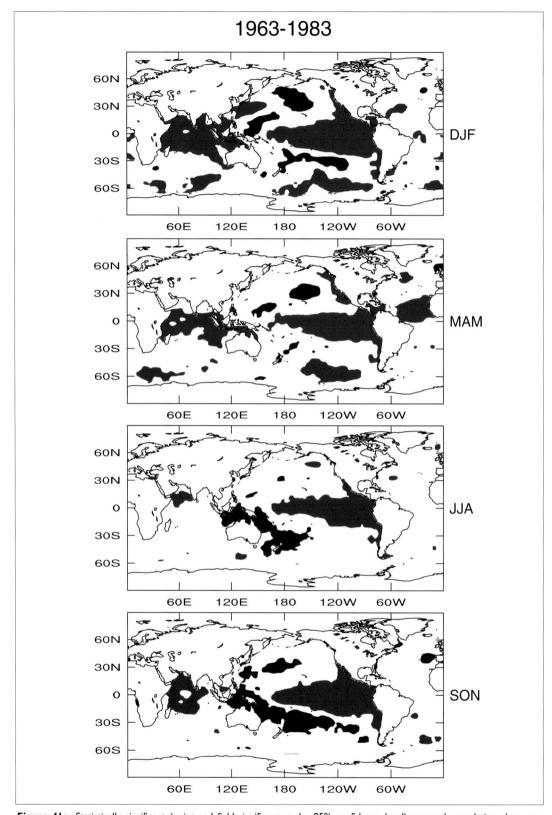

Figure 41e: Statistically significant (point and field significant at the 95% confidence level) seasonal correlations between eastern equatorial Pacific (EEP) (5°N– 5°S; 150°W– 20°W) and global SST during the 1963–1983 epoch. Red areas are significant positive correlations and blue areas significant negative correlations.

5. WIDER TERRESTRIAL AND MARINE ENVIRONMENTAL IMPACTS

There has been a rapid increase in research into the impacts of ENSO on aspects of marine and terrestrial ecology in the last 10–15 years, although such impacts (whether or not they were explicitly linked with ENSO events at the time) have been observed for centuries. In some cases, the impacts can be linked directly to the changes in rainfall, atmospheric circulation and oceanic variables that accompany El Niño and La Niña phases (as described in 'Oceanic, Atmospheric Hydrological Variable Responses to ENSO', p. 61), but in most instances the influence of ENSO is indirect and modulated or even obscured by other factors. Given the vast amount of research on ENSO impacts, this section aims to provide only a very basic overview of ENSO impacts in the terrestrial and marine environments. Readers interested in such aspects are encouraged to examine the references cited. A recent general overview of such material can be found in Glantz (1996).

In the terrestrial environment, much of the ENSO impacts research has focused on the economically and socially important agricultural sector. Through their often substantial influence on interannual variations in the occurrence of drought and flood periods, on intra-annual anomalies in the seasonality of rainfall during El Niño and La Niña years, and on associated modifications in the temperature regime, ENSO events can have a significant direct impact on agricultural production. Examples abound in Africa (Ogallo, 1987), India (Kiladis and Sinha, 1991), China (Wang and Mearns, 1987), parts of South-East Asia (Malingreau, 1987), Australia (Allan, 1991) and the Americas (Gasques and Magalhaes, 1987; Wilhite et al., 1987), amongst others; in particular, aspects of the impacts of the 1982–83 El Niño event on cereal crops in these and other areas are reviewed in Glantz et al. (1987). In general, El Niño-related droughts (and above-average rainfall or flooding linked with La Niña), coupled with high temperatures in drought periods and unseasonal rainfall distribution through the year, produce well below-average yields of rice, maize/corn, wheat and other cereals in affected regions. However, not all droughts or floods are ENSO-related, nor does the progress of the impacts of a particular ENSO phase depend entirely on the phase alone; local or regional factors can assume great significance. This has important implications when attempts are made to forecast for agricultural decision-making and crop-yield estimation, although there is a need for such forecasts (and an ever-improving ability to produce them on probabilistic principles).

In Australia, Nicholls (1985b, 1986a, 1988b) and Rimmington and Nicholls (1993) have shown clear associations between cereal production and ENSO-related rainfall anomalies. These are particularly significant for sorghum yields. In addition, Kuhnel (1993) has found regionally varying links between ENSO and sugar-cane yields in parts of Queensland. The importance of ENSO for the total rural economy of Australia has been illustrated graphically by the most recent period of prolonged drought in northeastern Australia, coinciding with the 1990–1995 climatic anomaly or 'persistent' El Niño sequence. For most cereal crops in Australia, reliable forecasts of major El Niño or La Niña phases in the winter–spring period would be most beneficial to the agricultural sector (McKeon et al., 1990; McKeon and White, 1992; Stone and McKeon, 1992; White and Howden, 1994; Bryceson and White, 1994). In general, such forecasts are most useful on an industry-wide basis rather than at the individual farm scale, as the statistical relationships on which they are based do not necessarily apply at smaller spatial scales. Examples of efforts to examine the economic and societal impacts of ENSO forecasts in the rural sector are found in O'Brien (1992), Cane *et al.*, (1994), Salafsky (1994), Chowdhury (1995), Golnaraghi and Kaul (1995) and Keppenne (1995).

Agriculture is not the only aspect of the terrestrial biosphere that is affected directly or indirectly by ENSO; rather more indirect, but nevertheless important, links have been suggested between the occurrence of large-scale wildfires in many parts of the world and the vegetation, temperature and precipitation/moisture regimes conducive to wildfire occurrence. Wetter than normal conditions (favourable to fuel-load accumulation) often prevail during the year preceding an El Niño event, to be followed by the characteristic spring–summer drought and high temperatures during the El Niño year (see some examples in Figure 36). A direct association has been suggested between fire activity in southeastern Australia and El Niño events (Skidmore, 1987). The February 1983 Ash Wednesday fires in southern Australia, directly linked with the 1982–83 El Niño event, probably caused more loss of life and property than any other fire in Australia this half-century, although the January 1994 Sydney fires, which could be linked with the climatic anomaly or 'persistent' El Niño sequence of 1990–1995, provide a more recent case and there were other extensive ENSO-related fires in 1944 and 1967. In the southeastern USA, there is a suggestion that fire activity decreases during El Niño phases (Simard *et al.*, 1985), and increases during the drier La Niña years. For forest and rangeland managers throughout the tropics and subtropics, ENSO-related drought, particularly prolonged drought that can impact seedling success and natural regeneration, and the attendant wildfire danger pose an ongoing and considerable management and conservation problem.

Outbreaks of both human and agricultural pests and diseases have also been linked with ENSO, particularly through the relationships between large-scale rainfall anomalies and increases/decreases in conditions favourable for the development and spread of various crop blights and the distributions of breeding sites and populations of disease-carriers such as mosquitoes (Nicholls, 1986b). Drake and Farrow (1988), amongst others, have suggested a link between insect life-cycles and ENSO, and Nicholls (1991b) reviews aspects of the relationships between ENSO and climate-related human health problems, many related to famine in drought-afflicted regions or to insect- or water-borne diseases in flooded areas. In Australia, most of the human-health impacts are linked with expanded insect-carrier populations in wetter, La Niña, years, causing diseases such as Ross River Fever and Murray Valley Encephalitis to become more prevalent. There has even been an extension of research on ENSO impacts from health concerns into evolutionary biological fields. This is exemplified by efforts to link the evolution of finches in the Galapagos Islands (Gibbs and Grant, 1987) and Australian vegetation (Nicholls, 1991a) to the occurrence of ENSO.

Rather more attention seems to have been given to the marine environmental impacts of ENSO than to those in the terrestrial environment. Marine biological productivity in some of the most productive, and economically and ecologically significant parts of the global oceans is directly related to ocean current systems and marine temperature regimes that form part of, or are influenced by, ENSO (Glantz, 1984). Associations between ENSO and fisheries and other aspects of marine ecology in many parts of the world are reviewed in Glantz and Thompson (1982), Glantz and Feingold (1990), Glynn (1990a) and Sharp and McLain (1993).

Jordan (1991) focuses attention on the southeastern Pacific region along the South American coast, where the pelagic fishery, and interdependent important seabird colonies and a substantial guano industry, are closely dependent on the upwelling of cold, nutrient-rich waters in the coastal zone. This system is significantly disturbed during El Niño phases, as warmer waters intrude into the area and the wind-driven upwelling is disrupted by the large-scale changes that occur in the pressure and wind systems across the Pacific basin in these years (see 'The Oceans: El Niño and La Niña' p. 4). Primary productivity falls in the southeastern Pacific during El Niño years because fertilisation in the euphotic zone is poor, the lowest trophic levels experience a decline in food quality and availability, many fish species are displaced or subject to high mortality rates, reproduction patterns are disturbed, and recruitment rates decline. Marine organisms that are directly affected by the suppression of upwelling and higher than normal ocean temperatures in the eastern Pacific during El Niño phases include phyto- and zooplankton, pelagic and coastal fish and benthic populations (kelp, abalones, lobsters, subtidal and intertidal species). Also directly affected are the seabird (Smith, 1990, Duffy, 1990) and pinniped (seal) (Limberger, 1990, Trillmich and Ono, 1991) populations that are dependent on these marine organisms, as well as other species such as marine iguanas on the Galapagos Islands (Laurie, 1990). As an example, the combination of the 1972–73 El Niño event and overfishing saw the collapse of the anchoveta fishery along the west coast of South America (see Jordan, 1991) to levels from which it has not yet recovered.

Coral is another marine organism that is affected by ENSO. During major El Niño phases, the prolonged period of ocean warming in the EEP, together with salinity changes, reduced zooplankton and nutrient supply and changed light intensity, causes coral bleaching (loss of the symbiotic fauna that normally inhabit the coral and are responsible for its colouring) and subsequent death (Glynn, 1990b). Coral mortalities of between 70 and 95 per cent were reported for much of the area during the 1982–83 event. Coral bleaching and mortality may also occur due to strong upwelling events and ocean temperature reduction, as happened in 1985 during La Niña conditions (Glynn, 1990b), but although this source of coral damage is apparently more frequent than that due to El Niño events it is also less damaging overall. ENSO-related coral mortality has also been reported for other parts of the world, including the Java Sea, the south-west Indian Ocean, southern Japan and the Caribbean (Coffroth et al., 1990). Another cause of coral death reported from Papua New Guinea during the 1982–1983 El Niño was exposure, with sea levels being reduced by as much as 20–30 cm in some areas (Godfrey, 1996, pers.comm.). Other impacts of ENSO on corals include changes in the potential for species introductions, range extensions and establishment as the circulation pattern in the Pacific changes during an ENSO event (Richmond, 1990). A recent review of coral bleaching can be found in Goreau and Hayes (1994).

The sensitivity of corals to the marine environmental disturbances introduced by ENSO means that coral reefs provide the potential for reconstructing a particularly useful record of past ENSO events (Cole and Fairbanks, 1990; Cole et al., 1992, 1993; Shen et al., 1992a, b; Dunbar and Cole, 1993, 1995; Quinn et al., 1993; Dunbar et al., 1994; Gagan and Chivas, 1995; Slowey and Crowley, 1995, Wellington and Dunbar, 1995). This may be achieved through the analysis of stable isotopes of carbon and oxygen in the coral skeletons (Druffel et al., 1990; Druffel and Griffin, 1993; Wellington and Dunbar, 1995), although such analyses may fail to identify major events such as that of 1982–83 when large-scale coral mortality occurred and skeleton-building ceased. It is also possible to use trace elements in the coral as more indirect indicators of climatic variability and ENSO. Shen and Sandford (1990) suggest that the concentrations of various trace elements may indicate lateral movement of water masses or signal the input of fresh water from rainfall anomalies into the reef zone. Similar ideas have been pursued by Isdale (1984), Isdale and Kotwicki (1987) and Kotwicki and Isdale (1991), who have reconstructed environmental conditions in the region of the Great Barrier Reef from analyses of coral cores. Marine environmental reconstruction from coral cores is an active research field in palaeo studies, and considerable attention is now being devoted to the ocean areas around Indonesia, in the Indian and Pacific basins and to the north of Australia (Figure 42). The area of the tropical western Pacific is an important one for the understanding of long-term variability in ENSO. Review of the potential resolution of high quality climatic data from corals are found in Lough et al. (1994) and Cook (1995).

Less direct marine environmental impacts of ENSO include variations in the shrimp fishery in the Gulf of Mexico, probably due to adjustments in regional weather patterns in response to ENSO and other large-scale atmospheric forcings (White and Downton, 1991), and mass mortalities and breeding failures of seabirds along the south-west coast of Africa (La Cock, 1986).

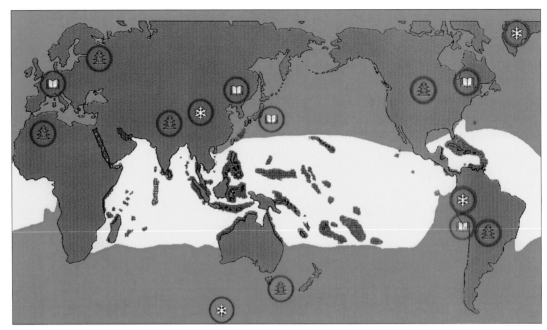

Figure 42: Distribution of potential coral coring sites (red shading) across the tropical–subtropical oceans relative to other historical documentary (book symbol), tree-ring (tree symbol) and ice core (ice symbol) proxy/palaeoenvironmental records (Lough and Barnes, 1995, pers. comm.).

In the southwestern Pacific, there are clear links between ENSO and interannual variability in shelf waters around Australia (Hsieh and Hamon, 1991). Aspects of the life cycles and food chains of marine organisms in Australian waters, such as tuna, prawns, green turtles, rock lobster and coral, have been linked with ENSO-related disturbances to oceanic variables such as temperature, salinity and nutrient levels. Much of this work is reviewed in Allan (1991).

6. Appendixes

A: Significant Scientific Contributors to the Early Evolution of ENSO Research

SCIENTIST	BIRTH–DEATH	LIFE HISTORY/ACHIEVEMENTS/OBITUARY
Todd, Sir Charles	7/7/1826–29/1/1910	*Quart. J. Roy. Meteor. Soc.,* **34,** (1910), 189–190 *Symons's Meteorological Magazine,* **45,** (1910), 21–22 *The Origins of Australian Meteorology,* Gibbs, W.J., (1975), Aust. Govt. Pub. Ser., Canberra, 14–15.
Blanford, Henry Francis	1834–23/1/1893	*The American Meteorological Journal,* **10,** (1893), 74–76 *Symons's Meteorological Magazine,* **28,** (1893), 1 *Proc. Asiatic Soc. Bengal,* (March, 1893), 72–74 *Nature,* **47,** (1893), 322–323 *The American Meteorological Journal,* **10,** (1894), 391–393 *Quart. J. Roy. Meteor. Soc.,* **20,** (1894), 104–105 *Hundred Years of Weather Service (1875–1975),* India Meteorological Department (1976), Deputy Director General of Observatories, Poona, India, 207 pp.
Lockyer, Sir Joseph Norman	17/5/1836–16/8/1920	*Meteorological Magazine,* **55,** (1920), 181–182 *Mon. Wea. Rev.,* **48,** (1920), 659 *Nature,* **105,** (1920), 781–784 *Quart. J. Roy. Meteor. Soc.,* **55,** (1929), 318–320 *Meteorological Magazine,* **64,** (1929), 278–281 *Science and controversy: A biography of Sir Norman Lockyer,* Meadows, A.J. (1972), MIT Press, 331 pp.
Hildebrandsson, Hugo Hildebrand	19/8/1838–29/7/1925	*Quart. J. Roy. Meteor. Soc.,* **46,** (1920), 197–199 *Meteorological Magazine,* **60,** (1925), 205–206 *Quart. J. Roy. Meteor. Soc.,* **51,** (1925), 425–428 *Mon. Wea. Rev.,* **53,** (1925), 311
Eliot, Sir John	25/5/1839–18/3/1908	*Symons's Meteorological Magazine,* **43,** (1908), 78 *Mon. Wea. Rev.,* **36,** (1908), 71–72 *Quart. J. Roy. Meteor. Soc.,* **35,** (1909), 111 *Hundred Years of Weather Service (1875–1975),* India Meteorological Department (1976), Deputy Director General of Observatories, Poona, India, 207 pp.

Bigelow,   1851–2/3/1924   *Bull. Amer. Meteor. Soc.,* **5**, (1924), 60–61
Frank Hagar   *Mon. Wea. Rev.,* **52**, (1924), 165–166
    Meteorological Magazine, **59**, (1924), 167

Lockyer,   1868–1934   *Science and controversy: A biography of Sir*
William James Stewart   *Norman Lockyer,* Meadows, A.J. (1972), MIT Press, 331 pp.

Walker,   14/6/1868–4/11/1958   *Quart. J. Roy. Meteor. Soc.,* **50**, (1924), 224
Sir Gilbert Thomas   *Meteorological Magazine,* **59**, (1924), 108
    Quart. J. Roy. Meteor. Soc., **54**, (1928), 125
    Quart. J. Roy. Meteor. Soc., **60**, (1934), 184–185
    Quart. J. Roy. Meteor. Soc., **85**, (1959), 186
    Nature, **182**, (1958), 1706
    Weather, **14**, (1959), 67–68
    WMO Bulletin, **8**, (1959), 38–39
    Indian J. Meteor. & Geophys., **10**, (1959), 1–2, 113–120
    Biogr. Mem. Fell. Roy. Soc., **8**, (1962), 166–174
    Hundred Years of Weather Service (1875–1975), India Meteorological Department (1976), Deputy Director General of Observatories, Poona, India, 207 pp.

Mossman,   7/11/1870–19/7/1940   *Quart. J. Roy. Meteor. Soc.,* **66**, (1940), 445
Robert Cockburn   *Proc, Roy. Soc. Edinburgh,* **60**, (1939–1940), 402

Bliss,   1888–4/8/1955   *Quart. J. Roy. Meteor. Soc.,* **82**, (1956), 121
Ernest William

Normand,   10/9/1889–25/10/1982   *WMO Bulletin,* **32**, (1983), 91–102
Sir Charles William Blyth   *Hundred Years of Weather Service (1875–1975),* India Meteorological Department (1976), Deputy Director General of Observatories, Poona, India, 207 pp.

Berlage,   24/10/1896–2/3/1968   *Bull. Amer. Meteor. Soc.,* **49**, (1968), 747
Hendrik Petrus   *KNMI Library* faxed material and photo

Bjerknes,   2/11/1897–7/7/1975   *Bull. Amer. Meteor. Soc.,* **56**, (1975), 1089–1090
Jacob Aall Bonnevie   *WMO Bulletin,* **24**, (1975), 263–264

Quinn,   28/9/1918–12/1/1994   *Corvallis Gazette–Times, Corvallis Ore,* Obituaries,
William Hewes   15/1/1994,Oregon, USA.

Troup,   18/8/1927–10/6/1983   See Appendix B.
Alexander James

B: ALEXANDER JAMES 'SANDY' TROUP

No biographical information on Alexander James 'Sandy' Troup has been published in the standard meteorological or climatological literature, and this short appraisal of his life and academic efforts is intended to redress this situation.

Sandy Troup's contribution to El Niño Southern Oscillation (ENSO) research is best remembered, and continues to be perpetuated, through the wide use of his normalised Tahiti minus Darwin Southern Oscillation Index (SOI) in its various forms. Contrary to statements in some later ENSO studies and general perception, this 'abbreviated' SOI was devised by Sandy Troup in the period after his major paper on the Southern Oscillation phenomenon in the *Quart. J. Roy. Meteor. Soc.* in 1965. Interestingly, Troup's SOI was termed 'abbreviated' because the final step of its calculation (detailed in 'Calculation and filtering of Niño 3, Niño 4 and SOI series', p. 57) involved multiplication by a factor of 10. This was done because Sandy was keen to round off the index so not to have any decimal points. Sandy Troup's index was not documented in the literature until 1974, in a conference paper by Barrie Pittock which is cited in the reference section of this book.

Born of British parents on the 18th of August 1927, in what was then Ceylon (now Sri Lanka), where his father managed a tea estate, he was sent with his brother to be educated in England just prior to the Second World War. As a consequence of the war, he was evacuated to Australia. He attended Xavier College, Melbourne, where he graduated dux of his school, and then undertook tertiary studies in the Physics Department of Melbourne University. During holidays, Sandy stayed with his uncle, Len Deacon, who had a major influence in this career choice.

After the war Len Deacon was joined by Bill Priestley, who had become the head of the Commonwealth Scientific, Industrial and Research Organisation (CSIRO) Section of Meteorological Physics at Highett in Melbourne. Deacon advised Sandy of a vacancy at the Section, and he applied, was accepted, and joined the group at Highett around 1951.

Sandy Troup's initial research efforts were in support of Priestley's meridional eddy flux studies and Berson's work on the northwestern Australian summer heat low. In the period from the mid 1950s until the early 1960s, he became involved in a major CSIRO study of cold fronts and sea breezes, in which concerted efforts were made to increase local observing station density and improve vertical resolution of these phenomena.

During the middle part of his career, Sandy spent a period away from the now CSIRO Division of Meteorological Physics. This time was spent assisting Neil Streten in synoptic analyses of satellite imagery and Henry Phillpot with examinations of Southern Hemisphere circumpolar circulation at the then International Antarctic Analysis Centre (IAAC) in the Australian Bureau of Meteorology.

Research on global teleconnections and the Southern Oscillation, and what was later to be termed the ENSO phenomenon, occupied the latter half of his research career. Like others of his time, and those that followed, he was attracted to the subject by the earlier work of Sir Gilbert Walker. In the course of his research he made contact with Hendrik Berlage, one of the other prominent ENSO researchers of the time. Both of these men are now remembered for their vital roles in repromoting Southern Oscillation research and linking that phenomenon with oceanic processes and especially El Niño events.

Sandy Troup died on the 10th of June 1983 after a long illness.

7. REFERENCES

Abbe, C. 1901. The physical basis of long-range weather forecasts. *Mon. Wea. Rev.,* **29**, 551–561.

Aceituno, P. 1988. On the functioning of the Southern Oscillation in the South American sector. Part I: Surface climate. *Mon. Wea. Rev.,* **116**, 505–524.

Aceituno, P. 1989. On the functioning of the Southern Oscillation in the South American sector. Part II: Upper-air circulation. *J. Climate,* **2**, 341–355.

Aceituno, P. 1992. El Niño, the Southern Oscillation, and ENSO: Confusing names for a complex ocean–atmosphere interaction. *Bull. Amer. Meteor. Soc.,* **73**, 483–485.

Ad Hoc Study Group on Oceanic Interdecadal Climate Variability, 1992. Oceanic Interdecadal Climate Variability. *IOC Tech. Ser.* **40**, 40 pp.

Allan, R.J. 1985. The Australasian summer monsoon, teleconnections, and flooding in the Lake Eyre basin. *South Australian Geographical Papers No.* **2**, Roy. Geogr. Soc. Australasia (South Australian Branch), 47 pp.

Allan, R.J. 1988. El Niño Southern Oscillation influences in the Australasian region. *Prog. Phys. Geogr.,* **12**, 4–40.

Allan, R.J. 1989. ENSO and climatic fluctuations in Australasia. In Donnelly, T.H. and Wasson, R.J. (eds), *CLIMANZ III Pro. Symp.,* University of Melbourne, pp. 49–61.Allan, R.J. 1991. Australasia. In Glantz, M., Katz, R. and Nicholls, N. (eds), *ENSO teleconnections linking worldwide climate anomalies: Scientific basis and societal impacts.* Cambridge University Press, Cambridge, UK, pp. 73–120.

Allan, R.J. 1993. Historical fluctuations in ENSO and teleconnection structure since 1879: Near-global patterns. *Quat. Aust.,* **11**, 17–27.

Allan, R.J. and D'Arrigo, R.D. 1996. 'Persistent' ENSO sequences: How unusual was the recent El Niño? *The Holocene* (submitted).

Allan, R.J. and Pariwono, J.I. 1990. Ocean–Atmosphere interactions in low latitude Australasia. *J. Climatol.,* **10**, 145–179.

Allan, R.J., Beck, K. and Mitchell, W.M. 1990. Sea level and rainfall correlations in Australia: Tropical links. *J. Climate,* **3**, 838–846.

Allan, R.J., Nicholls, N., Jones, P.D. and Butterworth, I.J. 1991. A further extension of the Tahiti–Darwin SOI, early ENSO events and Darwin pressure. *J. Climate,* **4**, 743–749.

Allan, R.J., Lindesay, J.A. and Reason, C.J.C. 1995. Multidecadal variability in the climate system over the Indian Ocean region during the austral summer. *J. Climate,* **8**, 1853–1873.

Allan, R.J., Beard, G., Close, A., Herczeg, A.L., Jones, P.D. and Simpson, H.J. 1996. Mean sea level pressure indices of the El Niño–Southern Oscillation: relevance to stream discharge in southeastern Australia. *CSIRO Division of Water Resources Report* 96/1.

Allen, B.J. 1989. Frost and drought through time and space, part I: The climatological record. *Mountain Research and Development,* **9**, 252–278.

Allen, B.J., Brookfield, H. and Byron, Y. 1989. Frost and drought through time and space, Part II: The written, oral and proxy records and their meaning. *Mountain Research and Development,* **9**, 279–305.

Allen, M.R. and Smith, L.A. 1994. Investigating the origins and significance of low-frequency modes of climate variability. *Geophys. Res. Lett.,* **21**, 883–886.

Angstrom, A. 1935. Teleconnections of climate changes in present time. *Geog. Ann.,* **17**, 242–258.

Annamalai, H. 1995. Intrinsic problems in the seasonal prediction of the Indian summer monsoon rainfall. *Meteorol., Atmos. Phys.,* **55**, 61–76.

Archibald, E.D. 1883. Indian Meteorology II. *Nature,* **28**, 428–430.

Archibald, E.D. 1896. The long period weather forecasts of India. *Nature,* **55**, 85–88.

Bacastow, R.B. 1976. Modulation of atmospheric carbon dioxide by the Southern Oscillation. *Nature,* **261**, 116–118.

Bacastow, R.B., Adams, J.A., Keeling, C.D., Moss, D.J., Whorf, T.P. and Wong, C.S. 1980. Atmospheric carbon dioxide, Southern Oscillation, and the weak 1975 El Niño. *Science,* **210**, 66–68.

Baldwin, M.P. and O'Sullivan, D. 1995. Stratospheric effects of ENSO-related tropospheric circulation anomalies. *J. Climate,* **8**, 649–667.

Balmaseda, M.A., Anderson, D.L.T. and Davey, M.K. 1994a. Seasonal dependence of ENSO prediction skill. *Climate Research Tech. Note* CRTN No. **51**, Hadley Centre for Climate Prediction and Research, Meteorological Office, Bracknell, Berkshire, UK, 10 pp.

Balmaseda, M.A., Anderson, D.L.T. and Davey, M.K. 1994b. Seasonal dependence of ENSO prediction skill. *Ocean Modelling,* **104**, 2–6.

Balmaseda, M.A., Anderson, D.L.T. and Davey, M.K. 1994c. ENSO prediction using a dynamical ocean model coupled to statistical atmospheres. *Tellus,* **46A**, 497–511.

Banglin, Z. 1994. A data-adaptive filter of the Tahiti–Darwin Southern Oscillation Index and the associate scheme of filling data gaps. *Adv. Atmos. Sci.,* **11**, 447–458.

Barnett, T.P. 1983. Interaction of the monsoon and Pacific trade wind system at interannual time scales. Part I: The equatorial zone. *Mon. Wea. Rev.,* **111**, 756–773.

Barnett, T.P. 1984a. Prediction of the El Niño of 1982–83. *Mon. Wea. Rev.,* **112**, 1403–1407.

Barnett, T.P. 1984b. Interaction of the monsoon and Pacific trade wind system at interannual time scales. Part II: The tropical band. *Mon. Wea. Rev.,* **112**, 2380–2387.

Barnett, T.P. 1984c. Interaction of the monsoon and Pacific trade wind system at interannual time scales. Part III: A partial anatomy of the Southern Oscillation. *Mon. Wea. Rev.,* **112**, 2388–2400.

Barnett, T.P. 1985. Variations in near-global sea level pressure. *J. Atmos. Sci.,* **42**, 478–501.

Barnett, T.P. 1988. Variations in near-global sea level pressure: Another view. *J. Climate,* **1**, 225–230.

Barnett, T.P. 1991. The interaction of multiple time scales in the tropical climate system. *J. Climate,* **4**, 269–285.

Barnett, T.P. and Jones, P.D. 1992. Intercomparison of two different Southern Hemisphere sea level pressure datasets. *J. Climate,* **5**, 93–99.

Barnett, T.P., Graham, N., Cane, M., Zebiak, S., Dolan, S., O'Brien, J.J. and Legler, D. 1988. On the prediction of the El Niño of 1986–1987. *Science,* **241**, 192–196.

Barnett, T.P., Latif, M., Kirk, E. and Roeckner, E. 1991. On ENSO physics. *J. Climate,* **4**, 487–515.

Barnett, T.P., Del Genio, A.D. and Ruedy, R.A. 1992. Unforced decadal fluctuations in a coupled model of the atmosphere and ocean mixed layer. *J. Geophys. Res.,* **97**, 7341–7354.

Barnett, T.P., Bengtsson, L., Arpe, K., Flugel, M., Graham, N., Latif, M., Ritchie, J., Roeckner, E., Schlese.U., Schulzweida, U. and Tyree, M. 1994. Forecasting global ENSO-related climate anomalies. *Tellus,* **46A**, 381–397.

Barnett, T.P., Latif, M., Graham, N. and Flugel, M. 1995. On the frequency–wavenumber structure of the tropical ocean/atmosphere system. *Tellus,* **47A**, 998–1012.

Barnston, A.G. and Ropelewski, C.F. 1992. Prediction of ENSO episodes using canonical correlation analysis. *J. Climate,* **5**, 1316–1345.

Barnston, A.G., van den Dool, H.M., Zebiak, S.E., Barnett, T.P., Ji, M., Rodenhuis, D.R., Cane, M.A., Leetmaa, A., Graham, N.E., Ropelewski, C.F., Kousky, V.E., O'Lenic, E.A. and Livezey, R.E. 1994. Long-lead seasonal forecasts – Where do we stand? *Bull. Amer. Meteor. Soc.,* **75**, 2097–2114.

Barros, V.R. and Scasso, L.M. 1994. Surface pressure and temperature anomalies in Argentina in connection with the Southern Oscillation. *Atmosfera,* **7**, 159–171.

Basnett, T.A. and Parker, D.E. 1996. Development of the Global Mean Sea Level Pressure data set GMSLP Version 2. *Climate Research Tech. Note* Hadley Centre for Climate Prediction and Research, Meteorological Office, Bracknell, Berkshire, UK, (in press).

Battisti, D.S. and Hirst, A.C. 1989. Interannual variability in a tropical atmosphere/ocean model: Influence of the basic state, ocean geometry and nonlinearity. *J. Atmos. Sci.,* **46**, 1687–1712.

Battisti, D.S. and Sarachik, E.S. 1995. Understanding and predicting ENSO. *Rev. Geophys. Supp.,* 1367–1376.

Behrend, H. 1987. Teleconnections of rainfall anomalies and of the Southern Oscillation over the entire tropics and their seasonal dependence. *Tellus,* **39A**, 138–151.

Beltrando, G. and Camberlin, P. 1993. Interannual variability of rainfall in the eastern horn of Africa and indicators of atmospheric circulation. *Int. J. Climatol.,* **13**, 533–546.

Berlage, Jr., H.P. 1927. East-Monsoon forecasting in Java. *Verhandelingen Koninklijk Magnetisch en Meteorologisch Observatorium te Batavia, No.* **20**, 42 pp.

Berlage, Jr., H.P. 1934. Further research into the possibility of long range forecasting in Netherlands-India. *Verhandelingen Koninklijk Magnetisch en Meteorologisch Observatorium te Batavia, No.* **26**, 31 pp.

Berlage, Jr., H.P. 1939. *Observations made at secondary stations in Netherlands East–India. Vol.* **XXI**, Royal Magnetic and Meteorological Observatory at Batavia, 77 pp.

Berlage, Jr., H.P. 1940. *Observations made at secondary stations in Netherlands East–India. Vol.* **XXII**, Royal Magnetic and Meteorological Observatory at Batavia, 78 pp.

Berlage, Jr., H.P. 1941. *Observations made at secondary stations in Netherlands East–India. Vol.* **XXIII**, Royal Magnetic and Meteorological Observatory at Batavia, 80 pp.

Berlage, Jr., H.P. 1957. *Fluctuations of the general atmospheric circulation of more than one year, their nature and prognostic value.* Mededlingen en Verhandelingen No. 69, Koninklijk Meteorologische Instituut, Staatsdrukkerijs-Gravenhage, Netherlands, 152 pp.

Berlage, Jr., H.P. 1961. Variations in the general atmospheric and hydrospheric circulation of periods of a few years duration affected by variations of solar activity. *Annals of the New York Academy of Sciences,* **95,** 354–367.

Berlage, Jr., H.P. 1966. *The Southern Oscillation and world weather.* Mededlingen en Verhandelingen No. 88, Koninklijk Meteorologische Instituut, Staatsdrukkerijs-Gravenhage, Netherlands, 152 pp.

Berlage, Jr., H.P. and de Boer, H.J. 1959. On the extension of the Southern Oscillation throughout the world during the period July 1 1949 to July 1 1957. *Pure Appl. Geophys.,* **44,** 287–295.

Berlage, Jr., H.P. and de Boer, H.J. 1960. On the Southern Oscillation, its way of operation and how it affects pressure patterns in the higher latitudes. *Pure Appl. Geophys.,* **46,** 329–351.

Bigelow, F.H. 1903. Studies on the circulation of the atmospheres of the Sun and of the Earth. II.—Synchronism of the variations of the solar prominences with the terrestrial barometric pressures and the temperatures. *Mon. Wea. Rev.,* **31,** 509–516.

Bigg, G.R. 1990. El Niño and the Southern Oscillation. *Weather,* **45,** 2–8.

Bigg, G.R. 1992. Validation of trends in the surface wind field over the Mozambique Channel. *Int. J. Climatol.,* **12,** 829–838.

Bigg, G.R. 1993. Comparison of coastal wind and pressure trends over the tropical Atlantic: 1946–1987. *Int. J. Climatol.,* **13,** 411–421.

Bigg, G.R. 1995. The El Niño event of 1991–94. *Weather,* **50,** 117–124.

Bigg, G.R. and Inoue, M. 1992. Rossby waves and El Niño during 1935–46. *Quart. J. Roy. Meteor. Soc.,* **118,** 125–152.

Bigg, G.R., Whysall, K.D.B. and Cooper, N.S. 1987. Long-term trends and a major climate anomaly in the tropical Pacific wind-field. *Trop. Ocean–Atmos. Newsl.,* **37,** 1–2.

Bjerknes, J. 1961. El Niño study based on analyses of ocean surface temperatures 1935–1957. *Bull. Inter-Amer. Trop. Tuna Comm.,* **5,** 217–303.

Bjerknes, J. 1966. A possible response of the atmospheric Hadley Circulation to equatorial anomalies of ocean temperature. *Tellus,* **18,** 820–829.

Bjerknes, J. 1969. Atmospheric teleconnections from the equatorial Pacific. *Mon. Wea. Rev.,* **97,** 163–172.

Bjerknes, J. 1972. Large-scale atmospheric response to the 1964–65 Pacific equatorial warming. *J. Phys. Oceanogr.,* **2,** 212–217.

Blanford, H.F. 1880a. On the barometric see-saw between Russia and India in the sun-spot cycle. *Nature,* **25,** 477–482.

Blanford, H.F. 1880b. *Report on the meteorology of India in 1878.* Government Printer, Calcutta, 149 pp.

Blanford, H.F. 1884. On the connection of Himalayan snowfall with dry winds and seasons of drought in India. *Proc. Roy. Soc., London,* **37,** 3–22.

Blanford, H.F. 1891. The paradox of the sun-spot cycle in meteorology. *Nature,* **43,** 583–587.

Blazejewski, H., Cadet, D.L. and Marsal, O. 1986. Low-frequency sea surface temperature and wind variations over the Indian and Pacific Oceans. *J. Geophys. Res.,* **91,** C4, 5129–5132.

Bliss, E.W. 1926. The Nile flood and world weather. *Mem. Roy. Meteor. Soc.,* **1,** 79–85.

Bliss, E.W. 1930. A study of rainfall in the West Indies. *Quart. J. Roy. Meteor. Soc.,* **56,** 67–72.

Bliss, E.W. 1936. A forecast of Barbados rainfall. *Quart. J. Roy. Meteor. Soc.,* **62,** 45–48.

Boerema, J. 1939. *Observations made at secondary stations in the Netherlands Indies. Vol.* **XIXA,** Climatological tables, Royal Magnetical and Meteorological Observatory at Batavia, 85 pp.

Bottomley, M., Folland, C.K., Hsiung, J., Newell, R.E. and Parker, D.E. 1990. *Global Ocean Surface Temperature Atlas (GOSTA),* Meteorological Office and Massachusetts Institute of Technology, 313 plates, 20 pp.

Bradley, R.S., Diaz, H.F., Kiladis, G.N. and Eischeid, J.K. 1987. ENSO signal in continental temperature and precipitation records. *Nature,* **327,** 497–501.

Brankovic, C., Palmer, T.N. and Ferranti, L. 1994. Predictability of seasonal atmospheric variations. *J. Climate,* **7,** 217–237.

British Board of Trade, 1861. *Fifth Number of Meteorological Papers.* Eyre and Spottiswoode, London, 77–82.

Broecker, W.S. 1991. The great ocean conveyor. *Oceanography,* **4,** 79–89.

Brookfield, H. 1989. Frost and drought through time and space, Part III: What were conditions like when the high valleys were first settled? *Mountain Research and Development,* **9,** 306–321.

Brooks, C.E.P. 1918. The meteorology of Ocean Island during the period 1905–1916. *Quart. J. Roy. Meteor. Soc.,* **44,** 50–53.

Brooks, C.E.P. 1919. Notes on the meteorological journal at Wei Hai Wei, kept by commander A.E. House, 1910 to 1916. *Quart. J. Roy. Meteor. Soc.,* **45,** 29–41.

Brooks, C.E.P. 1928a. Periodicities in the Nile floods. *Mem. Roy. Meteor. Soc.,* **11,** 10–26.

Brooks, C.E.P. 1928b. Long-range weather forecasting. *Meteor. Mag.,* **63,** 81–86.

Brooks, C.E.P. 1938. Long-range forecasting. *Meteor. Mag.,* **73,** 257–262.

Brooks, C.E.P. and Braby, H.W. 1921. The clash of the trades in the Pacific. *Quart. J. Roy. Meteor. Soc.,* **47,** 1–13.

Budin, G.R. 1985. Interannual variability of Australian snowfall. *Aust. Meteor. Mag.,* **33,** 145–149.

Bryceson, K.P. and White, D.H., (eds), 1994. *Proceedings of a Workshop on Drought and Decision Support.* Bureau of Resource Sciences Proceedings. Bureau of Resource Sciences, Canberra, 63 pp.

Bureau of Meteorology, 1945–1953. *Meteorological summary for selected climatological stations. Australia.* Government Printer, Melbourne.

Bureau of Meteorology, 1954–1956. *Monthly climatological records.* Government Printer, Melbourne.

Burnett, A.W. 1993. Size variations and long-wave circulation within the January Northern Hemisphere circumpolar vortex: 1946–89. *J. Climate,* **6,** 1914–1920.

Busalacchi, A.J. and O'Brien, J.J. 1981. Interannual variability of the equatorial Pacific in the 1960s. *J. Geophys. Res.,* **86,** 10,901–10,907.

Busalacchi, A.J., Takeuchi, K. and O'Brien, J.J. 1983. Interannual variability of the equatorial Pacific – revisited. *J. Geophys. Res.,* **88,** 7551–7562.

Cadet, D.L. 1985. The Southern Oscillation over the Indian Ocean. *J. Climatol.,* **5,** 189–212.

Campbell, J. 1879. Meteorology of Bangkok, Siam. *Quart. J. Roy. Meteor. Soc.,* **5,** 82–89.

Cane, M.A. 1986. El Niño. *Ann. Rev. Earth Planet. Sci.,* **14,** 43–70.

Cane, M.A. and Zebiak, S.E. 1985. A theory for El Niño and the Southern Oscillation. *Science,* **228,** 1085–1087.

Cane, M.A., Zebiak, S.E. and Dolan, S.C. 1986. Experimental forecasts of El Niño. *Nature,* **321,** 827–832.

Cane, M.A., Eshel, G. and Buckland, R.W. 1994. Forecasting Zimbabwean maize yield using eastern equatorial Pacific sea surface temperature. *Nature,* **370,** 204–205.

Cardone, V.J., Greenwood, J.G. and Cane, M. 1990. On trends in historical marine wind data. *J. Climate,* **3,** 113–127.

Carranza, L. 1891. Contra-corriente maritima observada Piata y Pacasmayo. *Bol. Soc. Geogr, Lima,* **1,** 9, 344–345.

Carrillo, C.N. 1892. Disertacion sobre las Corrientes Oceanicas y Estudios de la Corriente Peruana o de Humboldt. Microficha. *Bol. Soc. Geogr. Lima,* **2,** 11, 52–110.

Carleton, A.M. 1989. Antarctic sea-ice relationships with indices of the atmospheric circulation of the Southern Hemisphere. *Clim. Dyn.,* **3,** 207–220.

Casey, T. 1995. Optimal linear combination of seasonal forecasts. *Aust. Meteor. Mag.,* **44,** 219–224.

Cayan, D.R. and Redmond, K.T. 1994. ENSO influences on atmospheric circulation and precipitation in the western United States. *Proc. Tenth Ann. Pacific Climate (PACLIM) Workshop, Tech. Rep.,* **36,** Interagency Ecological Studies Program for the Sacramento–San Joaquin Estuary, Asilomar, California, pp. 5–26.

Central Meteorological Observatory, 1954. *Climatic records of Japan and the far east area.* Tokyo, Japan, 679 pp.

Chan, J.C.L. 1985. Tropical cyclone activity in the northwest Pacific in relation to the El Niño/Southern Oscillation phenomenon. *Mon. Wea. Rev.,* **113,** 599–606.

Chan, J.C.L. 1994. Prediction of the interannual variations of tropical cyclone movement over regions of the western North Pacific. *Int. J. Climatol.,* **14,** 527–538.

Chan, J.C.L. 1995. Prediction of annual tropical cyclone activity over the western North Pacific and the South China Sea. *Int. J. Climatol.,* **15,** 1011–1019.

Chang, T. and Yasunari, T. 1982. Fluctuations of global surface pressure patterns during the past 100 years and their relation to the Asian monsoon. Part I. Northern summer (July). *J. Meteor. Soc. Japan,* **60,** 1132–1142.

Chao, Y. and Philander, S.G.H. 1993. On the structure of the Southern Oscillation. *J. Climate,* **6,** 450–469.

Chavez, F.P. 1986. The legitimate El Niño Current. *Trop. Ocean Atmos. Newsl.,* **34,** 1.

Chen, T-C., van Loon, H., Wu, K-D. and Yen, M-C. 1992. Changes in the atmospheric circulation over the North Pacific–North America area since 1950. *J. Meteor. Soc. Japan,* **70,** 1137–1146.

Chen, W.Y. 1982. Assessment of Southern Oscillation sea-level pressure indices. *Mon. Wea. Rev.,* **110,** 800–807.

Chen, Y-Q., Battisti, D.S. and Sarachik, E.S. 1995. A new ocean model for studying the tropical oceanic aspects of ENSO. *J. Phys. Oceanogr.,* **25,** 2065–2089.

Cheney, R.E., Douglas, B., Agreen, R., Miller, L., Porter, D. and Doyle, N. 1987. *GEOSAT altimeter GDR, users handbook.* Natl. Ocean Surv., Natl. Oceanic and Atmos. Admin., Silver Spring, Md, 29 pp.

Cheney, R.E., Douglas, B. and Miller, L. 1989. Evaluation of Geosat altimeter data with application to tropical Pacific sea level variability. *J. Geophys. Res.,* **94,** 4737–4747.

Cheney, R.E., Doyle, N., Douglas, B., Agreen, R., Miller, L., Timmerman, E. and McAdoo, D. 1991. *The complete GEOSAT altimeter GDR handbook.* Natl. Ocean Surv., Natl. Oceanic and Atmos. Admin., Silver Spring, Md, 79 pp.

Chongyin, L. 1990. Interaction between anomalous winter monsoon in East Asia and El Niño events. *Adv. Atmos. Sci.,* **7,** 36–46.

Chowdhury, A. 1995. Crop yield in India in relation to El Niño. *Mausam,* **46,** 127–132.

Clarke, A.J. and Liu, X. 1994. Interannual sea level in the Northern and Eastern Indian Ocean. *J. Phys. Oceanogr.,* **24,** 1224–1235.

Climate Analysis Center, 1986. *Climate Diagnostics Bulletin, No.* **86/4,** NWS, NOAA, Washington, D.C., 9–10.

Climate Prediction Center, 1994–96. *Climate Diagnostics Bulletin, No.* **94/1-96/4**,NWS, NOAA, Washington, D.C.,

Climatological Table for the British Empire, 1923–1939. *The Meteorological Magazine,* **58** to **74.**

Close, A. 1990. The impact of man on the natural flow regime. In Mackay, N. and Eastburn, E., (eds), *The Murray,* Murray Darling Basin Commission, Canberra, Australia, 61–74.

Coffroth, M.A., Lasker, H.R. and Oliver, J.K. 1990. Coral mortality outside of the eastern Pacific during 1982–1983: relationship to El Niño. In Glynn, P.W., (ed.), *Global Ecological Consequences of the 1982–83 El Niño–Southern Oscillation,* Elsevier Oceanography Series, Elsevier, Amsterdam, pp. 141–182.

Cole, J.E. and Fairbanks, R.G. 1990. The Southern Oscillation recorded in the ^{18}O of corals from Tarawa atoll. *Palaeoceanography,* **5,** 669–683.

Cole, J.E., Shen, G.T., Fairbanks, R.G. and Moore, M. 1992. Coral monitors of El Niño/Southern Oscillation dynamics across the equatorial Pacific. Chapter 18, in Diaz, H.F. and Markgraf, V., (eds), *El Niño: Historical review and palaeoclimatic aspects of the Southern Oscillation,* CUP, Cambridge, UK, pp. 349–375.

Cole, J.E., Fairbanks, R.G. and Shen, G.T. 1993. Recent variability in the Southern Oscillation: Isotopic results from a Tarawa atoll coral. *Science,* **260,** 1790–1793.

Collen, B. 1992. *South Pacific Historical Climate Network, Climate Station Histories, Part 1: Southwest Pacific Region.* New Zealand Meteorological Service, Wellington, 113 pp.

Conway, T.J., Tans, P.P., Waterman, L.S., Thoning, K.W., Kitzis, D.R., Masarie, K.A. and Zhang, N. 1994. Evidence for interannual variability of the carbon cycle from the National Oceanic and Atmospheric Administration/Climate Monitoring and Diagnostics Laboratory Global Air Sampling Network. *J. Geophys. Res.,* **99,** D11, 22, 831–22, 855.

Cook, E.R. 1995. Temperature histories from tree rings and corals. *Clim Dyn.,* **11,** 211–222.

Cooke, W.E. 1901. *Climate of Western Australia from observations made during the years 1876–1899.* Government Printer, 35 pp.

Cooper, N.S. and Whysall, K.D.B. 1989. Recent decadal climate variations in the tropical Pacific. *Int. J. Climatol.,* **9,** 221–242.

Cordery, I. and Opoku-Ankomah, Y. 1994. Temporal Variation of Relations Between Tropical Sea-Surface Temperatures and New South Wales Rainfall. *Aust. Meteor. Mag.,* **43,** 73–80.

Crowley, T.J. and Kim, K-Y. 1993. Towards development of a strategy for determining the origin of decadal-centennial scale climate variability. *Quat. Sci. Rev.,* **12,** 375–385.

Cubasch, U, Santer, B.D., Maier-Reimer, E. and Bottinger, M. 1990. Sensitivity of a global coupled ocean–atmosphere circulation model to a doubling of carbon dioxide. In Pitcher, E.J., (ed.), *Science and Engineering on Supercomputers,* Computational Mechanics Publications, Southampton, pp. 347–352.

Davey, M.K., Ineson, S. and Balmaseda, M.A. 1994. Simulation and hindcasts of tropical Pacific Ocean interannual variability. *Tellus,* **46A,** 433–447.

Davey, M.K., Anderson, D.L.T. and Lawrence, S. 1996. A simulation of variability of ENSO forecast skill. *J. Climate,* **9,** 240–246.

Delcroix, T., Picaut, J. and Eldin, G. 1991. Equatorial Kelvin and Rossby Waves evidenced in the Pacific Ocean through Geosat sea level and surface current anomalies. *J. Geophys. Res.,* **96,** C12, 3249–3262.

Delecluse, P., Servain, J., Levy, C., Arpe, K. and Bengtsson, L. 1994. On the connection between the 1984 Atlantic warm event and the 1982–1983 ENSO. *Tellus,* **46A,** 448–464.

Department of Meteorological Services, 1952. *Southern Rhodesia Climate Handbook. Supp. No. 1, Climatological tables for the period July, 1931 to June, 1951.* Government Printing and Stationary Department, 158 pp.

Deser, C. and Blackmon, M.L. 1993. Surface climate variations over the North Atlantic Ocean during winter: 1900–1989. *J. Climate,* **6,** 1743–1753.

Deser, C. and Wallace, J.M. 1987. El Niño events and their relation to the Southern Oscillation: 1925–1986. *J. Geophy. Res.,* **92,** 14,189–14,196.

Deser, C. and Wallace, J.M. 1990. Large-scale atmospheric circulation features of warm and cold episodes in the tropical Pacific. *J. Climate,* **3,** 1254–1281.

Dettinger, M.D. and Cayan, D.R. 1995. Large-scale atmospheric forcing of recent trends towards early snowmelt runoff in California. *J. Climate,* **8,** 606–623.

Deutschen Seewarte, 1887–1904. *Deutsche Ueberseeische Meteorologische Beobachtungen.* Heft I to XII, Hamburg.

Diaz, H.F. and Kiladis, G.N. 1992. Atmospheric teleconnections associated with the extreme phases of the Southern Oscillation. Chapter 2, in Diaz, H.F. and Markgraf, V. (eds) *El Niño: Historical review and palaeoclimatic aspects of the Southern Oscillation,* Cambridge University Press, Cambridge, UK, pp. 7–28.

Diaz, H.F. and Kiladis, G.N. 1995. Climatic variability on decadal to century time-scales. In Henderson-Sellers, A.,(ed.), Future Climates of the World: A Modelling Perspective. *World Survey of Climatology,* **16,** 191–244.

Diaz, H.F. and Markgraf, V., (eds), 1992. *El Niño: Historical review and palaeoclimatic aspects of the Southern Oscillation.* Cambridge University Press, Cambridge, UK, 476 pp.

Diaz, H.F. and Pulwarty, R.S. 1994. An analysis of the time scales of variability in centuries-long ENSO-sensitive records in the last 1000 years. *Climatic Change*, **26**, 317–342.

Diaz, H.F., Bradley, R.S. and Eischeid, J.K. 1989. Precipitation fluctuations over global land areas since the late 1800s. *J. Geophys. Res.*, **94**, 1195–1210.

Dines, W.H. 1916. Review of 'Correlation in seasonal variations of weather'. *Quart, J. Roy. Meteor. Soc.*, **42**, 129–132.

Dines, W.H. 1917. Correlation in seasonal variations of weather. *Quart, J. Roy. Meteor. Soc.*, **43**, 333–335.

Dong, K. 1988. El Niño and tropical cyclone frequency in the Australian region and the northwest Pacific. *Aust. Meteor. Mag.*, **36**, 219–226.

Donguy, J-R. 1994. Surface and subsurface salinity in the tropical Pacific Ocean. Relations with climate. *Prog. Oceanog.*, **34**, 45–78.

Doraiswamy Iyer, V. and Francis, K.A. 1941. The climate of Seychelles with special reference to its rainfall. *Mem. India Meteorol. Dep.*, **XXVII**, Part **III**, 59 pp.

Drake, V.A. and Farrow, R.A. 1988. The influence of atmsopheric structure and motions on insect migration. *Ann. Rev. Entomol.*, **33**, 183–210.

Drosdowsky, W. 1988. Lag relations between the Southern Oscillation and the troposphere over Australia. *BMRC Research Report No.* **13**, Bureau of Meteorology, Melbourne, 201 pp.

Drosdowsky, W. 1993a. An analysis of Australian seasonal rainfall anomalies: 1950–1987. I: Spatial patterns. *Int. J. Climatol.*, **13**, 1–30.

Drosdowsky, W. 1993b. An analysis of Australian seasonal rainfall anomalies: 1950–1987. II: Temporal variability and teleconnection patterns. *Int. J. Climatol.*, **13**, 111–149.

Drosdowsky, W. 1993c. Potential predictability of winter rainfall over southern and eastern Australia using Indian Ocean sea-surface temperature anomalies. *Aust. Meteor., Mag.*, **42**, 1–6.

Drosdowsky, W. 1996. Variability of the Australian summer monsoon at Darwin: 1957–1992. *J. Climate*, **9**, 85–96.

Drosdowsky, W. and Williams, M. 1991. The Southern Oscillation in the Australian region. Part I: Anomalies at the extremes of the oscillation. *J. Climate*, **4**, 619–638.

Druffel, E.R.M. and Griffin S. 1993. Large variations of surface ocean radiocarbon: Evidence of circulation changes in the southwestern Pacific. *J. Geophys. Res.*, **98**, 20,249–20,259.

Druffel, E.R.M., Dunbar, R.B., Wellington, G.M. and Minnis, S.A. 1990. Reef-building corals and identification of ENSO warming episodes. In Glynn, P.W., (ed.), *Global Ecological Consequences of the 1982–83 El Niño–Southern Oscillation*, Elsevier Oceanography Series, vol. 52, Elsevier, Amsterdam, pp. 233–254.

Duffy, D.C. 1990. Seabirds and the 1982–1984 El Niño/Southern Oscillation. In Glynn, P.W., (ed.), Global Ecological Consequences of the 1982–83 El Niño–Southern Oscillation, *Elsevier Oceanography Series,* Elsevier, Amsterdam, 395–416.

Dunbar, R.B. and Cole, J.E. 1993. Coral records of ocean–atmosphere variability. Report from the Workshop on Coral Paleoclimate Reconstruction, *NOAA Climate and Global Change Program Special Report No.* **10**, 38 pp.

Dunbar, R.B. and Cole, J.E. 1995. The tropical influence on global climate: Are surprises the rule? *EOS Trans. AGU*, **76**, 431.

Dunbar, R.B, Wellington, G.M., Colgan, M.W. and Glynn, P.W. 1994. Eastern Pacific sea surface temperature since 1600 A.D.: The ^{18}O record of climate variability in Galapagos corals. *Paleoceanography*, **9**, 291–315.

Efimov, V.V., Prusov, A.V. and Shokurov, M.V. 1995. Patterns of interannual variability defined by a cluster analysis and their relation with ENSO. *Quart. J. Roy. Meteor. Soc.*, **121**, 1651–1679.

Eguiguren, V. 1894. Las lluvias en Piura. *Bol. Soc. Geogr. de Lima*, **4**(7–9), 241–258.

Eliot, J. 1896. On the origin of the cold weather storms of the year 1893 in India, and the character of the air movement on the indian seas and the equatorial belt, more especially during the south-west monsoon period (as shown by the data of the Indian monsoon area charts for the year 1893). *Quart. J. Roy. Meteor. Soc.*, **XXII**, 1–37.

Eliot, J. 1897. Periodic variations of rainfall in India. *Nature*, **56**, 110–115.

Eliot, J. 1904. The meteorology of the Empire during the unique period 1892–1902. *Broad Views*, **1**, 191–201.

Ellery, R.L.J. 1873–1891. *Monthly record of results of observations in meteorology, terrestrial magnetism, etc., etc., taken at the Melbourne Observatory.* Government Printer, Victoria.

Elliott, W.P. and Angell, J.K. 1987. The relation between Indian monsoon rainfall, the Southern Oscillation, and hemispheric air and sea temperature: 1884–1984. *J. Climate App. Meteor.*, **26**, 943–948.

Elliott, W.P. and Angell, J.K. 1988. Evidence for changes in Southern Oscillation relationships during the last 100 years. *J. Climate*, **1**, 729–737.

Enfield, D.B. 1987. Progress in understanding EL Niño. *Endeavour*, **11**, 197–204.

Enfield, D.B. 1988. Is El Niño becoming more common ? *Oceanogr. Mag.*, **1**, 23–27, 59.

Enfield, D.B. 1989. El Niño, past and present. *Rev. Geophys.*, **27**, **1**, 159–187.

Enfield, D.B. 1992. Historical and prehistorical overview of El Niño/Southern Oscillation. Chapter 5, in Diaz, H.F. and Markgraf, V. (eds) *El Niño: Historical review and palaeoclimatic aspects of the Southern Oscillation,* Cambridge University Press, Cambridge, UK, pp. 95–117.

Enfield, D.B. and Cid, L.S. 1991. Low-frequency changes in El Niño–Southern Oscillation. *J. Climate,* **4,** 1137–1146.

Enomoto, H. 1991. Fluctuations of snow accumulation in the Antarctic and sea level pressure in the Southern Hemisphere in the last 100 years. *Climatic Change,* **18,** 67–87.

Evans, J.L. and Allan, R.J. 1992. El Niño/Southern Oscillation modification to the structure of the monsoon and tropical cyclone activity in the Australasian region. *Int. J. Climatol.,* **12,** 611–623.

Ezer, T., Mellor, G.L. and Greatbatch, R.J. 1995. On the interpentadal variability of the North Atlantic Ocean: Model simulated changes in transport, meridional heat flux and coastal sea level between 1955–1959 and 1970–1974. *J. Geophys. Res.,* **100, C6,** 10,559–10,566.

Feely, R.A., Wanninkhof, R., Cosca, C.E., McPhaden, M.J., Byrne, R.H., Miller, F.J., Chavez, F.P., Clayton, T., Campbell, D.M. and Murphy, P.P. 1994. The effect of tropical instability waves on CO_2 species distributions along the equator in the eastern equatorial Pacific during the 1992 ENSO event. *Geophys. Res. Lett.,* **21,** 277–280.

Fitzharris, B.B., Hay, J.E. and Jones, P.D. 1992. Behaviour of New Zealand glaciers and atmospheric circulation changes over the past 130 years. *The Holocene,* **2,** 97–106.

Fletcher, J.O., Radok, U. and Slutz, R. 1982. Climatic signals of the Antarctic Ocean. *J. Geophys. Res.,* **87,** 4269–4276.

Flohn, H. 1986. Indonesian droughts and their teleconnections. *Berliner Geographische Studien,* **20,** 251–265.

Flohn, H. and Fleer, H. 1975. Climatic teleconnections with the equatorial Pacific and the role of ocean/atmosphere coupling. *Atmosphere,* **13,** 96–109.

Foley, J.C. 1957. Droughts in Australia: Review of records from earliest years of settlement. *Bureau of Meteorology, Bull. No.* **43,** 281 pp.

Folland, C.K. and Parker, D.E. 1995. Correction of instrumental biases in historical sea surface temperature data. *Quart. J. Roy. Meteor. Soc.,* **121,** 319–367.

Folland, C.K., Colman, A., Parker, D.E. and Bevan, A. 1990. Low frequency variability of the oceans. Climate Research Tech. Note CRTN No. **4,** Hadley Centre for Climate Prediction and Research, Meteorological Office, Bracknell, Berkshire, UK, 8 pp.

Fouhy, E., Coutts, L., McGann, R., Collen, B. and Salinger, J. 1992. *South Pacific Historical Climate Network, Climate Station Histories, Part 2: New Zealand and Offshore Islands.* New Zealand Meteorological Service, Wellington, 216 pp.

Fraedrich, K. 1994. An ENSO impact on Europe? *Tellus,* **46A,** 541–552.

Fraedrich, K. and Muller, K. 1992. Climate anomalies in Europe associated with ENSO extremes. *Int. J. Climatol.,* **12,** 25–31.

Fraedrich, K., Muller, K. and Kuglin, R. 1992. Northern Hemisphere circulation regimes during the extremes of the El Niño/Southern Oscillation. *Tellus,* **44A,** 33–40.

Francey, R.J., Tans, P.P., Allison, C.E., Enting, I.E., White, J.W.C. and Trolier, M. 1995. Changes in the oceanic and terrestrial carbon uptake since 1982. *Nature,* **373,** 326–330.

Fu, C. 1986. A review of studies of El Niño–Southern Oscillation phenomenon associated with interannual climate variability. *Chinese J. Atmos. Sci.,* **11,** 237–253.

Fu, C. and Fletcher, J. 1988. Large signals of climatic variation over the ocean in the Asian monsoon region. *Adv. Atmos. Sci.,* **5,** 389–404.

Fu, C., Diaz, H.F. and Fletcher, J.O. 1986. Characteristics of the response of sea surface temperature in the central Pacific associated with warm episodes of the Southern Oscillation. *Mon. Wea. Rev.,* **114,** 1716–1738.

Gagan, M.K. and Chivas, A.R. 1995. Oxygen isotopes in Western Australian coral reveal Pinatubo aerosol-induced cooling in the western Pacific warm pool. *Geophys. Res. Lett.,* **22,** 1069–1072.

Gasques, J.G. and Magalhaes, A.R. 1987. Climate anomalies and their impacts in Brazil during the 1982–83 ENSO event. In Glantz, M.H., Katz, R.W. and Krentz, M., (eds), *The Societal Impacts Associated with the 1982–83 Worldwide Climate Anomalies.* Environmental and Societal Impacts Group, NCAR, Boulder, CO, 30–36.

Gibbs, H.L. and Grant, P.R. 1987. Oscillating selection on Darwin's finches. *Nature,* **327,** 511–513.

Gill, A.E. and Rasmusson, E.M. 1983. The 1982–83 climate anomaly in the equatorial Pacific. *Nature,* **306,** 229–234.

Glantz, M.H. 1984. El Niño – should it take the blame for disasters? *Mazingira,* 21–26.

Glantz, M.H. 1996. *Currents of Change: El Niño's impact on Climate and Society.* Cambridge University Press, 200 pp.

Glantz, M.H., and Feingold, L.E. (eds). 1990. Climate Variability, *Climate Change and Fisheries.* Environmental and Societal Impacts Group, NCAR, Boulder, CO, 139 pp.

Glantz, M.H., and Thompson, J.D. (eds) 1981. Resource Management and Environment Uncertainty: Lessons from Coastal Upwelling Fisheries. P. Adv. Env. Sci., vol. 11, Wiley-Interscience, 491 pp.

Glantz, M.H., Katz, R.W. and Krentz, M. (eds). 1987. *The Societal Impacts Associated with the 1982–83 Worldwide Climate Anomalies.* Environmental and Societal Impacts Group, NCAR, Boulder, CO, 105 pp.

Glantz, M., Katz, R. and Nicholls, N. (eds). 1991. *ENSO teleconnections linking worldwide climate anomalies: Scientific basis and societal impacts.* Cambridge University Press, Cambridge, UK, 535 pp.

Gloersen, P. 1995. Modulation of hemispheric sea-ice cover by ENSO events. *Nature, 373,* 503–506.

Glynn, P.W., (ed.), 1990a. *Global Ecological Consequences of the 1982–83 El Niño–Southern Oscillation,* Elsevier Oceanography Series, vol. 52, Elsevier, Amsterdam, 563 pp.

Glynn, P.W. 1990b. Coral mortality and disturbances to coral reefs in the tropical eastern Pacific. In Glynn, P.W., (ed.), *Global Ecological Consequences of the 1982–83 El Niño–Southern Oscillation,* Elsevier Oceanography Series, vol. 52, Elsevier, Amsterdam, pp. 55–126.

Godfrey, J.S. 1975. On ocean spindown I: A linear experiment. *J. Phys. Oceanogr., 5,* 399–409.

Godfrey, J.S., Alexiou, A., Ilahude, A.G., Legler, D.M., Luther, M.E., McCreary, Jr., J.P., Meyers, G.A., Mizumo, K., Rao, R.R., Shetye, S.R., Toole, J.H. and Wacongne, S. 1995. The role of the Indian Ocean in the global climate system: Recommendations regarding the Global Ocean Observing System. *Report of the Ocean Observing System Development Panel, OOSDP Background Report Number 6,* Texas A&M University, College Station, TX, USA, 89 pp.

Golnaraghi, M. and Kaul, R. 1995. Responding to ENSO. *Environment, 37,* 16–20, 38–43.

Gordon, A.L., Zebiak, S.E. and Bryan, K. 1992. Climate variability and the Atlantic Ocean. *EOS Trans. AGU, 73,* 161, 164–165.

Gordon, C. 1989. Tropical ocean–atmosphere interactions in a coupled model. *Phil. Trans. Roy. Soc. Lond., A329,* 207–223.

Gordon, H.B. and O'Farrell, S.P. 1996. Transient climate change in the CSIRO coupled model with a dynamic sea ice model. *Mon. Wea. Rev.* (in press).

Gordon, N.D. 1985. The Southern Oscillation: A New Zealand perspective. *J. Roy. Soc. New Zealand, 15,* 137–155.

Gordon, N.D. 1986. The Southern Oscillation and New Zealand weather. *Mon. Wea. Rev., 114,* 371–387.

Goreau, T.J. and Hayes, R.L, 1994. Coral bleaching and ocean "hot spots". *Ambio, 23,* 176–180.

Graham, N.E. 1994. Decadal–scale climate variability in the tropical and North Pacific during the 1970s and 1980s: Observations and model results. *Clim. Dyn., 10,* 135–162.

Graham, N.E. and Barnett, T.P. 1995. ENSO and ENSO-related predictability. Part II: Northern Hemisphere 700-mb height predictions based on a hybrid coupled ENSO model. *J. Climate, 8,* 544–549.

Graham, N.E. and White, W.B. 1988. The El Niño cycle: A natural oscillator of the Pacific Ocean–atmosphere system. *Science, 240,* 1293–1302.

Graham, N.E. and White, W.B. 1990. The role of the western boundary in the ENSO cycle: Experiments with coupled models. *J. Phys. Oceanogr., 20,* 1935–1948.

Graham, N.E. and White, W.B. 1991. Comments on 'On the role of off-equatorial oceanic Rossby Waves during ENSO'. *J. Phys. Oceanogr., 21,* 453–465.

Graham, N.E., Barnett, T.P., Wilde, R., Ponater, M. and Schubert, S. 1994. On the roles of tropical and midlatitude SSTs in forcing interannual to interdecadal variability in the winter Northern Hemisphere circulation. *J. Climate, 7,* 1416–1441.

Gray, W.M. and Sheaffer, J.D. 1991. El Niño and QBO influences on tropical cyclone activity. In Glantz, M., Katz, R. and Nicholls, N., (eds), *ENSO teleconnections linking worldwide climate anomalies: Scientific basis and societal impacts.* Cambridge University Press, Cambridge, UK, 257–284.

Gray, W.M., Sheaffer, J.D. and Knaff, J.A. 1992. Influence of the stratospheric QBO on ENSO variability. *J. Meteor. Soc. Japan, 70,* 975–994.

Greatbatch, R.J. and Xu, J. 1993. On the transport of volume and heat through sections across the North Atlantic: Climatology and the pentads 1955–1959, 1970–1974. *J. Geophys. Res., 98,* C6, 10,125–10,143.

Greatbatch, R.J., Fanning, A.F. and Goulding, A.D. 1991. A diagnosis of interpentadal circulation changes in the North Atlantic. *J. Geophys. Res., 96,* 22,009–22,023.

Gregory, S. 1991. Interrelationships between Indian and northern Australian summer monsoon rainfall values. *Int. J. Climatol., 11,* 55–62.

Griffiths, R.F. 1910. *Meteorological observations made at the Adelaide Observatory during the Year 1907.* Government Printer, Adelaide, South Australia, 250 pp.

Gu, D. and Philander, S.G.H. 1995. Secular changes of annual and interannual variability in the tropics during the past century. *J. Climate, 8,* 864–876.

Gunther, F.R. 1936. A report on oceanographical investigation in the Peru coastal current. *Discovery Rep., 13,* 107–276.

Hackert, E.C. and Hastenrath, S. 1986. Mechanisms of Java rainfall anomalies. *Mon. Wea. Rev., 114,* 745–757.

Halpert, M.S. and Ropelewski, C.F. 1992. Surface Temperature Patterns Associated with the Southern Oscillation. *J. Climate, 5,* 577–593.

Hameed, S., Sperber, K.R. and Cess, R.D. 1989. Genesis of the Southern Oscillation with the atmosphere: Illustration with global general circulation models. *Geophys. Res. Lett., 16,* 691–844.

Hamilton, K. and Garcia, R.R. 1986. El Niño/Southern Oscillation events and their associated midlatitude teleconnections 1531–1841. *Bull. Amer. Meteor. Soc., 67,* 1354–1361.

Handler, P. 1984. Possible association of stratospheric aerosols and El Niño type events. *Geophys. Res. Lett., 11,* 1121–1124.

Handler, P. 1986. Possible association between the climatic effects of stratospheric aerosols and sea surface temperatures in the eastern tropical Pacific Ocean. *J. Climatol., 6,* 31–41.

Handler, P. and Andsager, K. 1990. Volcanic aerosols, El Niño and the Southern Oscillation. *Int. J. Climatol.,* **10**, 413–424.

Hastenrath, S. 1991. *Climate dynamics of the tropics.* Kluwer, Hingham, Mass., 488 pp.

Hastenrath, S. 1995. Recent advances in tropical climate prediction. *J. Climate,* **8**, 1519–1532.

Hastenrath, S. and Greischar, L. 1993. Further work on the prediction of northeast Brazil rainfall anomalies. *J. Climate,* **6**, 743–758.

Hastenrath, S. and Wolter, K. 1992. Large-scale patterns and long-term trends of circulation variability associated with Sahel rainfall anomalies. *J. Meteor. Soc. Japan,* **70**, 1045–1055.

Hastenrath, S., Nicklis, A. and Greischar, L. 1993. Atmospheric-hydrospheric mechanisms of climate anomalies in the western equatorial Indian Ocean. *J. Geophys. Res.,* **98**, C11, 20,219–20,235.

Hastenrath, S., Greischar, L. and van Heerden, J. 1995. Prediction of the summer rainfall over South Africa. *J. Climate,* **8**, 1511–1518.

Hastings, P.A. 1990. Southern Oscillation influences on tropical cyclone activity in the Australia/south-west Pacific region. *Int. J. Climatol.,* **10**, 291–298.

Henry, A.J. 1925. Meteorological data for Midway Island, North Pacific Ocean. *Mon. Wea. Rev.,* **53**, 357.

Hildebrandsson, H.H. 1897. Quelques recherches sur les centres d'action de l'atmosphère. *Kunglica Svenska Vetenskaps-akademiens Handlingar,* **29**, 36 pp

Hirono, M. 1988. On the trigger of El Niño Southern Oscillation by the forcing of early El Chichon volcanic aerosols. *J. Geophys. Res.,* **93**, 5365–5384.

Hirst, A.C. 1989. Recent advances in the theory of ENSO. *Newslett. Aust. Meteor. Oceanogr. Soc.,* **2**, 101–113.

Hirst, A.C. and Godfrey, J.S. 1993. The role of Indonesian throughflow in a global ocean GCM. *J. Phys. Oceanogr.,* **23**, 1057–1086.

Hirst, A.C. and Godfrey, J.S. 1994. The response to a sudden change in Indonesian throughflow in a global ocean GCM. *J. Phys. Oceanogr.,* **24**, 1895–1910.

Hisard, P. 1992. The centenary of the 1891 – El Niño report by Carranza (1892): First encounters of the 'ENSO phenomenon' by Krusenstern and by Humboldt. In Ortlieb, L. and Machare, J., (eds), *Paleo-ENSO Records Intern. Symp., Extended Abstracts,* ORSTOM (Lima) and CONCYTEC, Lima, pp. 133–141.

Hocquenghem, A.M. and Ortlieb, L. 1992. Eventos El Niño y Iluvias anormales en la costa del Peru: siglos XVI–XIX. *Bull. Inst. Fr. Etud. Andines, Lima,* **21**(1), 197–278.

Holland, G.J. 1986. Interannual variability of the Australian summer monsoon at Darwin 1952–82. *Mon. Wea. Rev.,* **114**, 594–604.

Horel, J.D. and Wallace, J.M. 1981. Planetary-scale atmospheric phenomena associated with the Southern Oscillation. *Mon. Wea. Rev.,* **109**, 813–829.

Hsieh, W.W. and Hamon, B.V. 1991. The El Niño–Southern Oscillation in south-eastern Australian waters. *Aust. J. Mar. Freshwater Res.,* **42**, 263–275.

Hughes, T. 1992. Abrupt climatic change related to unstable ice-sheet dynamics: Towards a new paradigm. *Palaeogeogr., Palaeoclim., Palaeoecol. (Global and Planetary Change Section),* **97**, 203–234.

Hulme, M. 1992. Rainfall changes in Africa: 1931–1960 to 1961–1990. *Int. J. Climatol.,* **12**, 685–699.

Hunt, B.G., Zebiak, S.E. and Cane, M.A. 1994. Experimental predictions of climatic variability for lead times of twelve months. *Int. J. Climatol.,* **14**, 507–526.

Hunt, H.A. 1910–1913. *Australian Monthly Weather Report and Meteorological Abstract. Vol.* **1** to **4**. Government Printer, Melbourne.

Hunt, H.A. 1911. *Results of rainfall observations made in Victoria during 1840–1910.* Commonwealth Government Printer, Melbourne, 55 pp.

Hunt, H.A. 1914. *Results of rainfall observations made in Queensland.* Commonwealth Government Printer, Melbourne, 285 pp.

Hunt, H.A. 1916. *Results of rainfall observations made in New South Wales during 1909–1914.* Commonwealth Government Printer, Melbourne, 224 pp.

Hunt, H.A. 1918. *Results of rainfall observations made in South Australia and the Northern Territory.* Commonwealth Government Printer, Melbourne, 421 pp.

Hunt, H.A. 1929. *Results of rainfall observations made in Western Australia.* Commonwealth Government Printer, Melbourne, 387 pp.

Hurrell, J.W. 1995. Decadal trends in the North Atlantic Oscillation: Regional temperatures and precipitation. *Science,* **269**, 676–679.

Hurrell, J.W. and van Loon, H. 1994. A modulation of the atmospheric annual cycle in the Southern Hemisphere. *Tellus,* **46A**, 325–338.

Hutchinson, P. 1992. The Southern Oscillation and prediction of 'Der' season rainfall in Somalia. *J. Climate,* **5**, 525–531.

India Meteorological Department, 1976. *Hundred years of weather service (1875–1975).* Dy. Director General of Observatories, Poona, 207 pp.

Ineson, S. and Gordon, C. 1989. Parametrization of the upper ocean mixed layer in a tropical ocean GCM. *Dynamical Climatology Tech. Note No. 74,* Meteorological Office, Bracknell, Berkshire, UK, 31 pp..

Inoue, M. and Bigg, G.R. 1995. Trends in wind and sea-level pressure in the tropical Pacific Ocean for the period 1950–1979. *Int. J. Climatol.,* **15**, 35–52.

Isdale, P. 1984. Fluorescent bands in massive corals record centuries of coastal rainfall. *Nature*, **310**, 578–579.

Isdale, P. and Kotwicki, V. 1987. Lake Eyre and the Great Barrier Reef: A palaeohydrological ENSO connection. *J. Roy. Geogr. Soc. Austr. S. Aust. Branch*, **87**, 44–55.

Jacobs, G.A., Hurlburt, H.E., Kindle, J.C., Metzger, E.J., Mitchell, J.L., Teague, W.J. and Wallcraft, A.J. 1994. Decade-scale trans-Pacific propagation and warming effects of an El Niño anomaly. *Nature*, **370**, 360–363.

Ji, M., Kumar, A. and Leetmaa, A. 1994a. A multiseason climate forecast system at the National Meteorological Center. *Bull. Amer. Meteor. Soc.*, **75**, 569–577.

Ji, M., Kumar, A. and Leetmaa, A. 1994b. An experimental coupled forecast system at the National Meteorological Center. Some early results. *Tellus*, **46A**, 398–418.

Jinghua, Y., Longxun, C. and Gu, W. 1988. The propagation characteristics of interannual low-frequency oscillations in the tropical air-sea system. *Adv. Atmos. Sci.*, **5**, 405–420.

Jingxi, L., Yihui, D. and Shiying, G. 1990. Interannual low-frequency oscillations of meridional winds over the equatorial Indian–Pacific Oceans. *Acta Meteor. Sin.*, **4**, 586–597.

Jordan, R.S. 1991. Impact of ENSO events on the southeastern Pacific region with special reference to the interaction of fishing and climate variability. In Glantz, M., Katz, R. and Nicholls, N., (eds), *ENSO teleconnections linking worldwide climate anomalies: Scientific basis and societal impacts*. Cambridge University Press, Cambridge, UK, 401–430.

Jose, A.M. 1990. El Niño-related drought events in the Philippines during the past decade (1980–1989). Paper presented at the *1990 Philippine Water Congress and Exhibitions*, Baguio, 18 pp.

Jose, A.M. 1992. A preliminary assessment of the 1991–1992 El Niño-related drought in the Philippines. CAB T.P. No., **92.1**, PAGASA, 11 pp.

Joseph, P.V., Liebmann, B. and Hendon, H.H. 1991. Interannual variability of the Australian summer monsoon onset: Possible influence of Indian summer monsoon and El Niño. *J. Climate*, **4**, 529–538.

Joseph, P.V., Eischeid, J.K. and Pyle, R.J. 1994. Interannual variability of the onset of the Indian summer monsoon and its association with atmospheric features, El Niño, and sea surface temperature anomalies. *J. Climate*, **7**, 81–105.

Jones, P.D. 1988. The influence of ENSO on global temperatures. *Climate Monitor*, **17**, 80–89.

Jones, P.D. 1991. Southern Hemisphere sea-level pressure data: An analysis and reconstructions back to 1951 and 1911. *Int. J. Climatol.*, **11**, 585–607.

Jones, P.D. and Wigley, T.M.L. 1988. Antarctic gridded sea level pressure data: An analysis and reconstruction back to 1957. *J. Climate*, **1**, 1199–1220.

Julian, P.R. and Chervin, R.M. 1978. A study of the Southern Oscillation and Walker Circulation phenomenon. *Mon. Wea. Rev.*, **106**, 1433–1451.

Jury, M.R. and Pathack, B.M.R. 1991. A study of climate and weather variability over the southwest Indian Ocean. *Meteorol. Atmos. Phys.*, **47**, 37–48.

Jury, M.R. and Pathack, B.M.R. 1993. Composite climatic patterns associated with extreme modes of summer rainfall over southern Africa: 1975–1984. *Theor. Appl. Climatol.*, **47**, 137–145.

Jury, M.R., Parker, B. and Waliser, D. 1994a. Evolution and variability of the ITCZ in the SW Indian Ocean: 1988–90. *Theor. Appl. Climatol.*, **48**, 187–194.

Jury, M.R., McQueen, C. and Levey, K. 1994b. SOI and QBO signals in the African region. *Theor. Appl. Climatol.*, **50**, 103–115.

Jury, M.R., Parker, B.A., Raholijao, N. and Nassor, A. 1995. Variability of summer rainfall over Madagascar: Climatic determinants at interannual scales. *Int. J. Climatol.*, **15**, 1323–1332.

Kahya, E. and Dracup, J.A. 1993. U.S. streamflow patterns in relation to the El Niño/Southern Oscillation. *Water Resour. Res.*, **29**, 2491–2503.

Kahya, E. and Dracup, J.A. 1994. The influences of Type 1 El Niño and La Niña events on streamflows in the Pacific southwest of the United States. *J. Climate*, **7**, 965–976.

Kane, R.P. 1994. Interannual variability of some trace elements and surface aerosol. *Int. J. Climatol.*, **14**, 691–704.

Karoly, D.J. 1989. Southern Hemisphere circulation features associated with El Niño–Southern Oscillation events. *J. Climate*, **2**, 1239–1252.

Karoly, D.J., Hope, P. and Jones, P.D. 1996. Decadal variations of the Southern Hemisphere circulation. *Int. J. Climatol.* (in press).

Kawamura, R. 1994. A rotated EOF analysis of global sea surface temperature variability with interannual and interdecadal scales. *J. Phys. Oceanogr.*, **24**, 707–715.

Kawamura, R., Sugi, M. and Sato, N. 1995a. Interdecadal and interannual variability in the northern extratropical circulation simulated with the JMA global model. Part I: Wintertime leading mode. *J. Climate*, **8**, 3006–3019.

Kawamura, R., Sugi, M. and Sato, N. 1995b. Interdecadal and interannual variability in the northern extratropical circulation simulated with the JMA global model. Part II: Summertime leading mode. *J. Climate*, **8**, 3020–3027.

Keen, R.A. 1982. The role of cross-equatorial tropical cyclone pairs in the Southern Oscillation. *Mon. Wea. Rev.*, **110**, 1405–1416.

Keppenne, C.L. 1995. An ENSO signal in soybean futures prices. *J. Climate*, **8**, 1685–1689.

Kerr, R.A. 1994. Did the tropical Pacific drive the world's warming? *Science*, **266**, 544–545.

Kessler, W.S. 1991. Can reflected extra-equatorial Rossby Waves drive ENSO? *J. Phys. Oceanogr.*, **21**, 444–452.

Kessler, W.S. and McPhaden, M.J. 1995. The 1992-1993 El Niño in the central Pacific. *Deep-Sea Res. II*, **42**, 295–333.

Kestin, T.S., Karoly, D.J., Yano, J-I. and Rayner, N.A. 1996. Time-frequency analysis of ENSO signals. *J. Climate* (submitted).

Khandekar, M.L. 1991. Eurasian snow cover, Indian monsoon and El Niño/Southern Oscillation – A synthesis. *Atmosphere-Ocean*, **29**, 636–647.

Kidson, E. 1925. Some periods in Australian weather. *Bureau of Meteorology, Bull. No.* **17**, 5–33.

Kidson, J.W. 1975. Tropical eigenvector analysis and the Southern Oscillation. *Mon. Wea. Rev.*, **103**, 187–196.

Kiladis, G.N. and Diaz, H.F. 1986. An analysis of the 1877–78 ENSO episode and comparison with 1982–83. *Mon. Wea. Rev.*, **114**, 1035–1047.

Kiladis, G.N. and Diaz, H.F. 1989: Global climatic anomalies associated with extremes in the Southern Oscillation. *J. Climate*, **2**, 1069–1090.

Kiladis, G.N. and Sinha, S.K. 1991. ENSO monsoon and drought in India. In Glantz, M., Katz, R. and Nicholls, N., (eds), *ENSO teleconnections linking worldwide climate anomalies: Scientific basis and societal impacts*. Cambridge University Press, Cambridge, UK, 431–458.

Kiladis, G.N. and van Loon, H. 1988. The Southern Oscillation. Part VII: Meteorological anomalies over the Indian and Pacific sectors associated with the extremes of the oscillation. *Mon. Wea. Rev.*, **116**, 120–136.

Kirchner, I. and Graf, H-F. 1995. Volcanos and El Niño: signal separation in Northern Hemisphere winter. *Clim. Dyn.*, **11**, 341–358.

Klaben, J., Hense, A. and Romer, U. 1994. Climate anomalies north of 55°N associated with tropical climate extremes. *Int. J. Climatol.*, **14**, 829–842.

Kleeman, R. 1991. A simple model of the atmospheric response to ENSO sea surface temperature anomalies. *J. Atmos. Sci.*, **48**, 3–18.

Kleeman, R. 1993. On the dependence of hindcast skill on ocean thermodynamics in a coupled ocean–atmosphere model. *J. Climate*, **6**, 2012–2033.

Kleeman, R. and Power, S.B. 1994. Limits to predictability in a coupled ocean–atmosphere model due to atmospheric noise. *Tellus*, **46A**, 529–540.

Knutson, T.R. and Manabe, S. 1994. Impact of increasing CO_2 on simulated ENSO-like phenomena. *Geophys. Res. Lett.*, **21**, 2295–2298.

Knutson, T.R. and Manabe, S. 1995. Time-mean response over the tropical Pacific to increased CO_2 in a coupled ocean–atmosphere model. *J. Climate*, **8**, 2181–2199.

Kotwicki, V. and Isdale, P. 1991. Hydrology of Lake Eyre: El Niño link. *Palaeogeogr., Palaeoclimatol., Palaeoecol.*, **84**, 87–98.

Kousky, V.E. and Kayano, M.T. 1993. A comprehensive atmospheric index of the Southern Oscillation. Paper presented at the *6th Conference on Climate Variations*, AMS, Nashville, Tenn., pp. 264–268.

Kousky, V.E., Kagano, M.T. and Cavalcanti, I.F.A. 1984. A review of the Southern Oscillation: Oceanic–atmospheric circulation changes and related rainfall anomalies. *Tellus* **36A**, 490–504.

Krishnamurti, T.N., Chu, S.H. and Iglesias, W. 1986. On the sea level pressure of the Southern Oscillation. *Arch. Met. Geoph. Biokl. Ser. A*, **34**, 385–425.

Kuhnel, I. 1993. Periodicity and strength of the ENSO climatic signal and its consequences for sugarcane production in Queensland, *Proceedings of the Australian Society of Sugar Cane Technologists*, 261–267.

Kuhnel, I., McMahon, T.A., Findlayson, B.L., Haines, A., Whetton, P.H. and Gibson, T.T. 1990. Climatic influences on streamflow variability: A comparison between southeastern Australia and southeastern United States of America. *Water Resour. Res.*, **26**, 2483–2496.

Kumar, A., Hoerling, M., Ji, M., Leetmaa, A. and Sardeshmukh, P. 1996. Assessing a GCM's suitability for making seasonal predictions. *J. Climate*, **9**, 115–129.

Kutzbach, G. 1987. Concepts of monsoon physics in historical perspective: The Indian Monsoon (Seventeenth to early Twentieth Century). In Fein, J.S. and Stephens, P.L., (eds), *Monsoons*, Wiley, New York, pp. 159–209.

La Cock, G.D. 1986. The Southern Oscillation, environmental anomalies, and mortality of two southern African seabirds. *Climatic Change*, **8**, 173–184.

Lagerloef, G.S.E. 1995. Interdecadal variations in the Alaska Gyre. *J. Phys. Oceanogr.*, **25**, 2242–2258.

Lall, U. and Mann, M. 1995. The Great Salt Lake: A barometer of low-frequency climatic variability. *Water Resour. Res.*, **31**, 2502–2515.

Lander, M.A. 1994. An exploratory analysis of the relationship between tropical storm formation in the western North Pacific and ENSO. *Mon. Wea. Rev.*, **122**, 636–651.

Landsea, C.W., Gray, W.M., Mielke, Jr., P.W. and Berry, K.J. 1992. Long-term variations of western Sahelian monsoon rainfall and intense U.S. landfalling hurricanes. *J. Climate*, **5**, 1528–1534.

Latif, M. and Barnett, T.P. 1994a. Causes of decadal climate variability over the North Pacific and North America. *Max-Planck-Institut fur Meteorologie, Report No.* **141**, 17 pp.

Latif, M. and Barnett, T.P. 1994b. Causes of decadal climate variability over the North Pacific and North America. *Science,* **266**, 634–637.

Latif, M. and Barnett, T.P. 1995. Interactions of the tropical oceans. *J. Climate,* **8**, 952–964.

Latif, M. and Graham, N.E. 1992. How much predictive skill is contained in the thermal structure of an oceanic GCM *J. Phys. Oceanogr.,* **22**, 951–962.

Latif, M. and Neelin, J.D. 1994. El Niño/Southern Oscillation. *Max-Planck-Institut fur Meteorologie, Report No.* **129**, 25 pp.

Latif, M. and Villwock, A. 1990. Interannual variability as simulated in coupled ocean–atmosphere models. *J. Marine Systems,* **1**, 51–60.

Latif, M., Biercamp, J., von Storch, H., McPhaden, M.J. and Krik, E. 1990. Simulation of ENSO related surface wind anomalies with an atmospheric GCM forced by observed SST. *J. Climatol.,* **3**, 509–521.

Latif, M., Sterl, A., Assenbaum, M., Junge, M.M. and Maier-Reimer, E. 1994a. Climate variability in a coupled GCM. Part II: The Indian Ocean and monsoon. *J. Climate,* **7**, 1449–1462.

Latif, M., Stockdale, T., Wolff, J., Burgers, G., Maier-Reimer, E., Junge, M.M., Arpe, K. and Bengtsson, L. 1994b. Climatology and variability in the ECHO coupled GCM. *Tellus,* **46A**, 351–366.

Latif, M., Barnett, T.P., Cane, M.A., Flugel, M., Graham, N.E., von Storch, H., Xu, J-S. and Zebiak, S.E. 1994c. A review of ENSO prediction studies. *Clim. Dyn.,* **9**, 167–179.

Latif, M., Kleeman, R. and Eckert, C. 1995. Greenhouse warming, decadal variability, or El Niño? An attempt to understand the anomalous 1990s. *Max-Planck-Institut fur Meteorologie, Report No.* **175**, 51 pp.

Lau, K-M. 1985. Subseasonal-scale oscillation, bimodal climatic state and the El Niño/Southern Oscillation. In Nihoul, J.C.J., (ed.), *Coupled ocean–atmosphere models.,* Amsterdam, Elsevier Oceanography Series **40**, 29–40.

Lau, K-M. and Chan, P.H. 1985. Aspects of the 40–50 Day Oscillation during the northern winter as inferred from satellite-observed outgoing longwave radiation. *Mon. Wea. Rev.,* **113**, 1889–1909.

Lau, N-C, Philander, S.G.H. and Nath, M.J. 1992. Simulation of ENSO-like phenomena with a low-resolution coupled GCM of the global ocean and atmosphere. *J. Climate,* **5**, 284–307.

Laurie, W.A. 1990. Effects of the 1982–83 El Niño–Southern Oscillation event on marine iguana (Amblyrhynchus cristatus Bell, 1825) populations on Galapagos. In Glynn, P.W., (ed.), *Global Ecological Consequences of the 1982–83 El Niño–Southern Oscillation,* Elsevier Oceanography Series, vol. 52, Elsevier, Amsterdam, pp. 361–380.

Leighly, J.B. 1933 Marquesan meteorology. *Uni. Calif. Pub. Meteor.,* **6**, 147–172.

Lejenas, H. 1995. Long term variations of atmospheric blocking in the Northern Hemisphere. *J. Meteor. Soc. Japan,* **73**, 79–88.

Lenderink, G. and Haarsma, R.J. 1994. Variability and multiple equilibria of the thermohaline circulation associated with deep-water formation. *J. Phys. Oceanogr.,* **24**, 1480–1493.

Levitus, S. 1982. *Climatological Atlas of the World Ocean.* US Dep. Comm., Rockville, MA, NOAA Prof. Pap., 13, 173 pp.

Levitus, S., Antonov, J.I. and Boyer, T.P. 1994. Interannual variability of temperature at a depth of 125 meters in the North Atlantic Ocean. *Science,* **266**, 96–99.

Liang, X-Z., Samel, A.N. and Wang, W-C. 1995. Observed and GCM simulated decadal variability of monsoon rainfall in east China. *Clim. Dyn.,* **11**, 103–114.

Limberger, D. 1990. El Niño effect on South American pinniped species. In Glynn, P.W., (ed.), *Global Ecological Consequences of the 1982–83 El Niño–Southern Oscillation,* Elsevier Oceanography Series, vol. 52, Elsevier, Amsterdam, pp. 417–432.

Lindesay, J.A. 1988. South African rainfall, the Southern Oscillation and a southern hemisphere semi-annual cycle. *J. Climatol.,* **8**, 17–30.

Lindesay, J.A. and Vogel, C.H. 1990. Historical evidence for Southern Oscillation-southern African rainfall relationships. *Int. J. Climatol.,* **10**, 679–689.

Lindesay, J.A. and Jury, M.R. 1991. Atmospheric circualtion controls and characteristics of a flood event in central South Africa. *Int. J. Climatol.,* **11**, 609–627.

Lindesay, J.A. and Allan, R.J. 1993. Summer rainfall modulation in southern Africa and Australia with interannual fluctuations over the Indian Ocean. *Prepr. 4th Int. Conf. Sth. Hemisph. Meteorol. Oceanogr.,* Amer. Met. Soc., Boston, pp. 532–533.

Liu, W.T., Tang, W. and Fu, L-L. 1995. Recent warming event in the Pacific may be an El Niño. *EOS Trans. AGU,* **76**, 429 and 437.

Liu, Y. and Ding, Y. 1992. Influence of El Niño on weather and climate in China. *Acta Meteor. Sin.,* **6**, 117–131.

Lobell, M.G. 1942. Some observations on the Peruvian coastal current. *EOS Trans. AGU,* **23**, 332–336.

Lockwood, J.G. 1984. The Southern Oscillation and El Niño. *Prog. Phys. Geogr.,* **8**, 102–110.

Lockyer, N. 1904. Simultaneous solar and terrestrial changes. *Nature,* **69**, 351–357.

Lockyer, N. 1908. *Monthly Mean Values of Barometric Pressure for 73 Selected Stations over the Earth's surface.* Solar Physics Committee, London, 97 pp.

Lockyer, N. and Lockyer, W.J.S. 1900a. On solar changes of temperature and variations in rainfall in the region surrounding the Indian Ocean. I. *Nature*, **63**, 107–109.

Lockyer, N. and Lockyer, W.J.S. 1900b. On solar changes of temperature and variations in rainfall in the region surrounding the Indian Ocean. II. *Nature*, **63**, 128–133.

Lockyer, N. and Lockyer, W.J.S. 1902a. On some phenomena which suggest a short-period of solar and meteorological changes. *Proc. Roy. Soc., London*, **70**, 500.

Lockyer, N. and Lockyer, W.J.S. 1902b. On the similarity of the short-period pressure variation over large areas. *Proc. Roy. Soc., London*, **71**, 134–135.

Lockyer, N. and Lockyer, W.J.S. 1904. The behaviour of the short-period atmospheric pressure variation over the earth's surface. *Proc. Roy. Soc., London*, **73**, 457–470.

Lockyer, W.J.S. 1906. Barometric variations of long duration over large areas. *Proc. Roy. Soc., London, A*, **78**, 43–60.

Lockyer, W.J.S. 1909. *A discussion of Australian Meteorology*. Solar Physics Committee, London, 117 pp.

Lockyer, W.J.S. 1910. Does the Indian climate change? *Nature*, **84**, 178.

Lough, J.M. 1991. Rainfall variations in Queensland, Australia: 1891–1986. *Int. J. Climatol.*, **11**, 745–768.

Lough, J.M. 1992. Variations of sea-surface temperatures off north-eastern Australia and associations with rainfall in Queensland: 1956–1987. *Int. J. Climatol.*, **12**, 765–782.

Lough, J.M. 1993. Variations of some seasonal rainfall characteristics in Queensland, Australia: 1921–1987. *Int. J. Climatol.*, **13**, 391–409.

Lough, J.M. 1994. Climate variation and El Niño–Southern Oscillation events on the Great Barrier Reef: 1958 to 1987. *Coral Reefs*, **13**, 181–195.

Lough, J.M. and Fritts, H.C. 1985. The Southern Oscillation and tree rings: 1600–1961. *J. Clim. Appl. Meteor.*, **24**, 952–966.

Lough, J.M. and Fritts, H.C. 1990. Historical aspects of El Niño/Southern Oscillation – Information from tree rings. In Glynn, P.W., (ed.), *Global Ecological Consequences of the 1982–83 El Niño–Southern Oscillation*, Elsevier Oceanography Series, vol. 52, Elsevier, Amsterdam, pp. 285–321.

Lough, J.M., Barnes, D.J. and Taylor, R.B. 1994. The potential of massive corals for the study of high-resolution climate variation in the past millennium. *Proc. NATO Advanced Research Workshop on Climate Variations and forcing mechanisms of the last 2,000 years*, Il Ciocco, Italy, 14 pp.

Lukas, R., Webster, P.J., Ji, M. and Leetma, A. 1995. The large-scale context for the TOGA Coupled Ocean–Atmosphere Response Experiment. *Meteorol. Atmos. Phys.*, **56**, 3–16.

Luksch, U. and von Storch, H. 1992. Modeling the low-frequency sea surface temperature variability in the North Pacific. *J. Climate*, **5**, 893–906.

Lunkeit, F., Sausen, R. and Oberhuber, J.M. 1996. Climate simulations with the global coupled atmosphere–ocean model ECHAM2/OPYC. *Clim. Dyn.*, **12**, 195–212.

Lyons, H.G. 1905. On the relation between variations of atmospheric pressure in north-east Africa and the Nile flood. *Proc. Roy. Soc., A76*, 66–86.

Mabres, A., Woodman, R. and Zeta, R. 1993. Some additional historical notes on the chronology of the El Niño. *Bull. Inst. Fr. Etud. Andines, Lima*, **22(1)**, 395–406.

Machare, J. and Ortlieb, L., (eds), 1993. Records of El Niño phenomena and ENSO events in South America. *Bull. Inst. Fr. Etud. Andines, Lima*, **22(1)**, 406 pp.

Malingreau, J-P. 1987. The 1982–83 drought in Indonesia: assessment and monitoring. In Glantz, M.H., Katz, R.W. and Krentz, M., (eds), *The Societal Impacts Associated with the 1982–83 Worldwide Climate Anomalies*. Environmental and Societal Impacts Group, NCAR, Boulder, CO, pp. 11–18.

Mann, M.E. and Park, J. 1993. Spatial correlations of interdecadal variation in global surface temperatures. *Geophys. Res. Lett.*, **20**, 1055–1058.

Mann, M.E. and Park, J. 1994a. Globally correlated variability in surface temperatures. Paper presented at the *6th Conference on Climate Variations*, AMS, Nashville, Tenn., pp. 297–301.

Mann, M.E. and Park, J. 1994b. Global-scale modes of surface temperature variability on interannual to century timescales. *J. Geophys. Res.*, **99**, 25819–25833.

Mann, M.E., Lall, U. and Saltzman, B. 1995a. Decadal-to-centennial-scale climate variability: Insights into the rise and fall of the Great Salt Lake. *Geophys. Res. Lett.*, **22**, 937–940.

Mann, M.E., Park, J. and Bradley, R.S. 1995b. Global interdecadal and century-scale climate oscillations during the past five centuries. *Nature*, **378**, 266–270.

Mantua, N.J. and Battisti, D.S. 1994. Evidence for the delayed oscillator mechanism for ENSO: The "observed" oceanic Kelvin mode in the far western Pacific. *J. Phys. Oceanogr.*, **24**, 691–699.

Mantua, N.J. and Battisti, D.S. 1995. Aperiodic variability in the Zebiak-Cane coupled ocean–atmosphere model: Air–sea interactions in the western equatorial Pacific. *J. Climate*, **8**, 2897–2927.

Marengo, J.A. 1995. Variations and change in South American streamflow. *Climatic Change*, **31**, 99–117.

Marengo, J.A. and Hastenrath, S. 1993. Case studies of extreme climatic events in the Amazon Basin. *J. Climate*, **6**, 617–627.

Mariategui, J., de Vildoso, A. Ch. and Velez, J. 1985. *Bibliography of "El Niño" phenomenon from 1891 to 1985*. Instituto del Mar del Peru. Boletin, Volumen Extraordinario, 136 pp.

Marotzke, J. and Willebrand, J. 1991. Multiple equilibria of the thermohaline circulation. *J. Phys. Oceanogr., 21*, 1372–1385.

Martin, L., Fournier, M., Mourguiart, P., Sifeddine, A., Turcq, B., Absy, M.L. and Flexor, J-M. 1993. Southern Oscillation signal in South American palaeoclimatic data of the last 7000 years. *Quat. Res., 39*, 338–346.

Mason, S.J. 1990. Temporal variability of sea surface temperatures around Southern Africa: a possible forcing mechanism for the 18-year rainfall oscillation? *S. Afr. J. Sci., 86*, 243–252.

Mason, S.J. 1995. Sea-surface temperature – South African rainfall associations, 1910–1989. *Int. J. Climatol., 15*, 119–135.

Mason, S.J. and Lindesay, J.A. 1993. A note on the modulation of Southern Oscillation–southern African rainfall associations with the Quasi-Biennial Oscillation. *J. Geophys. Res., 98*, D5, 8847–8850.

Mason, S.J. and Tyson, P.D. 1992. The modulation of sea surface temperature and rainfall associations over southern Africa with solar activity and the Quasi-Biennial Oscillation. *J. Geophys. Res., 97*, D5, 5847–5856.

Maung Po, E. 1942. The foreshadowing of the rainfall of Burma. *Quart. J. Roy. Meteor. Soc., 68*, 217–228.

McBride, J.L. and Nicholls, N. 1983. Seasonal relationships between Australian rainfall and the Southern Oscillation. *Mon. Wea. Rev., 111*, 1998–2004.

McCreary, Jr., J.P. and Anderson, D.L.T. 1991. An overview of coupled ocean–atmosphere models of El Niño and the Southern Oscillation. *J. Geophys. Res., 96*, 3125–3150.

McKeon, G.M. and White, D.H. 1992. El Niño and better land management. *Search, 23*, 197–200.

McKeon, G.M., Day, K.A., Howden, S.M., Mott, J.J., Orr, D.M., Scattini, W.J. and Weston, E.J. 1990. Management for pastoral production in northern Australian savannas. *J. Biogeogr., 17*, 1–18.

McPhaden, M.J. 1993. TOGA-TAO and the 1991–93 El Niño–Southern Oscillation event. *Oceanography, 6*, 36–44.

Meadows, A.J. 1972. *Science and controversy: A biography of Sir Norman Lockyer*. MIT Press, Cambridge, Mass., 331 pp.

Mears, E.C. 1943. The Callao painter. *Sci. Mon., 57*, 331–336.

Mears, E.C. 1944. The ocean current called 'The Child'. *Smithson. Inst. Ann. Rep.*, 245–251.

Mechoso, C.R. and Iribarren, G.P. 1992. Streamflow in southeastern South America and the Southern Oscillation. *J. Climate, 5*, 1535–1539.

Mechoso, C.R., Fisher, M., Ghil, M., Halpern. M. and Spahr, M. 1990. Coupling experiments of an atmospheric and an oceanic GCM. *Proc. Int. TOGA Scientific Conference*, WMO, *Tech. Doc. No. 379*, WCRP-43, WCRP, Geneva, 239 pp.

Mechoso, C.R., Robertson, A.W., Barth, N., Davey, M.K., Delecluse, P., Gent, P.R., Ineson, S., Kirtman, B., Latif, M., Le Treut, H., Nagai, T., Neelin, J.D., Philander, S.G.H., Polcher, J., Schopf, P.S., Stockdale, T., Suarez, M.J., Terray, L., Thual, O. and Tribbia, J.J. 1995. The seasonal cycle over the tropical Pacific in coupled ocean–atmosphere general circulation models. *Mon. Wea. Rev., 123*, 2825–2838.

Meehl, G.A. 1990a. Development of global coupled ocean–atmosphere general circulation models. *Clim. Dyn., 5*, 19–33.

Meehl, G.A. 1990b. Seasonal cycle forcing of El Niño–Southern Oscillation in a global, coupled ocean–atmosphere GCM. *J. Climate, 3*, 72–98.

Meehl, G.A. 1991. The Southern Oscillation in a coupled GCM: Implications for climate sensitivity and climate change. In Schlesinger, M.E., (ed.), *Greenhouse-Gas-Induced Climatic Change: A critical appraisal of simulations and observations*, Elsevier Science Publishers B.V., Amsterdam, pp. 111–128.

Meehl, G.A. 1993. A coupled air–sea biennial mechanism in the tropical Indian and Pacific regions: Role of the ocean. *J. Climate, 6*, 31–41.

Meehl, G.A. 1994. Coupled land–ocean–atmosphere processes and south Asian monsoon variability. *Science, 266*, 263–267.

Meehl, G.A., Branstator, G.W. and Washington, W.M. 1993. Tropical Pacific interannual variability and CO_2 climate change. *J. Climate, 6*, 42–63.

Mehta, V.M. 1991. Meridional oscillations in an idealised ocean–atmosphere system. Part 1: Uncoupled modes. *Clim. Dyn., 6*, 49–65.

Mehta, V.M. 1992. Meridionally propagating interannual-to-interdecadal variability in a linear ocean–atmosphere model. *J. Climate, 5*, 330–342.

Meteorological Council, 1890. *Meteorological observations at the foreign and colonial stations of the Royal Engineers and the Army Medical Department, 1852–1886*. H.M. Stationary Office, London, vol. 1.

Meyers, P.G. and Weaver, A.J. 1992. Low-frequency internal oceanic variability under seasonal forcing. *J. Geophys. Res., 97*, 9541–9563.

Meyers, S.D. and O'Brien, J.J. 1995. Pacific Ocean influences atmospheric carbon dioxide. *EOS Trans. AGU, 76*, 533 and 537.

Michaelsen, J. and Thompson, L.G. 1992. A comparison of proxy records of El Niño/Southern Oscillation. Chapter 17, in Diaz, H.F. and Markgraf, V. (eds), *El Niño: Historical review and palaeoclimatic aspects of the Southern Oscillation*, Cambridge University Press, Cambridge, UK, pp. 323–348.

Miller, L., Cheney, R. and Lillibridge, J. 1993a. Blending ERS-1 altimetry and tide-gauge data. *EOS Trans. AGU*, **74**, 185 and 197.

Miller, A.J., Barnett, T.P. and Graham, N.E. 1993b. A comparison of some tropical ocean models: Hindcast skill and El Niño evolution. *J. Phys. Oceanogr.*, **23**, 1567–1591.

Miller, A.J., Cayan, D.R., Barnett, T.P., Graham, N.E. and Oberhuber, J.M. 1994a. Interdecadal variability of the Pacific Ocean: Model response to observed heat flux and wind stress anomalies. *Clim. Dyn.*, **9**, 287–302.

Miller, A.J., Cayan, D.R., Barnett, T.P., Graham, N.E. and Oberhuber, J.M. 1994b. The 1976–77 climate shift of the Pacific Ocean. *Oceanography*, **7**, 21–26.

Mitchell, W.M. 1994. Using mean sea level anomalies as an indicator of regional climate variability. *Agr. Sys. Info. Tech.*, **6**, 19–21.

Mo, K.C. and Wang, X.L. 1995. Sensitivity of the systematic error of extended range forecasts to sea surface temperature anomalies. *J. Climate*, **8**, 1533–1543.

Montgomery, R.B. 1940. Report on the work of G.T. Walker. *Mon. Wea. Rev., Supp. No.* **39**, 22 pp.

Moore, A. 1995. Tropical interannual variability in a global coupled GCM: Sensitivity to mean climate state. *J. Climate*, **8**, 807–828.

Moron, V., Bigot, S. and Roucou, P. 1995. Rainfall variability in subequatorial America and Africa and relationships with the main sea-surface temperature modes (1951–1990). *Int. J. Climatol.*, **15**, 1297–1322.

Moss, M.E., Pearson, C.P. and McKerchar, A.I. 1994. The Southern Oscillation index as a predictor of the probability of low streamflows in New Zealand. *Water Resour. Res.*, **30**, 2717–2723.

Mossman, R.C. 1913. Southern Hemisphere seasonal correlations. *Symons's Meteorological Magazine*, **48**, 2–7, 44–47, 82–85, 104–105, 119–124, 160–163, 200–207, 226–229.

Mossman, R.C. 1923. On Indian monsoon rainfall in relation to South American weather 1875–1914. *Mem. India Meteorol. Dep.*, **XXIII**, Part **VI**, 157–242.

Mullan, A.B. 1995. On the linearity and stability of Southern Oscillaton–climate relationships for New Zealand. *Int. J. Climatol.*, **15**, 1365–1386.

Murata, A.M. and Fushimi, K. 1996. Temporal and spatial variations in atmospheric and oceanic CO_2 in the western North Pacific from 1990 to 1993: Possible link to the 1991/92 ENSO event. *J. Meteor. Soc. Japan*, **74**, 1–20.

Murphy, R.C. 1926. Oceanic and climatic phenomena along the west coast of South America during 1925. *Geogr. Rev.*, **16**, 26–54.

Murphy, R.C. 1932. The Humboldt and Niño Currents. *Geogr. Rev.*, **22**, 148–150.

Murray, J.W., Leinen, M.W., Feely, R.A., Toggweiler, J.R. and Wanninkhof, R. 1992. EQPAC: A process study in the central equatorial Pacific. *Oceanography*, **5**, 134–142.

Murray, J.W., Barber, R.T., Roman, M.R., Bacon, M.P. and Feely, R.A. 1994. Physical and biological controls on carbon cycling in the equatorial Pacific. *Science*, **266**, 58–65.

Murray, J.W., Johnson, E. and Garside, C. 1995. A U.S. JGOFS process study in the equatorial Pacific (EqPac): Introduction. *Deep-Sea Res. II*, **42**, 275–293.

Nagai, T., Tokioka, T., Endoh, M. and Kitamura, Y. 1992. El Niño–Southern Oscillation simulated in an MRI atmosphere–ocean coupled general circualtion model. *J. Climate*, **5**, 1202–1233.

Nagai, T., Kitamura, Y., Endoh, M. and Tokioka, T. 1995. Coupled atmosphere–ocean model simulations of El Niño–Southern Oscillation with and without an active Indian Ocean. *J. Climate*, **8**, 3–14.

National Climate Centre, 1988. *Seasonal Outlooks (Based on El Niño/Southern Oscillation (ENSO) relationships)*. Bureau of Meteorology, 34 pp.

National Environmental Satellite Data and Information Service, 1951–1995. *Monthly Climatic Data for the World*. U.S. Dept. Comm., Washington, D.C.

Neelin, J.D., Latif, M., Allaart, M.A.F., Cane, M.A., Cubasch, U., Gates, W.L., Gent, P.R., Ghil, M., Gordon, C., Lau, N-C., Mechoso, C.R., Meehl, G.A., Oberhuber, J.M., Philander, S.G.H., Schopf, P.S., Sperber, K.R., Sterl, A., Tokioka, T., Tribbia, J. and Zebiak, S.E. 1992. Tropical air–sea interaction in general circulation models. *Clim. Dyn.*, **7**, 73–104.

Neelin, J.D., Latif, M. and Jin, F-F. 1994. Dynamics of coupled ocean–atmosphere models: The tropical problem. *Ann. Rev. Fluid Mech.*, **26**, 617–659.

Newell, R.E., Selkirk, R. and Ebisuzaki, W. 1982. The Southern Oscillation: Sea surface temperature and wind relationships in a 100-year data set. *J. Climatol.*, **2**, 357–373.

Nicholls, N. 1979. A possible method for predicting seasonal tropical cyclone activity in the Australian region. *Mon. Wea. Rev.*, **107**, 1221–1224.

Nicholls, N. 1984a. The Southern Oscillation and Indonesian sea surface temperature. *Mon. Wea. Rev.*, **112**, 424–432.

Nicholls, N. 1984b. The Southern Oscillation, sea-surface-temperature, and interannual fluctuations in Australian tropical cyclone activity. *J. Climatol.*, **4**, 661–670.

Nicholls, N. 1985a. Towards the prediction of major Australian droughts. *Aust. Meteor. Mag.*, **33**, 161–166.

Nicholls, N. 1985b. Impact of the Southern Oscillation on Australian crops. *J. Climatology,* **5,** 553–560.

Nicholls, N. 1986a. Use of the Southern Oscillation to predict Australian sorghum yield. *Agricultural and Forest Meteorology,* **38,** 9–15.

Nicholls, N. 1986b. A method for predicting Murray Valley Encephalitis in southeast Australia using the Southern Oscillation. *Aust. J. Exp. Bio. Med. Sci.,* **64,** 587–594.

Nicholls, N. 1988a. More on early ENSOs: Evidence from Australian documentary sources. *Bull. Amer. Meteor. Soc.,* **69,** 4–6.

Nicholls, N. 1988b. El Niño–Southern Oscillation impact prediction. *Bull. Amer. Meteor. Soc.,* **69,** 173–176.

Nicholls, N. 1988c. Low latitude volcanic eruptions and the El Niño–Southern Oscillation. *J. Climatol.,* **8,** 91–95.

Nicholls, N. 1989. Sea surface temperatures and Australian winter rainfall. *J. Climate,* **2,** 965–973.

Nicholls, N. 1990. Low latitude volcanic eruptions and the El Niño–Southern Oscillation: A reply. *Int. J. Climatol.,* **10,** 425–429.

Nicholls, N. 1991a. The El Niño–Southern Oscillation and Australian vegetation. *Vegetatio,* **91,** 23–36.

Nicholls, N. 1991b. Teleconnections and health. In Glantz, M., Katz, R. and Nicholls, N., (eds), *ENSO teleconnections linking worldwide climate anomalies: Scientific basis and societal impacts.* Cambridge University Press, Cambridge, UK, pp. 493–509.

Nicholls, N. 1991c. Advances in long-term weather forecasting. In Muchow, R.C. and Bellamy, J.A., (eds), *Climatic risk in crop production: Models and management for the semiarid tropics and subtropics.* CAB International, pp. 427-444.

Nicholls, N. 1992. Recent performance of a method for forecasting Australian seasonal tropical cyclone activity. *Aust. Meteor. Mag.,* **40,** 105–110.

Nicholls, N. 1995. All-India summer monsoon rainfall and sea surface temperatures around northern Australia and Indonesia. *J. Climate,* **8,** 1463–1467.

Nicholls, N. and Kariko, A. 1993. East Australian Rainfall Events: Interannual Variations, Trends, and Relationships with the Southern Oscillation. *J. Climate,* **6,** 1141–1152.

Nicholson, S.E. 1993. An overview of African rainfall fluctuations of the last decade. *J. Climate,* **6,** 1463–1466.

Nicholson, S.E. and Entekhabi, D. 1986. Quasi-periodic behaviour of rainfall variability in Africa. *Arch. Met. Geoph. Biokl. Ser. A,* **34,** 311–348.

Nicholson, S.E. and Palao, I.M. 1993. A re-evaluation of rainfall variability in the Sahel. Part I. Characteristics of rainfall fluctuations. *Int. J. Climatol.,* **13,** 371–389.

Nigam, S. 1994. On the dynamical basis for the Asian summer monsoon rainfall–El Niño relationship. *J. Climate,* **7,** 1750–1771.

Nigam, S. and Shen, H-S. 1993. Structure of oceanic and atmospheric low-frequency variability over the tropical Pacific and Indian Oceans. Part I: COADS observations. *J. Climate,* **6,** 657–676.

Nitta, T. 1992. Interannual and decadal scale variations of atmospheric temperature and circulations. In Ye, D., Zeng, Q., Zhang, R., Matsuno, T. and Huang, R., (eds), *Climate Variability, Proceedings of International Workshop on Climate Variabilities (IWCV),* Beijing, China, pp. 15–22.

Nitta, T. and Kachi, M. 1994. Interdecadal variations in precipitation over the tropical Pacific and Indian Oceans. *J. Meteor. Soc. Japan,* **72,** 823–830.

Nitta, T. and Yamada, S. 1989. Recent warming of tropical sea surface temperature and its relationship to the Northern Hemisphere circulation. *J. Meteor. Soc. Japan,* **67,** 375–383.

Nitta, T. and Yoshimura, J. 1993. Trends and interannual and interdecadal variations of global land surface air temperature. *J. Meteor. Soc. Japan,* **71,** 367–374.

Normand, C.W.B. 1932. Some problems of moderm meteorology, No. 6 Present position of seasonal weather forecasting. *Quart. J. Roy. Meteor. Soc.,* **58,** 3–10.

Normand, C.W.B. 1953. Monsoon seasonal forecasting. *Quart. J. Roy. Meteor. Soc.,* **79,** 463–473.

Oak Ridge National Laboratory, 1992. The global historical climatology network: Long-term monthly temperature, precipitation, sea level pressure, and station pressure data. *Env. Sci. Div., Pub. No.* **3912,** *ORNL/CDIAC-***53,** 99 pp.

Oberhuber, J.M., Lunkeit, F. and Sausen, R. 1990. CO_2 doubling experiments with a coupled global isopycnic ocean–atmosphere circulation model. *Rep. d. Met. Inst. Univ. Hamburg,* Hamburg, FRG.

O'Brien, J.J. 1992. *Workshop on the Economic Impact of ENSO Forecasts on the American, Australian and Asian Continents, Vol. 1: Executive Summary and Panel Reports.* Florida State University, Tallahassee, Florida, USA, 85 pp.

O'Brien, J.J., Busalacchi, A. and Kindle, J. 1981. Ocean models of El Niño. In Glantz, M. and Thompson, J.D., (eds), *Resource management and environmental uncertainty: Lessons from coastal upwelling fisheries.* P. Adv. Env. Sci. vol. 11, Wiley-Interscience, pp. 159–212.

Ogallo, L. 1987. Impacts of the 1982–83 ENSO event on eastern and southern Africa. In Glantz, M.H., Katz, R.W. and Krentz, M., (eds), *The Societal Impacts Associated with the 1982–83 Worldwide Climate Anomalies.* Environmental and Societal Impacts Group, NCAR, Boulder, CO, pp. 55–61.

Oladipo, E.O. 1995. Some statistical characteristics of drought area variations in the savanna region of Nigeria. *Theor. Climatol.,* **50,** 147–155.

O'Neill, G. 1995. Coping with climate. *ECOS,* **84,** 11–26.

Oort, A.H., Pan, Y.H., Reynolds, R.W. and Ropelewski, C.F. 1987. Historical trends in the surface temperature over the oceans based on the COADS. *Clim. Dyn.*, **2**, 29–38.

Opoku-Ankomah, Y. and Cordery, I. 1993. Temporal variation of relations between New South Wales rainfall and the Southern Oscillation. *Int. J. Climatol.*, **13**, 51–64.

Ortlieb, L. 1994. Major historical rainfalls in central Chile and the chronology of ENSO events during the XVI–XIX centuries. *Revista Chilena Historia Natural*, 67, 463–485..

Ortlieb, L. and Machare, J., (eds), 1992. *Paleo-ENSO Records Intern. Symp., Extended Abstracts*, ORSTOM (Lima) and CONCYTEC, Lima, 333 pp.

Ortlieb, L. and Machare, J. 1993. Former El Niño events: records from western South America. *Global and Planetary Change*, 7, 181–202.

Palmer, T.N. and Anderson, D.L.T. 1994. The prospects for seasonal forecasting – A review paper. *Quart. J. Roy. Meteor. Soc.*, **120**, 755–793.

Pan, Y.H. and Oort, A.H. 1990. Correlation analyses between sea surface temperature anomalies in the eastern equatorial Pacific and the world ocean. *Clim. Dyn.*, **4**, 191–205.

Pant, G.B., Kumar, K.R., Parthasarathy, B. and Borgaonkar, H.P. 1988. Long-term variability of the Indian summer monsoon and related parameters. *Adv. Atmos. Sci.*, **5**, 469–481.

Pariwono, J.I., Bye, A.T. and Lennon, G.W. 1986. Long period variations of sea level in Australasia. *Geophys. J. Roy. Astron. Soc.*, **87**, 43–54.

Parker, D.E. 1983. Documentation of a Southern Oscillation Index. *Meteor. Mag.*, **112**, 184–188.

Parker, D.E. 1984. The statistical effects of incomplete sampling of coherent data series. *J. Climatol.*, **4**, 445–449.

Parker, D.E. and Folland, C.K. 1994. Global data required for monitoring climate change. *Climate Research Tech. Note* CRTN No. **46**, Hadley Centre for Climate Prediction and Research, Meteorological Office, Bracknell, Berkshire, UK, 52 pp.

Parker, D.E., Jones, P.D., Folland, C.K. and Bevan, A. 1994. Interdecadal changes of surface temperature since the late nineteenth century. *J.Geophys. Res.*, **99**, 14,373–14,399.

Parker, D.E., Folland, C.K., Bevan, A., Ward, M.N., Jackson, M. and Maskell, K. 1995a. Marine surface data for analysis of climatic fluctuations on inter-annual to century timescales. In Martinson, D.G., Bryan, K., Ghil, M., Hall, M.M., Karl, T.R., Sarachik, E.S., Sorooshian, S. and Talley, L.D. (eds) *Natural climate variability on decade to century timescales*, National Academy Press, Washington, DC, pp. 214–250 (and colour figs pp.222–228).

Parker, D.E., Jackson, M. and Horton, E.B. 1995b. The GISST2.2 sea surface temperature and sea-ice climatology. *Climate Research Tech. Note* CRTN No. **63**, Hadley Centre for Climate Prediction and Research, Meteorological Office, Bracknell, Berkshire, UK, 35 pp.

Parthasarathy, B. and Yang, S. 1995. Relationships between regional Indian summer monsoon rainfall and Eurasian snow cover. *Adv. Atmos. Sci.*, **12**, 143–150.

Parthasarathy, B., Kumar, K.R. and Munot, A.A. 1991. Evidence of secular variations in Indian monsoon rainfall–circulation relationships. *J. Climate*, **4**, 927–938.

Parthasarathy, B., Kumar, K.R. and Munot, A.A. 1992. Surface pressure and summer monsoon rainfall over India. *Adv. Atmos. Sci.*, **9**, 359–366.

Pazan, S.E. and White, W.B. 1986. Off-equatorial influence upon Pacific equatorial dynamic height variability during the 1982–1983 El Niño/Southern Oscillation event. *J. Geophys. Res.*, **91**, C7, 8437–8449.

Pazan, S.E., White, W.B., Inoue, M. and O'Brien, J.J. 1986. Off-equatorial influence upon Pacific equatorial dynamic height variability during the 1982–83 El Niño/Southern Oscillation event. *J. Geophys. Res.*, **91**, C7, 8437–8449.

Peiji, L. 1995. Comments on 'An apparent relationship between Himalayan snow cover and summer monsoon rainfall over India'. *Acta Meteor. Sin.*, **9**, 360–367.

Penland, C. and Magorian, T. 1993. Prediction of Niño 3 sea surface temperatures using linear inverse modeling. *J. Climate*, **6**, 1067–1076.

Penland, C. and Matrosova, L. 1994. A balance condition for stochastic numerical models with application to the El Niño–Southern Oscillation. *J. Climate*, **7**, 1352–1372.

Perigaud, C. and Dewitte, B. 1996. El Niño–La Niña events simulated with Cane and Zebiak's model and observed with satellite or in situ data. Part I: Model data comparison. *J. Climate*, **9**, 66–84.

Pezet, F.A. 1896. The counter-current 'El Niño' on the coast of northern Peru. *Geogr. J. (London)*, 7, 603–606.

Philander, S.G.H. 1983. El Niño Southern Oscillation phenomena. *Nature*, **302**, 295–301.

Philander, S.G.H. 1985. El Niño and La Niña. *J. Atmos. Sci.*, **42**, 2652–2662.

Philander, S.G.H. 1986. Predictability of El Niño. *Nature*, **321**, 810–811.

Philander, S.G.H. 1989. El Niño and La Niña. *Amer. Sci.*, **77**, 451–459.

Philander, S.G.H. 1990. *El Niño, La Niña, and the Southern Oscillation*. Academic Press, New York, 293 pp.

Philander, S.G.H. 1992a. El Niño. *Oceanus*, **35**, 56–61.

Philander, S.G.H. 1992b. Ocean–atmosphere interactions in the tropics: A review of recent theories and models. *J. Appl. Meteor.*, **31**, 938–945.

Philander, S.G.H. and Rasmusson, E.M. 1985. The Southern Oscillation and El Niño. *Adv. Geophys.*, **28A**, 197–215.

Philander, S.G.H., Pacanowski, R.C., Lau, N-C. and Nath, M.J. 1992. Simulation of ENSO with a global atmospheric GCM coupled to a high-resolution, tropical Pacific Ocean GCM. *J. Climate*, **5**, 308–329.

Pierce, D.W., Barnett, T.P. and Mikolajewicz, U. 1995. Comparing roles of heat and freshwater flux in forcing thermohaline oscillations. *J. Phys. Oceanogr.*, **25**, 2046–2064.

Pisciottano, G., Diaz, A., Cazes, G. and Mechoso, C.R. 1994. El Niño–Southern Oscillation impact on rainfall in Uruguay. *J. Climate*, **7**, 1286–1302.

Pittock, A.B. 1974. Global interactions in stratosphere and troposphere. *Proc. Int. Conf. Structure, Composition and General Circulation of the Upper and Lower Atmospheres and Possible Anthropogenic Perturbations, Vol.* **II**, IUGG, IAMAP, The Australian Academy of Science, US Dept. of Transportation, pp. 716–726.

Pittock, A.B. 1984. On the reality, stability, and usefulness of Southern Hemisphere teleconnections. *Aust. Met. Mag.*, **32**, 75–82.

Polonsky, A.B. 1994. Comparative study of the Pacific ENSO event of 1991–2 and the Atlantic ENSO-like event of 1991. *Aust. J. Mar. Freshwater Res.*, **45**, 705–725.

Ponte, R.M., Rosen, R.D. and Boer, G.J. 1994. Angular momentum and torques in a simulation of the atmosphere's response to the 1982–83 El Niño. *J. Climate*, **7**, 538–550.

Portman, D.A. and Gutzler, D.S. 1996. Explosive volcanic eruptions, the El Niño–Southern Oscillation, and U.S. climate variability. *J. Climate*, **9**, 17–33.

Post and Telegraph Department, 1889–1896. *Queensland Meteorological Record 1889–1896.* Government Printer, Brisbane.

Postmentier, E.S., Cane, M.A. and Zebiak, S.E. 1989. Tropical Pacific climate trends since 1960. *J. Climate*, **2**, 731–736.

Potter, K.W. 1981. Illustration of a new test for detecting a shift in mean precipitation series. *Mon. Wea. Rev.*, **109**, 2040–2045.

Privalsky, V.E. and Jensen, D.T. 1995. Assessment of the influence of ENSO on annual global air temperatures. *Dyn. Atmos. Oceans*, **22**, 161–178.

Qu, T., Meyers, G. and Godfrey, J.S. 1994. Ocean dynamics in the region between Australia and Indonesia and its influence on the variation of sea surface temperature in a global general circulation model. *J. Geophys. Res.*, **99**, 18,433–18,445.

Quayle, E.T. 1929. Long range rainfall forecasting from tropical (Darwin) air pressures. *Proc. Roy. Soc., Victoria*, **41**, 160–164.

Quinn, T.M., Taylor, F. and Crowley, T.J. 1993. A 173 year stable isotope record from a tropical South Pacific coral. *Quat. Sci. Rev.*, **12**, 407–418.

Quinn, W.H. 1974. Monitoring and predicting El Niño invasions. *J. Appl. Meteor.*, **13**, 825–830.

Quinn, W.H. 1992. A study of Southern Oscillation-related climatic activity for AD 622—1900 incorporating Nile River flood data. Chapter 6, in Diaz H.F. and Markgraf, V., (eds), *El Niño: Historical and Paleoclimatic Aspects of the Southern Oscillation*, CUP, Cambridge, 119—149.

Quinn, W.H. 1993: The large-scale ENSO event, the El Niño and other important regional features. *Bull. Inst. Fr. Etud. Andines, Lima*, **22**(1), 13–34.

Quinn, W.H. and Burt, W.V. 1970. Prediction of abnormally heavy precipitation over the Equatorial Pacific Dry Zone. *J. Appl. Meteor.*, **9**, 20–28.

Quinn, W.H. and Burt, W.V. 1972. Use of the Southern Oscillation in weather prediction. *J. Appl. Meteor.*, **11**, 616–628.

Quinn, W.H. and Neal, V.T. 1983. Long-term variations in the Southern Oscillation, El Niño, and Chilean subtropical rainfall. *Fish. Bull.*, **81**, 363–374.

Quinn, W.H. and Neal, V.T. 1992. The historical record of El Niño events. In *Climate since A.D. 1500*, Bradley, R.S. and Jones, P.D., (eds), Routledge, London, 623–648.

Quinn, W.H. and Zopf, D.O. 1975. The Southern Oscillation, equatorial Pacific anomalies and El Niño. *Geofisica Internacional*, **15**, 327–353.

Quinn, W.H., Zopf, D.O., Short, K.S. and Kuo Yang, R.T.W. 1978. Historical trends and statistics of the Southern Oscillation, El Niño, and Indonesian droughts. *Fish. Bull.*, **76**, 663–678.

Quinn, W.H., Neal, V.T. and Antunez de Mayolo, S.E. 1987. El Niño Occurrences over the Past Four and a Half Centuries. *J. Geophys. Res.*, **92**, 14,449–14,461.

Ramage, C.S. 1968. Role of a tropical 'maritime continent' in the atmosphere circulation. *Mon. Wea. Rev.*, **96**, 365–370.

Ramage, C.S. 1975. Preliminary discussion of the meteorology of the 1972–73 El Niño. *Bull. Amer. Meteor. Soc.*, **56**, 234–242.

Ramage, C.S. 1983. Teleconnections and the seige of time. *J. Climatol.*, **3**, 223–231.

Ramage, C.S. 1986. El Niño. *Sci. Amer.*, **254**, 55–61.

Ramage, C.S. and Hori, A.M. 1981. Meteorological aspects of El Niño. *Mon. Wea. Rev.*, **109**, 1827–1835.

Rasmusson, E.M. 1984. El Niño: The ocean/atmosphere connection. *Oceanus*, **27**, 4–12.

Rasmusson, E.M. 1985. El Niño and variations in climate. *Amer. Sci.*, **73**, 168–177.

Rasmusson, E.M. 1987. The prediction of drought: a meteorological perspective. *Endeavour, New Series*, **11**, 175–182.

Rasmusson, E.M. and Arkin, P.A. 1985. Interannual climate variability associated with El Niño/Southern Oscillation. In Nihoul, J.C.J., (ed.), *Coupled ocean–atmosphere models.*, Amsterdam, Elsevier Oceanography Series 40, 697–725.

Rasmusson, E.M. and Arkin, P.A. 1993. A global view of large-scale precipitation variability. *J. Climate*, 6, 1495–1522.

Rasmusson, E.M. and Carpenter, T.H. 1982. Variations in tropical sea surface temperature and surface wind fields associated with the Southern Oscillation/El Niño. *Mon. Wea. Rev.*, 110, 354–384.

Rasmusson, E.M. and Wallace, J.M. 1983. Meteorological aspects of the El Niño/Southern Oscillation. *Science*, 222, 1195–1202.

Rasmusson, E.M., Wang, X. and Ropelewski, C.F. 1990. The biennial component of ENSO variability. *J. Mar. Sys.*, 1, 71–96.

Rawson, A.E. 1908. The anticyclonic belt of the Southern Hemisphere. *Quart. J. Roy. Meteor. Soc.*, 34, 165–188.

Rayner, N.A., Folland, C.K., Parker, D.E. and Horton, E.B. 1995. A new global sea-ice and sea surface temperature (GISST) data set for 1903–1994 for forcing climate models. *Hadley Centre for Climate Prediction and Research Note* No. 69, Hadley Centre for Climate Prediction and Research, Meteorological Office, Bracknell, Berkshire, UK, 14 pp.

Reason, C.J.C., Allan, R.J. and Lindesay, J.A. 1996. Dynamical response of the oceanic circulation and temperature to interdecadal variability in the surface winds over the Indian Ocean. *J. Climate*, 9, 97–114 .

Rebert, J.P. and Donguy, J.R. 1988. The Southern Oscillation Index since 1882. Time series of Ocean Measurements 4, *IOC Tech. Ser.* 33, UNESCO, 49–53.

Reiter, E.R. 1978. Long-term variability in the tropical Pacific, its possible causes and effects. *Mon. Wea. Rev.*, 106, 324–330.

Reiter, E.R. 1979. Trade-wind variability, Southern Oscillation and Quasi-Biennial Oscillation. *Arch. Met. Geoph. Biokl. Ser.*, A28, 113–126.

Reiter, E.R. 1983. Teleconnections with tropical precipitation surges. *J. Atmos. Sci.*, 40, 1631–1647.

Reseau Mondial, 1910–1934. *British Meteorological and Magnetic Year Book Part* V, H.M. Stationary Office.

Revell, C.G. and Goulter, S.W. 1986a. South Pacific tropical cyclones and the Southern Oscillation. *Mon. Wea. Rev.*, 114, 1138–1145.

Revell, C.G. and Goulter, S.W. 1986b. Lagged relations between the Southern Oscillation and numbers of tropical cyclones in the South Pacific region. *Mon. Wea. Rev.*, 114, 2669–2670.

Reynolds, R.W., Folland, C.K. and Parker, D.E. 1989. Biases in satellite derived sea-surface-temperatures. *Nature*, 341, 728–731.

Reynolds, R.W. and Smith, T.M. 1994. Improved global sea surface temperature analyses using optimum interpolation. *J. Climate*, 7, 929–948.

Rhodesia and Nyasaland Meteorological Service, 1951. *Climatological summaries Northern Rhodesia July, 1938– June, 1948.* 53 pp.

Richey, J.E., Nobre, C. and Deser, C. 1989. Amazon River discharge and climate variability: 1903 to 1985. *Science*, 246, 101–103.

Richmond, R.H. 1990. The effects of the El Niño/Southern Oscillation on the dispersal of corals and other marine organisms. In Glynn, P.W., (ed.), *Global Ecological Consequences of the 1982–83 El Niño–Southern Oscillation*, Elsevier Oceanography Series, vol. 52, Elsevier, Amsterdam, pp. 127–140.

Rimmer, T. and Hossack, A.W.W. 1939. Forecasting summer rain in Queensland. *Uni. Qld. Pap., Dep. Phys.*, 1, 15 pp.

Rimmington, G.M. and Nicholls, N. 1993. Forecasting wheat yields in Australia with the Southern Oscillation Index. *Aust. J. Agr. Res.*, 44, 625–632.

Rind, D. and Chandler, M. 1991. Increased ocean heat transports and warmer climate. *J. Geophys. Res.*, 96, 7437–7461.

Robb, J. 1880. Notes on the meteorology of Zanzibar, East Africa. *Quart. J. Roy. Meteor. Soc.*, 6, 30–39.

Robock, A., Taylor, K.E., Stenchikov, G.L. and Liu, Y. 1995. GCM evaluation of a mechanism for El Niño triggering by the El Chicon ash cloud. *Geophys. Res. Lett.*, 22, 2369–2372.

Rocha, A. and Simmonds, I. 1996a. Interannual variability of southern African summer rainfall. Part I: Relationships with air–sea interaction processes. *Int. J. Climate* (in press).

Rocha, A. and Simmonds, I. 1996b. Interannual variability of southern African summer rainfall. Part II: Modelling the impact of sea surface temperatures on rainfall and circulation. *Int. J. Climate* (in press).

Rodionov, S.N. 1994. Association between winter precipitation and water level fluctuations in the Great Lakes and atmospheric circulation patterns. *J. Climate*, 7, 1693–1706.

Roeckner, E., Oberhuber, J.M., Bacher, A., Christoph, M. and Kirchner,I. 1995. ENSO variability and atmospheric response in a global coupled atmosphere–ocean GCM. *Max-Planck-Institut fur Meteorologie, Report No.* 178, 33pp.

Ronghui, H. 1994. Interactions between the 30–60 Day Oscillation, the Walker Circulation and the convective activities in the tropical western Pacific and their relations to the Interannual Oscillation. *Adv. Atmos. Sci.*, 11, 367–384.

Ropelewski, C.F. and Halpert, M.S. 1986. North American precipitation and temperature patterns associated with the El Niño/Southern Oscillation (ENSO). *Mon. Wea. Rev.*, 114, 2352–2362.

Ropelewski, C.F. and Halpert, M.S. 1987. Global and regional scale precipitation patterns associated with the El Niño/Southern Oscillation. *Mon. Wea. Rev.*, 115, 1606–1626.

Ropelewski, C.F. and Halpert, M.S. 1989. Precipitation patterns associated with the high index phase of the Southern Oscillation. *J. Climate*, **2**, 268–284.

Ropelewski, C.F. and Jones, P.D. 1987. An extension of the Tahiti–Darwin Southern Oscillation Index. *Mon. Wea. Rev.*, **115**, 2161–2165.

Ropelewski, C.F., Halpert, M.S. and Wang, X. 1992. Observed tropospheric biennial variability and its relationship to the Southern Oscillation. *J. Climate*, **5**, 594–614.

Rosen, R.D., Eubanks, T.M., Dickey, J.O. and Steppe, J.A. 1984. An El Niño signal in atmospheric angular momentum and earth rotation. *Science*, **225**, 411–414.

Russell, H.C. 1871–1886. *Results of Meteorological Observations made in New South Wales during 1871–1886*. Government Printer, Sydney.

Russell, H.C. 1904. *Results of Meteorological Observations in New South Wales during 1900, 1901 and 1902*. Government Printer, Sydney, 216 pp.

Russell, H.C. 1905. *NSW Meteorological Observations 1903*. Government Printer, Sydney.

Russell, H.C. 1906. *Results of Meteorological Observations made in New South Wales during 1891–1895*. Government Printer, Sydney, 484 pp.

Rutllant, J. and Fuenzalida, H. 1991. Synoptic aspects of the central Chile rainfall variability associated with the Southern Oscillation. *Int. J. Climatol.*, **11**, 63–76.

Ryan, B.F., Watterson, I.G. and Evans, J.L. 1992. Tropical cyclone frequencies inferred from Gray's yearly genesis parameter: Validation of GCM tropical climates. *Geophys. Res. Lett.*, **19**, 1831–1834.

Salafsky, N. 1994. Drought in the rain forest: Effects of the 1991 El Niño–Southern Oscillation event on a rural economy in west Kalimantan, Indonesia. *Climatic Change*, **27**, 373–396.

Saravanan, R. and McWilliams, J.C. 1995. Multiple equilibria, natural variability, and climate transitions in an idealized ocean–atmosphere model. *J. Climate*, **8**, 2296–2323.

Sarker, R.P. and Thapliyal, V. 1988. Climatic change and variability. *Mausam*, **39**, 127–138.

Schell, I.I. 1947. Dynamic persistence and its applications to long-range foreshadowing. *Harvard Meteor. Stud.*, **8**, 79 pp

Schell, I.I. 1956. On the nature and origin of the Southern Oscillation. *J. Meteorol.*, **13**, 592–598.

Schell, I.I. 1968. Hendrik Petrus Berlage 1896–1968. *Bull. Amer. Meteor. Soc.*, **49**, 747.

Schimel, D. and Sulzman. E. 1995. Variability in the earth climate system: Decadal and longer timescales. *Rev. Geophys., Supp.*, 873–882.

Schlesinger, M.E. and Ramankutty, N. 1994. An oscillation in the global climate system of period 65–70 years. *Nature*, **367**, 723–726.

Schmitz, Jr., W.J. 1995. On the interbasin-scale thermohaline circulation. *Rev. Geophys.*, **33**, 151–173.

Schneider, E.K. and Kinter III, J.L. 1994. An examination of internally generated variability in long climate simulations. *Clim. Dyn.*, **10**, 181–204.

Schneider, E.K., Huang, B. and Shukla, J. 1995. Ocean wave dynamics and El Niño. *J. Climate*, **8**, 2415–2439.

Schneider, N., Barnett, T.P., Latif, M. and Stockdale, T. 1996. Warm pool physics in a coupled GCM. *J. Climate*, **9**, 219–239.

Schopf, P.S. and Suarez, M.J. 1988. Vacillations in a coupled ocean–atmosphere model. *J. Atmos. Sci.*, **45**, 549–566.

Schott, G. 1931. The Peru (Humboldt) Current and its northern vicinity in normal and abnormal conditions. *Ann. Hydrogr. Marit. Meteor.*, **59**, 161–169.

Schove, D.J. 1961. The major pressure oscillation, 1875 to 1960. *Pure Appl. Geophys.*, **49**, 255–261.

Schove, D.J. 1963. Models of the Southern Oscillation in the 300–100 mb layer and the basis of seasonal forecasting. *Pure Appl. Geophys.*, **55**, 249–261.

Schove, D.J. and Berlage, H.P. 1965. Pressure anomalies in the Indian Ocean area, 1796–1960. *Pure Appl. Geophys.*, **61**, 219–231.

Schweigger, E. 1945. La 'legitima' Corriente del Niño. *Bol. Comp. Admin. Guano*, **21**, 255–316.

Schweigger, E. 1961. Temperature anomalies in the eastern Pacific Ocean and their forecasting. *Sociedad Geografica de Lima, Boletin*, **78** (3/4), 3–50.

Seleshi, Y., Demaree, G.R. and Vannitsem, S. 1992. Statistical analysis of long-term monthly and annual Ethiopian precipitation series and their relationship with ENSO events. In Ye, D., Zeng, Q., Zhang, R., Matsuno, T. and Huang, R., (eds), *Climate Variability, Proceedings of International Workshop on Climate Variabilities (IWCV)*, Beijing, China, 80–92.

Sellick, N.P. 1932. Seasonal foreshadowing by correlation. *Quart. J. Roy. Meteor. Soc.*, **58**, 226–228.

Shaowu, W. 1992. Reconstruction of El Niño event chronology for the last 600 year period. *Acta Meteor. Sin.*, **6**, 47–57.

Sharp, G.D. and McLain, D.R. 1993. Fisheries, El Niño–Southern Oscillation and upper-ocean temperature records: An eastern Pacific example. *Oceanography*, **6**, 13–22.

Shen, G.T., and Sandford, C.L. 1990. Trace elements indicators of climate variability in reef-building corals. In Glynn, P.W. (ed.) *Global Ecological Consequences of the 1982–83 El Niño–Southern Oscillation*, Elsevier Oceanography Series, vol. 52, Elsevier, Amsterdam, pp. 255–284.

Shen, G.T., Linn, L.J., Campbell, T.M., Cole, J.E. and Fairbanks, R.G. 1992a. A chemical indicator of trade wind reversal in corals from the western tropical Pacific. *J. Geophys. Res.*, **97**, C8, 12,689–12,697.

Shen, G.T., Cole, J.E., Lea, D.W., Linn, L.J., McConnaughey, T.A. and Fairbanks, R.G. 1992b. Surface ocean variability at Galapagos from 1936–1982: Calibration of geochemical tracers in corals. *Paleoceanography*, 7, 563–588.

Shen, S. and Lau, K-M. 1995. Biennial Oscillation associated with the East Asian summer monsoon and tropical sea surface temperatures. *J. Meteor. Soc. Japan*, 73, 105–124.

Shinoda, M. and Kawamura, R. 1996. Relationships between rainfall over semi-arid southern Africa, geopotential heights, and sea surface temperatures. *J. Meteor. Soc. Japan*, 74, 21–36.

Shiotani, M. 1992. Annual, Quasi-Biennial, and El Niño–Southern Oscillation (ENSO) time-scale variations in equatorial total ozone. *J. Geophys. Res.*, 97, D7, 7625–7633.

Simard, A.J., Haines, D.A. and Main, W.A. 1985. Relations between El Niño–Southern Oscillation and anomalies in wildfire activity in the US. *Agr. For. Meteor.*, 36, 93–104.

Simmonds, I. 1990. A modelling study of winter circulation and precipitation anomalies associated with Australian region ocean temperatures. *Aust. Meteor. Mag.*, 38, 151–161.

Simmonds, I. and Jacka, T.H. 1995. Relationships between the interannual variability of Antarctic sea ice and the Southern Oscillation. *J. Climate*, 8, 637–647.

Simmonds, I. and Rocha, A. 1991. The association of Australian winter climate with ocean temperatures to the west. *J. Climate*, 4, 1147–1161.

Simpson, H.J., Cane, M.A., Lin, S.K., Zebiak, S.E. and Herczeg, A.L. 1993a. Forecasting annual discharge of River Murray, Australia, from a geophysical model of ENSO. *J. Climate*, 6, 386–390.

Simpson, H.J., Cane, M.A., Herczeg, A.L., Zebiak, S.E. and Simpson, J.H. 1993b. Annual River Discharge in Southeastern Australia Related to El Niño–Southern Oscillation Forecasts of Sea Surface Temperatures. *Water Resour. Res.*, 20, 11, 3671–3680.

Simpson, H.J., Herczeg, A.L., Smith, A., Smith, I.N., Allan, R.J., Zebiak, S., Blumenthal, B., Drosdowsky, W., Nicholls, N. and Close, A. 1996. Mean sea surface temperature anomalies in the low latitude Pacific and Indian Oceans relevant to stream discharge in SE Australia. *CSIRO Tech. Rep.* (in press).

Singh, S.V., Kripalani, R.H. and Sikka, D.R. 1992. Interannual variability of the Madden-Julian Oscillations in Indian summer monsoon rainfall. *J. Climate*, 5, 973–978.

Sittel, M.C. 1994a. Marginal probablities of the extremes of ENSO events for temperature and precipitation in the southeastern United States. *Tech. Rep.* 94–1, Center for Ocean-Atmospheric Studies, Florida, USA.

Sittel, M.C. 1994b. Differences in the means of ENSO extremes for maximum temperature and precipitation in the southeastern United States.. *Tech. Rep.* 94–2, Center for Ocean-Atmospheric Studies, Florida, USA.

Skidmore, A.K. 1987. Predicting bushfire activity in Australia from El Niño–Southern Oscillation events. *Aust. For.*, 50, 231–235.

Slowey, N.C. and Crowley, T.J. 1995. Interdecadal variability of Northern Hemisphere circulation recorded by Gulf of Mexico corals. *Geophys. Res. Lett.*, 22, 2345–2348.

Smith, I.N. 1994a. Assessments of categorial rainfall predictions. *Aust. Meteor. Mag.*, 43, 143–151.

Smith, I.N. 1994b. Indian Ocean sea-surface temperature patterns and Australian winter rainfall. *Int. J. Climatol.*, 14, 287–305.

Smith, I.N. 1995. A GCM simulation of global climate interannual variability: 1950–1988. *J. Climate*, 8, 709–718.

Smith, N.G. 1990. The Gulf of Panama and El Niño events: the fate of two refugee boobies from the 1982–83 event. In Glynn, P.W., (ed.), *Global Ecological Consequences of the 1982–83 El Niño–Southern Oscillation*, Elsevier Oceanography Series, vol. 52, Elsevier, Amsterdam, 381–393.

Smith, N.R. 1993. Ocean modelling in a global ocean observing system. *Rev. Geophys.*, 31, 281–317.

Smith, S.R. and Stearns, C.R. 1993. Antarctic pressure and temperature anomalies surrounding the maximum in the Southern Oscillation Index. *J. Geophys. Res.*, 98, D7, 13,071–13,083.

Smith, T.M., Reynolds, R.W. and Ropelewski, C.F. 1994. Optimal averaging of seasonal sea surface temperatures and associated confidence intervals. *J. Climate*, 7, 949–964.

Smith, T.M., Reynolds, R.W., Livezey, R.E. and Stokes, D.C. 1996. Reconstruction of historical sea surface temperatures using empirical orthogonal functions. *J Climate*, 9, 1403–1420.

Smithsonian Institution, 1927. World Weather Records, –1920. *Misc. Coll., Vol.* 79, Washington, D.C.

Smithsonian Institution, 1934. World Weather Records, 1921–1930. *Misc. Coll., Vol.* 90, Washington, D.C.

Smithsonian Institution, 1947. World Weather Records, 1931–1940. *Misc. Coll., Vol.* 105, Washington, D.C.

Solow, A.R. 1995. Testing for change on the frequency of El Niño events. *J. Climate*, 8, 2563–2566.

Solow, A.R. and Nicholls, N. 1990. The relationship between the Southern Oscillation and tropical cyclone frequency in the Australian region. *J. Climate*, 3, 1097–1101.

South African Journal of Science, 1987. Stemming the flood. *S. Afr. J. Sci.*, 83, 343–346.

Sperber, K.R. and Hameed, S. 1991. Southern Oscillation simulation in the OSU coupled upper ocean–atmosphere GCM. *Clim. Dyn.,* **6**, 83–97.

Sperber, K.R., Hameed, S., Gates, W.L. and Potter, G.L. 1987. Southern Oscillation simulated in a global climate model. *Nature,* **329**, 140–142.

Sperber, K.R., Hameed, S. and Gates, W.L. 1992. Surface currents and equatorial thermocline in a coupled upper ocean–atmosphere GCM. *Clim. Dyn.,* **7**, 121–131.

Stephenson, D.B. and Royer, J-F. 1995a. GCM simulation of the Southern Oscillation from 1979–88. *Clim. Dyn.,* **11**, 115–128.

Stephenson, D.B. and Royer, J-F. 1995b. Low-frequency variability of total ozone mapping spectrometer and general circulation model total ozone stationary waves associated with the El Niño/Southern Oscillation for the period 1979–1988. *J. Geophys. Res.,* **100**, D4, 7337–7346.

Stearns, S.D. and D.R. Hush, 1990. *Digital Signal Analysis.* Prentice-Hall, 440 pp.

Stockdale, T., Anderson, D., Davey, M., Delecluse, P., Kattenburg, A., Kitamura, Y., Latif, M. and Yamagata, T. 1993. Intercomparison of tropical ocean GCMs. *WCRP* **79**, World Climate Programme, 43 pp.

Stockdale, T., Latif, M., Burgers, G. and Wolff, J-O. 1994. Some sensitivities of a coupled ocean–atmosphere GCM. *Tellus,* **46A**, 367–380.

Stockton, C.W. 1990. Climatic variability on the scale of decades to centuries. *Climatic Change,* **16**, 173–183.

Stone, R.C. and Auliciems, A. 1992. SOI phase relationships with rainfall in eastern Australia. *Int. J. Climatol.,* **12**, 625–636.

Stone, R.C. and McKeon, G.M. 1992. Prospects for using weather prediction to reduce pasture establishment risk. *Trop. Grasslands,* **27**, 406–413.

Stone, R.C., Nicholls, N. and Hammer, G.L. 1996. Frost in NE Australia: Trends and influences of phases of the Southern Oscillation. *J. Climate* (in press).

Street-Perrott, F.A. and Perrott, R.A. 1990. Abrupt climate fluctuations in the tropics: The influence of Atlantic Ocean circulation. *Nature,* **343**, 607–612.

Streten, N.A. 1973. Some characteristics of satellite-observed bands of persistent cloudiness over the Southern Hemisphere. *Mon. Wea. Rev.,* **101**, 486–495.

Streten, N.A. 1975. Satellite derived inferences to some characteristics of the South Pacific atmospheric circulation associated with the Niño event of 1972–73. *Mon. Wea. Rev.,* **103**, 989–995.

Streten, N.A. 1981. Southern Hemisphere sea surface temperature variability and apparent associations with Australian rainfall. *J. Geophys. Res.,* **86**, 485–497.

Streten, N.A. 1983. Extreme distributions of Australian annual rainfall in relation to sea surface temperature. *J. Climatol.,* **3**, 143–153.

Streten, N.A. 1987. Sea surface temperature anomalies associated with the transition to a year of widespread low rainfall over southern Africa. *Aust. Meteor. Mag.,* **35**, 91–93.

Strong, A.E. 1986. The effect of El Chichon on the 82/83 El Niño (abstract). *EOS Trans. AGU,* **67**, 880.

Suppiah, R. 1989. Relationships between the Southern Oscillation and the rainfall of Sri Lanka. *Int. J. Climatol.,* **9**, 601–618.

Suppiah, R. 1996. Spatial and temporal variations in the relationships between the Southern Oscillation phenmenon and rainfall of Sri Lanka. *J. Climatol.* (in press).

Suarez, M.J. and Schopf, P.S. 1988. A delayed action oscillator for ENSO. *J. Atmos. Sci.,* **45**, 3283–3287.

Suzuki, H. 1973. Recent and wurm climates of the west coast of South America. *Bull. Dept. Geogr. Uni. Tokyo,* **5**, 3–32.

Tanimoto, Y., Iwasaka, N., Hanawa, K. and Toba, Y. 1993. Characteristic variations of sea surface temperature with multiple time scales in the North Pacific. *J. Climate,* **6**, 1153–1160.

Taylor, G.I., (1962). Gilbert Thomas Walker 1868–1958. *Biogr. Mem. Fell. Roy. Soc.,* **8**, 166–174.

Tett, S. 1995. Simulation of El Niño–Southern Oscillation-like variability in a global AOGCM and its response to CO_2 increase. *J. Climate,* **8**, 1473–1502.

Thapliyal, V. and Kulshrestha, S.M. 1991. Climate changes and trends over India. *Mausam,* **42**, 333–338.

Thompson, L.G. 1992. Ice core evidence from Peru and China. The historical record of El Niño events. In Bradley, R.S. and Jones, P.D. (eds), *Climate since A.D. 1500,* Routledge, London, pp. 517–548.

Thompson, L.G. 1993. Reconstructing the paleo ENSO records from tropical and subtropical ice cores. *Bull. Inst. Fr. Etud. Andines, Lima,* **22**(1), 65–83.

Thompson, L.G., Mosley-Thompson, E. and Arnao, B.M. 1984. El Niño–Southern Oscillation events recorded in the stratigraphy of the tropical Quelccaya ice cap, Peru. *Science,* **234**, 361–364.

Thompson, L.G., Mosley-Thompson, E. and Thompson, P.A. 1992. Reconstructing interannual climate variability from tropical and subtropical ice-core records. Chapter 16, In Diaz, H.F. and Markgraf, V. (eds) *El Niño: Historical review and palaeoclimatic aspects of the Southern Oscillation,* Cambridge University Press, Cambridge, UK, 295–322.

Todd, C. 1879–1910. *Meteorological observations made at the Adelaide Observatory for the years 1879 to 1906.* Government Printer, Adelaide, South Australia.

Todd, C. 1888. *The Australasian.* 1456.

Todd, C. 1893. Meteorological work in Australia: A review. *Report of the Fifth Meeting of the Australiasian Association for the Advancement of Science,* Adelaide, pp. 246–270.

Tokioka, T., Endoh, M. and Nagai, T. 1984. A description of the Meteorological Research Institute atmospheric general circulation model (MRI-GCM-I). *Tech. Rep. of the MRI*, 13, 1–249.

Tokioka, T., Endoh, M. and Nagai, T. 1988. *Development of a numerical model for studies on the climate variations. 1. A coupled atmosphere–ocean model.* Japanese Climate Study (JAPACS), Research and Development Bureau, Science and Technology Agency, 175–181.

Tokioka, T., Noda, A., Kitoh, A., Nikaidou, Y., Nakagawa, S., Motoi, T. and Yukimoto, S. 1995. A transient CO_2 experiment with the MRI CGCM – quick report. *J. Meteor. Soc. Japan*, 73, 817–826.

Tomita, T. and Yasunari, T. 1993. On the two types of ENSO. *J. Meteor. Soc. Japan*, 71, 273–283.

Tourre, Y.M. and White, W.B. 1995. ENSO signals in global upper-ocean temperature. *J. Phys. Oceanogr.*, 25, 1317–1332.

Trenberth, K.E. 1976. Spatial and temporal variations of the Southern Oscillation. *Quart. J. Roy. Meteor. Soc.*, 102, 639–653.

Trenberth, K.E. 1984. Signal versus noise in the Southern Oscillation. *Mon. Wea. Rev.*, 112, 326–332.

Trenberth, K.E. and Hoar, T.J. 1996. The 1990–1995 El Niño–Southern Oscillation event. *Geophys. Res. Lett.*, 23, 57–60.

Trenberth, K.E. and Hurrell, J.W. 1994. Decadal atmosphere–ocean variations in the Pacific. *Clim. Dyn.*, 9, 303–319.

Trenberth, K.E. and Shea, D.J. 1987. On the evolution of the Southern Oscillation. *Mon. Wea. Rev.*, 115, 3078–3096.

Trillmich, F. and Ono, K.A., (eds), 1991. *Pinnipeds and El Niño: Responses to Environmental Stress.* Springer-Verlag, Berlin, 293 pp.

Troup, A.J. 1965. The Southern Oscillation. *Quart. J. Roy. Meteor. Soc.*, 91, 490–506.

Tsimplis, M.N. and Woodworth, P.L. 1994. The global distribution of the seasonal sea level cycle calculated from coastal tide guage data. *J. Geophys. Res.*, 99, C8, 16,031–16,039.

Tu, C-W. 1936. China rainfall and world weather: A preliminary study and its applications to seasonal forecasting. *Mem. Roy. Meteor. Soc.*, 4, 99–117.

Tu, C-W. 1937a. Atmospheric circulation and world temperature. *Mem. Nat. Res. Inst. Meteorol.*, XI, 2, reprinted in *Collected Scientific Papers Meteorology, 1919–1949.* Academia Sinica, Peking, China, 1954, pp. 325–347.

Tu, C-W. 1937b. China weather and world oscillation with applications to long-range forecasting of floods and droughts of China during the summer. *Mem. Nat. Res. Inst. Meteorol.*, XI, 4, reprinted in *Collected Scientific Papers Meteorology, 1919–1949.* Academia Sinica, Peking, China, 1954, pp. 349–391.

Tyson, P.D. 1986. *Climatic change and variability in southern Africa.* Oxford University Press, 220 pp.

Ulbrich, U., Graf, H-F. and Kirchner, I. 1995. The impact of El Niño and volcanic forcing on the atmospheric energy cycle and the zonal mean atmospheric circulation. *Beitr. Phys. Atmosph.*, 68, 59–74.

Unal, Y.S. and Ghil, M. 1995. Interannual and interdecadal oscillation patterns in sea level. *Clim. Dyn.*, 11, 255–278.

US Weather Bureau, 1959. World Weather Records, 1941–1950. US Dept. Comm., Washington, DC.

US Weather Bureau, 1967. World Weather Records, 1951–1960 (Vols. 1–6). US Dept. Comm., Washington, D.C.

van Bemmelen, W. 1913. *Observations made at secondary stations in Netherlands East-India. Vol.* I, Royal Meteorological Observatory at Batavia, 20 pp.

van Heerden, J., Terblanche, D.E. and Schulze, G. 1988. The Southern Oscillation and South African summer rainfall. *Int. J. Climatol.*, 8, 577–597.

van Loon, H. and Henry, S.L. 1986. Comments on warm events in the Southern Oscillation and local rainfall over Southeast Asian. *Mon. Wea. Rev.*, 114, 1419–1423.

van Loon, H. and Shea, D.J. 1985. The Southern Oscillation. Part IV: The precursors south of 15°S to the extremes of the oscillation. *Mon. Wea. Rev.*, 113, 2063–2074.

van Loon, H., Kidson, J.W. and Mullan, A.B. 1993. Decadal variation of the annual cycle in the Australian dataset. *J. Climate*, 6, 1227–1231.

Verma, R.K. 1994. Variability of Indian summer monsoon: Relationship with global SST anomalies. *Mausam*, 45, 205–212.

Vermeulen, J.H. and Jury, M.R. 1992. Tropical cyclones in the south-west Indian Ocean – track prediction and verification 1989–91. *Meteor. Mag.*, 121, 186–192.

Vernekar, A.D., Zhou, J. and Shukla, J. 1995. The effect of Eurasian snow cover on the Indian monsoon. *J. Climate*, 8, 248–266.

Vijayakumar, R. and Kulkarni, J.R. 1995. The variability of the interannual oscillations of the Indian summer monsoon rainfall. *Adv. Atmos. Sci.*, 12, 95–102.

Vincent, D.G. 1994. The South Pacific Conzergence Zone (SPCZ): A review. *Mon. Wea. Rev.*, 122, 1949–1970.

Vogel, C.H. 1989. A documentary-derived climatic chronology for South Africa, 1820–1900. *Climatic Change*, 14, 291–307.

von Storch, H. and Hasslemann, K. 1995. Climate variability and change. *Max-Planck-Institut fur Meteorologie, Report No.* 152, 26 pp.

von Storch, H., Schriever, D., Arpe, K., Branstator, G.W., Legnani, R. and Ulbrich, U. 1994. Numerical experiments on the atmospheric response to cold equatorial Pacific conditions ("La Niña") during northern summer. *The Global Atmosphere and Ocean System,* 2, 99–120.

von Storch, J-S. 1994. Interdecadal variability in a global coupled model. *Tellus,* 46A, 419–432.

Wakata, Y. and Sarachik, E.S. 1991. On the role of equatorial ocean modes in the ENSO cycle. *J. Phys. Oceanogr.,* 21, 434–443.

Wajsowicz, R.C. 1994. A relationship between interannual variations in the South Pacific wind stress curl, the Indonesian Throughflow, and the West Pacific warm water pool. *J. Phys. Oceanogr.,* 24, 2180–2187.

Wajsowicz, R.C. 1995. The response of the Indo–Pacific Throughflow to interannual variations in the Pacific wind stress. Part I: Idealized geometry and variations. *J. Phys. Oceanogr.,* 25, 1805–1826.

Walker, G.T. 1910a. On the meteorological evidence for supposed changes of climate in India. *Mem. India Meteorol. Dep.,* XXI, Part I, 21 pp.

Walker, G.T. 1910b. Correlation in seasonal variations of weather II. Sunspots and temperature. *Mem. India Meteorol. Dep.,* 21, Part II, 22–45.

Walker, G.T. 1915a. Correlation in seasonal variations of weather V. Sunspots and temperature. *Mem. India Meteorol. Dep.,* 21, Part XI, 61–90.

Walker, G.T. 1915b. Correlation in seasonal variations of weather V. Sunspots and pressure. *Mem. India Meteorol. Dep.,* 21, Part XII, 91–118.

Walker, G.T. 1922. Correlation in seasonal variations of weather, VII. The local distribution of monsoon rainfall. *Mem. India Meteorol. Dep.,* 23, 23–39.

Walker, G.T. 1923. Correlation in seasonal variations of weather, VIII. A preliminary study of world weather. *Mem. India Meteorol. Dep.,* 24, Part IV, 75–131.

Walker, G.T. 1924. Correlation in seasonal variations of weather, IX. A further study of world weather. *Mem. India Meteorol. Dep.,* 24, Part IX, 275–332.

Walker, G.T. (and Bliss, E.W.), 1928a. World weather III. *Mem. Roy. Meteor. Soc.,* 2, 97–134.

Walker, G.T. 1928b. World weather. *Quart. J. Roy. Meteor. Soc.,* 54, 79–87.

Walker, G.T. 1928c. World weather. *Mon. Wea. Rev.,* 54, 167–170.

Walker, G.T. 1928d. Ceara (Brazil) famines and the general air movement. *Beitr. z. Phys. der freien Atmosphare,* 14, 88–93.

Walker, G.T. 1930. Seasonal forecasting. *Quart. J. Roy. Meteor. Soc.,* 56, 359–364.

Walker, G.T. 1947. Arctic conditions and world weather. *Quart. J. Roy. Meteor. Soc.,* 73, 226–256.

Walker, G.T. and Bliss, E.W. 1930. World weather IV. *Mem. Roy. Meteor. Soc.,* 3, 81–95.

Walker, G.T. and Bliss, E.W. 1932. World weather V. *Mem. Roy. Meteor. Soc.,* 4, 53–84.

Walker, G.T. and Bliss, E.W. 1937. World weather VI. *Mem. Roy. Meteor. Soc.,* 4, 119–139.

Walker, N.D. 1990. Links between South African summer rainfall and temperature variability of the Agulhas and Benguela current systems. *J. Geophys. Res.,* 95, 3297–3319.

Walker, N.D. and Lindesay, J.A. 1989. Preliminary observations of oceanic influences on the February–March 1988 floods in central South Africa. *S. Afr. J. Sci.,* 85, 164–169.

Wallace, J.M. and Gutzler, D.S. 1981. Teleconnections in the geopotential height fields during the Northern Hemisphere winter. *Mon. Wea. Rev.,* 109, 784–812.

Walter, A. 1948. Observations of atmospheric pressure in East Africa. Part 1. Results from first order stations. *Mem. East African Meteor. Dept.,* 2, 1, 13–50.

Wang, B. 1992. An overview of the Madden-Julian Oscillation and its relation to monsoon and mid-latitude circulation. *Adv. Atmos. Sci.,* 9, 93–111.

Wang, B. 1995a. Transition from a cold to a warm state of the El Niño–Southern Oscillation cycle. *Meteorol. Atmos. Phys.,* 56, 17–32.

Wang, B. 1995b. Interdecadal changes in El Niño onset in the last four decades. *J. Climate,* 8, 267–285.

Wang, C. and Weisberg, R.H. 1994a. Equatorially trapped waves of a coupled ocean–atmosphere system. *J. Phys. Oceanogr.,* 24, 1978–1998.

Wang, C. and Weisberg, R.H. 1994b. On the "Slow Mode" mechanism in ENSO-related coupled ocean–atmosphere models. *J. Climate,* 7, 1657–1667.

Wang, S. and Mearns, L.O. 1987. The impact of the 1982–83 El Niño event on crop yields in China. In Glantz, M.H., Katz, R.W. and Krentz, M., (eds), *The Societal Impacts Associated with the 1982–83 Worldwide Climate Anomalies.* Environmental and Societal Impacts Group, NCAR, Boulder, CO, 43–49.

Wang, W-C. and Li, K. 1990. Precipitation fluctuation over semiarid region in northern China and the relationship with El Niño/Southern Oscillation. *J. Climate,* 3, 769–783.

Wang, X.L. and Ropelewski, C.F. 1995. An assessment of ENSO-scale secular variability. *J. Climate,* 8, 1584–1599.

Ward, M.N. and Folland, C.K. 1991. Prediction of seasonal rainfall in the north Nordeste of Brazil using eignvectors of sea-surface temperature. *Int. J. Climatol.,* 11, 711–743.

Ward, M.N., Maskell, K., Folland, C.K., Rowell, D.P. and Washington, R. 1994. A tropic-wide oscillation of boreal summer rainfall and patterns of sea-surface temperature. *Climate Research Tech. Note* CTRN No. 48, Hadley Centre for Climate Prediction and Research, Meteorological Office, Bracknell, Berkshire, UK, 29 pp.

Warren, H.N. 1940. *Results of rainfall observations made in Queensland (Supp. Vol.).* Government Printer, Melbourne, 432 pp.

Warren, H.N. 1948. *Results of rainfall observations made in New South Wales.* Government Printer, Melbourne, 546 pp.

Watt, W.S, 1936. *Results of rainfall observations made in Tasmania.* Government Printer, Melbourne, 143 pp.

Watt, W.S, 1940. *Results of rainfall observations made in Papua, mandated territory of New Guinea, Solomon Islands, New Hebrides, etc.,.* Government Printer, Melbourne, 74 pp.

Watterson, I.G., Evans, J.L. and Ryan, B.F. 1995. Seasonal and interannual variability of tropical cyclogenesis: Diagnostics from large-scale fields. *J. Climate,* **8**, 3052–3066.

WCRP, (1995). *CLIVAR SCIENCE PLAN: A study of climate variability and predictability.* WCRP-89, WMO/TD No. 690, ICSU, WMO, UNESCO, 157 pp.

Weather Bureau, 1942–1951. *Report for the Year.* Dept. Transport, Union of South Africa.

WeatherDisc Associates, 1994. World Monthly Surface Station Climatology (TD-9645). On *World WeatherDisc: Climate data for the planet Earth. – Version* **3.0**, Seattle, Washington.

Webster, P.J. 1995. The annual cycle and the predictability of the tropical coupled ocean–atmosphere system. *Meteorol. Atmos. Phys.,* **56**, 33–55.

Webster, P.J. and Yang, S. 1992. Monsoon and ENSO: Selectively interactive systems. *Quart. J. Roy. Meteor. Soc.,* **118**, 877–926.

Wellington, G.M. and Dunbar, R.B. 1995. Stable isotopes signature of ENSO in eastern Pacific reef corals. *Coral Reefs,* **14**, 1–21.

West, F. and Healy, T. 1993. The Southern Oscillation Index and climatic parameters, and their relationship to snow and ski conditions at Mt. Ruapehu, New Zealand. *Weather and Climate,* **13**, 22–29.

Whetton, P.H. and Rutherfurd, I. 1994. Historical ENSO teleconnections in the Eastern Hemisphere. *Climatic Change,* **28**, 221–253.

Whetton, P.H., Adamson, D. and Williams, M.A.J. 1990. Rainfall and river flow variability in Africa, Australia and East Asia linked to El Niño–Southern Oscillation events. In Bishop, P. (Ed) Lessons for Human Survival: Nature's Record from the Quaternary *Geol. Soc. Aust. Symp. Proc.,* **1**, 71–82.

Whetton, P.H., Allan, R.J. and Rutherfurd,I. 1996. Historical ENSO teleconnections in the Eastern Hemisphere: Comparison with latest El Niño series of Quinn. *Climatic Change,* **32**, 103–109.

White, D.H. and Howden, M., (eds), 1994. Climate and Risk. *Agr. Sys. Info. Tech.,* **6**, 75 pp.

White, M.E. and Downton, M.W. 1991. The shrimp fishery in the Gulf of Mexico: relation to climatic variability and global atmospheric patterns. In Glantz, M., Katz, R. and Nicholls, N., (eds), *ENSO teleconnections linking worldwide climate anomalies: Scientific basis and societal impacts.* Cambridge University Press, Cambridge, UK, pp. 459–492.

White, W.B. 1994. Slow El Niño–Southern Oscillation boundary waves. *J. Geophys. Res.,* **99**, 22,737–22,751.

White, W.B. and Tai, C-K. 1992. Reflection of interannual Rossby Waves at the maritime western boundary of the tropical Pacific. *J. Geophys. Res.,* **97**, C9, 14,305–14,322.

White, W.B., He, Y. and Pazan, S.E. 1989. Off-equatorial propagating Rossby Waves in the tropical Pacific during the 1982–83 and 1986–87 ENSO events. *J. Phys. Oceanogr.,* **19**, 1397–1406.

Whysall, K.D.B., Cooper, N.S. and Bigg, G.R. 1987. Long-term changes in the tropical Pacific surface wind field. *Nature,* **327**, 216–219.

Wilby, R. 1993. Evidence of ENSO in the synoptic climate of the British Isles since 1880. *Weather,* **48**, 234–239.

Wilhite, D.A., Wood, D.A. and Meyer, S.J. 1987. Climate-related impacts in the United States during the 1982–83 El Niño. In Glantz, M.H., Katz, R.W. and Krentz, M., (eds), *The Societal Impacts Associated with the 1982–83 Worldwide Climate Anomalies.* Environmental and Societal Impacts Group, NCAR, Boulder, CO, pp. 75–78.

Willett, H.C. and Bodurtha, Jr., F.T. 1952. *Bull. Amer. Meteor. Soc.,* **33**, 429–430.

Williams, M. 1987. Relations between the Southern Oscillation and the troposphere over Australia. *BMRC Research Report No.* **6**, Bureau of Meteorology, Melbourne, 201 pp.

Wishart, J. 1928. On errors in the multiple correlation coefficient due to random sampling. *Mem. Roy. Meteor. Soc.,* **11**, 29–37.

Woodruff, S.D., Slutz, R.J., Jenne, R.L. and Steurer, P.M. 1987. A comprehensive ocean–atmosphere data set. *Bull. Amer. Meteorol. Soc.,* **68**, 1239–1250.

Wooster, W. 1960. El Niño. *Rep. Calif. Op. Ocean. Invest.,* **7**, 43–45.

Wright, P.B. 1977. The Southern Oscillation – Patterns and mechanisms of the teleconnections and the persistence. *Hawaii Institute of Geophysics Report,* **HIG-77-13**.

Wright, P.B. 1984. Relationships between indices of the Southern Oscillation. *Mon. Wea. Rev.,* **112**, 1913–1919.

Wright, P.B. 1985. The Southern Oscillation: An ocean–atmosphere feedback system? *Bull. Amer. Meteor. Soc.,* **66**, 398–412.

Wright, P.B. 1986. Precursors of the Southern Oscillation. *J. Climatol.,* **6**, 17–30.

Wright, P.B. 1988. On the reality of climatic changes in wind over the Pacific. *J. Climatol.,* **8**, 521–527.

Wright, P.B. 1989. Homogenized Long-Period Southern Oscillation Indices. *Int. J. Climatol.,* **9**, 33–54.

Wright, P.B., Mitchell, T.P. and Wallace, J.M. 1985. Relationships between surface observations over the global oceans and the Southern Oscillation. *Data Report ERL PMEL-*12, Seattle, NOAA.

Wright, P.B., Wallace, J.M., Mitchell, T.P. and Deser, C. 1988. Correlation structure of the El Niño/Southern Oscillation phenomenon. *J. Climate,* **1**, 609–625.

Wu, D-H., Anderson, D.L.T. and Davey, M.K. 1994. ENSO prediction experiments using a simple ocean–atmosphere model. *Tellus,* **46A**, 465–480.

Wunsch, C. 1992. Decade-to-century changes in the ocean circulation. *Oceanography,* **5**, 99–106.

Wyrtki, K. 1973. Teleconnections in the equatorial Pacific Ocean. *Science,* **180**, 66–68.

Wyrtki, K. 1974. Equatorial currents in the Pacific 1950 to 1970 and their relations to the trade winds. *J. Phys. Oceanogr.,* **4**, 372–380.

Wyrtki, K. 1975a. Fluctuations of the dynamic topography in the Pacific Ocean. *J. Phys. Oceanogr.,* **5**, 450–459.

Wyrtki, K. 1975b. El Niño – The dynamic response of the equatorial Pacific Ocean to atmospheric forcing. *J. Phys. Oceanogr.,* **5**, 572–584.

Wyrtki, K. 1976. Predicting and observing El Niño. *Science,* **191**, 343–346.

Wyrtki, K. 1977. Sea level during the 1973 El Niño. *J. Phys. Oceanogr.,* **7**, 779–787.

Wyrtki, K. 1987. Indonesian Through Flow and the associated pressure gradient. *J. Geophys. Res.,* **92**, C12, 12,941–12,946.

Wyrtki, K., Constantine, K., Kilonsky, B.J., Mitchum, G., Miyamoto, B., Murphy, T., Nakahara, S. and Caldwell, P. 1988a. The Pacific Island Sea Level Network. *JIMAR Contrib.,* **88-0137**, *Data Rep.,* 002.

Wyrtki, K., Kilonsky, B.J., and Nakahara, S. 1988b. The IGOSS Sea Level Pilot Project in the Pacific. *JIMAR Contrib.,* **88-0150**, *Data Rep.,* 003.

Xiangong, Z., Jie, S. and Zhen, Z. 1989. The Southern Oscillation reconstruction and drought/flood in China. *Acta. Meteor. Sin.,* **3**, 290-301.

Xu, J-S. 1992. On the relationship between the stratospheric Quasi-Biennial Oscillation and the tropospheric Southern Oscillation. *J. Atmos. Sci.,* **49**, 725-734.

Xue, Y., Cane, M.A., Zebiak, S.E. and Blumenthal, M.B. 1994. On the prediction of ENSO: a study with a low-order Markov model. *Tellus,* **46A**, 512-528.

Yang, S. and Xu, L. 1994. Linkage between Eurasian winter snow cover and regional Chinese summer rainfall. *Int. J. Climatol.,* **14**, 739–750.

Yarnal, B. 1985. Extratropical teleconnections with El Niño/Southern Oscillation (ENSO) events. *Prog. Phys. Geog.,* **9**, 315–352.

Yarnal, B. and Kiladis, G. 1985. Tropical teleconnections associated with El Niño/Southern Oscillation (ENSO) events. *Prog. Phys. Geog.,* **9**, 524–558.

Yasunari, T. 1985. Zonally propagating modes of the global east-west circulation associated with the Southern Oscillation. *J. Meteor. Soc. Japan,* **63**, 1013–1029.

Yasunari, T. 1987a. Global structure of the El Niño/Southern Oscillation. Part I. El Niño Composites. *J. Meteor. Soc. Japan,* **65**, 67–80.

Yasunari, T. 1987b. Global structure of the El Niño/Southern Oscillation. Part II. Time evolution. *J. Meteor. Soc. Japan,* **65**, 81–102.

Yasunari, T. 1990. Impact of Indian monsoon on the coupled atmosphere/ocean system in the tropical Pacific. *Meteorol. Atmos. Phys.,* **44**, 29–41.

Yi, S. and Longxun, C. 1992. The characteristics of 30–60 Day Oscillation and its relation to the interannual oscillations. *Adv. Atmos. Sci.,* **9**, 323–336.

Yin, Z-Y. 1994. Moisture condition in the south-eastern USA and teleconnection patterns. *Int. J. Climatol.,* **14**, 947–967.

Yongqiang, L. and Yihui, D. 1992. Influence of El Niño on weather and climate in China. *Acta. Meteor. Sin.,* **6**, 117–131.

Yoshino, M. and Yasunari, T. 1986. Climatic anomalies of El Niño and anti-El Niño years and their socio-economic impacts in Japan. *Sci. Rep. Inst. Geosci., Uni. Tsukuba, Sect.* **A**, 7, 41–53.

Young, K.C. 1993. Detecting and Removing Inhomogeneities from Long-Term Monthly Sea Level Pressure Time Series. *J. Climate,* **6**, 1205–1220.

Zebiak, S.E. and Cane, M.A. 1987. A model El Niño–Southern Oscillation. *Mon Wea. Rev.,* **115**, 2262–2278.

Zerefos, C.S., Bais, A.F. and Ziomas, I.C. 1992. On the relative importance of Quasi-Biennial Oscillation and El Niño/Southern Oscillation in the revised Dobson total ozone records. *J. Geophys. Res.,* **97**, D7, 10,135–10,144.

Zhang, X-G. and Casey, T.M. 1992. Long term variation in the Southern Oscillation and the relationship with rainfall in Australia. *Aust. Meteor. Mag.,* **40**, 211–225.

Zhang, X-G., Song, J. and Zhao, Z. 1989. The Southern Oscillation reconstruction and drought/flood in China. *Acta Meteor. Sin.,* **3**, 290–301.

ANNOTATED MAPS OF CLIMATIC EVENTS AND FLUCTUATIONS:

1871–1994

8. DETAILED HISTORY OF SELECTED ENSO PHASES

Assessment of Event History Using ENSO Impact Maps: A Range of El Niño and La Niña Phases and their Impacts Since 1871

The near-global impacts of El Niño and La Niña events on a variety of weather- and climate-related phenomena are synthesised here in a series of maps. Years for which El Niño (Table 3) and La Niña (Table 4) impacts are shown have been drawn from the full time period covered in this book, and are characteristic of the extreme ENSO phases. The maps also illustrate aspects of the inherent variability of the effects of El Niño and La Niña events, and of the climate system generally.

The maps include information on parameters known to be influenced by ENSO: rainfall, river flow and temperature anomalies, and tropical cyclone numbers. A variety of sources has been used in the compilation of the impacts maps, including global precipitation and air temperature data for ENSO-sensitive regions from Ropelewski and Halpert (1986, 1987, 1989) and Halpert and Ropelewski (1992). Regional precipitation information was obtained for northeastern Brazil, Peru and Surinam (Berlage, 1966; Jones, 1995, per. comm.), the central equatorial Pacific (Wright, 1989), the Indian and Sri Lankan monsoon areas (Pant *et al.*, 1988; Suppiah, 1989; Parthasarathy *et al.*, 1991, 1992; Vijayakumar and Kulkarni, 1995), China (Tu, 1936, 1937a,b; Wang and Li, 1990; Whetton and Rutherfurd, 1994), Indonesia (Quinn *et al.*, 1978), southern Africa (Nicholson and Entekhabi, 1986; Vogel, 1989; Lindesay and Vogel, 1990; Hulme, 1992; Nicholson and Palao, 1993; Nicholson, 1993), and East Africa (Nicholson and Entekhabi, 1986; Hulme, 1992; Hutchinson, 1992; Nicholson, 1993). Rivers for which flow anomalies are shown are the Nile, Senegal and Orange rivers in Africa, the Krishna River in India (*South African Journal of Science*, 1987; Whetton *et al.*, 1990; Quinn, 1992, 1993), the Murray-Darling River system in Australia (Close, 1990; Allan *et al.*, 1996), and the Amazon and Parana rivers in South America (Aceituno, 1988; Richey *et al.*, 1989; Mechoso and Iribarren, 1992). Information on Atlantic hurricane frequency (Gray and Sheaffer, 1991; Landsea *et al.*, 1992) and occasional flooding of Lake Eyre in central Australia (Allan, 1985) is also included.[1]

The impacts maps use colour shading and a number of symbols (defined in the key) to indicate the sign and magnitude of rainfall and temperature anomalies, the sign of river discharge anomalies, the presence of Lake Eyre flooding, and the number of Atlantic hurricanes/tropical cyclones in a particular year. Rainfall anomalies are expressed as percentages, above or below the fiftieth percentile with the precise anomalies indicated next to each area on the maps; in cases where no percentage was available, although the area was known to have been drier/wetter than normal, shading indicates the relevant anomaly in sign only. The temperature anomalies are expressed as average standardised positive or negative increments. This means that both anomalies can only be interpreted as relative increases or decreases and not as absolute values in millimetres or degrees C.

[1] For references see pp. 91–116.

El Niño Impact Maps

Table 3: El Niño years for which impacts on precipitation, river discharge, air temperature and tropical cyclone frequency are shown			
1877	1923	1941	1966
1888	1925	1957	1967
1899	1930	1958	1972
1905	1940	1965	1982

KEY TO IMPACT MAPS

⌇ 24	Hurricane/cyclone frequency
↘	Above-average river flow
⇘	Below-average river flow
✱	Lake Eyre flooding

1877

Rainfall

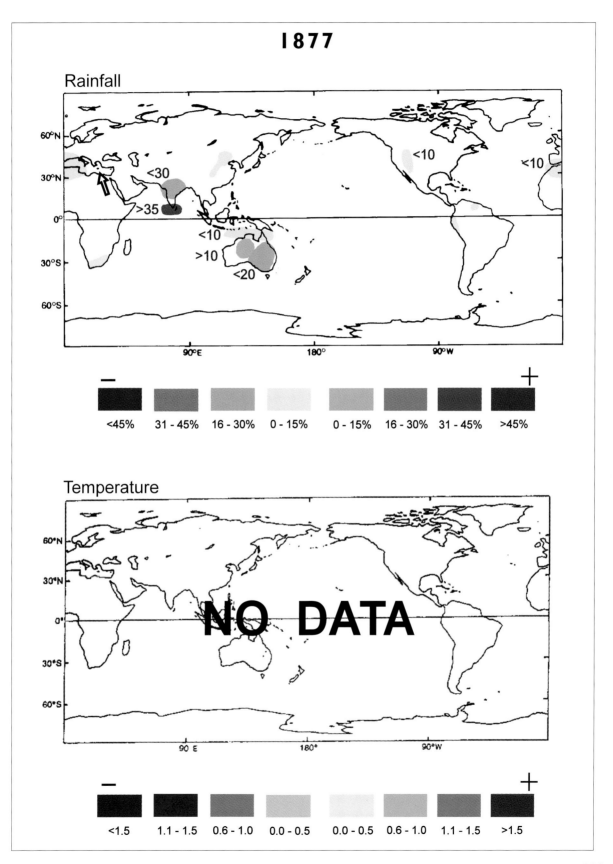

Temperature

NO DATA

1888

Rainfall

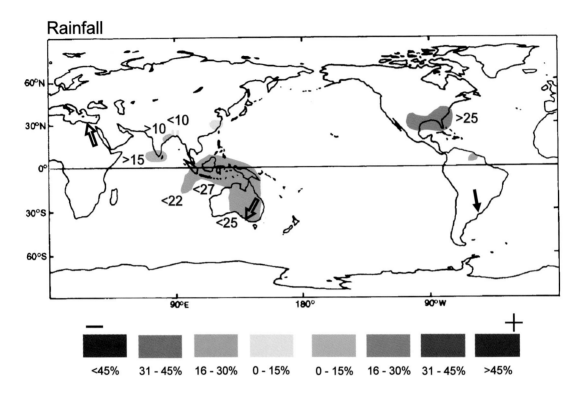

| <45% | 31 - 45% | 16 - 30% | 0 - 15% | 0 - 15% | 16 - 30% | 31 - 45% | >45% |

Temperature

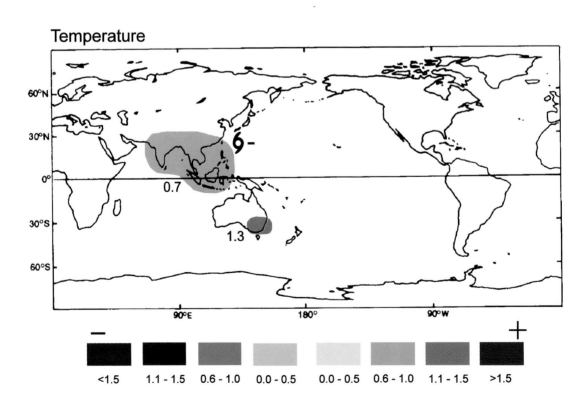

| <1.5 | 1.1 - 1.5 | 0.6 - 1.0 | 0.0 - 0.5 | 0.0 - 0.5 | 0.6 - 1.0 | 1.1 - 1.5 | >1.5 |

1899

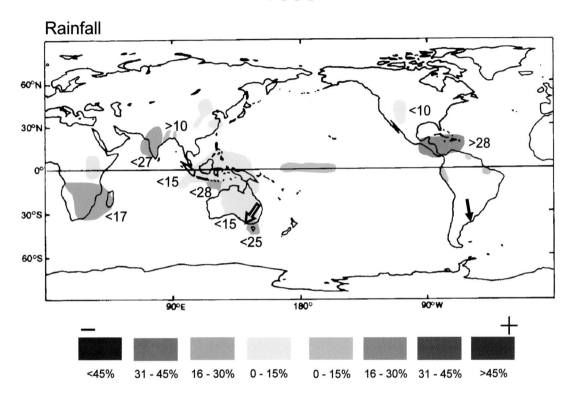

Rainfall

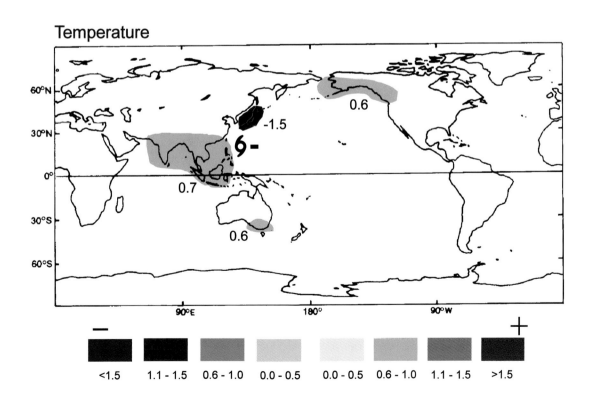

Temperature

1905

Rainfall

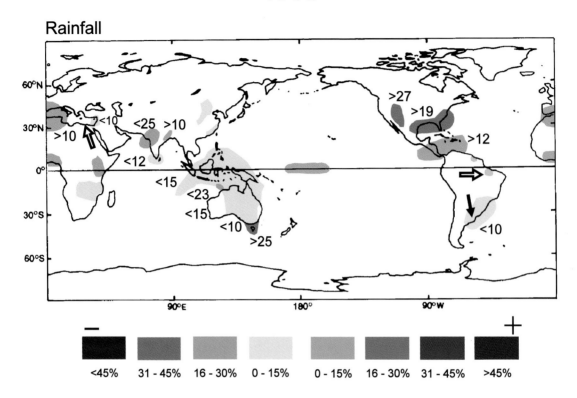

| <45% | 31 - 45% | 16 - 30% | 0 - 15% | 0 - 15% | 16 - 30% | 31 - 45% | >45% |

Temperature

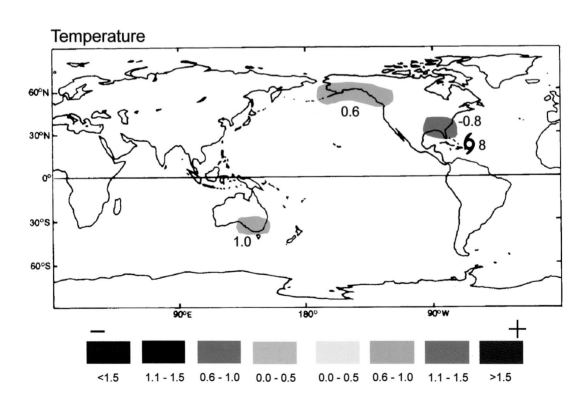

| <1.5 | 1.1 - 1.5 | 0.6 - 1.0 | 0.0 - 0.5 | 0.0 - 0.5 | 0.6 - 1.0 | 1.1 - 1.5 | >1.5 |

1923

Rainfall

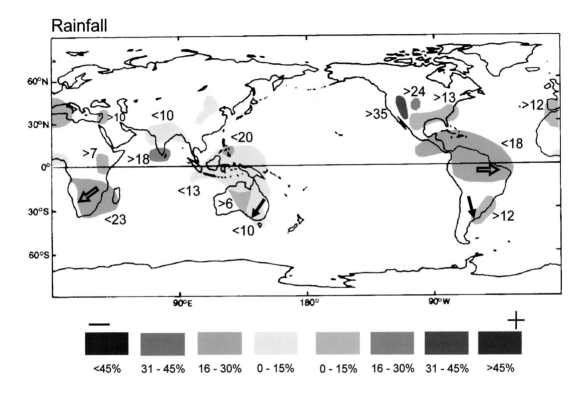

<45%	31 - 45%	16 - 30%	0 - 15%	0 - 15%	16 - 30%	31 - 45%	>45%

Temperature

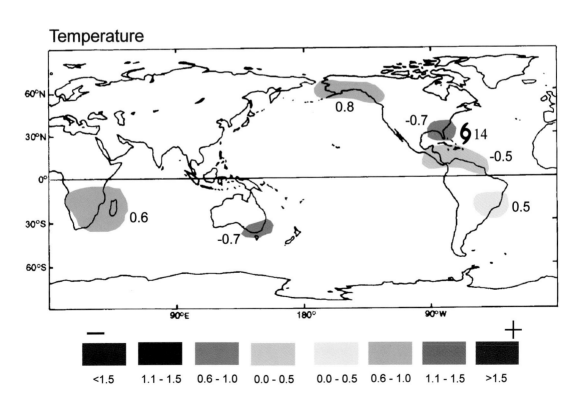

<1.5	1.1 - 1.5	0.6 - 1.0	0.0 - 0.5	0.0 - 0.5	0.6 - 1.0	1.1 - 1.5	>1.5

1925

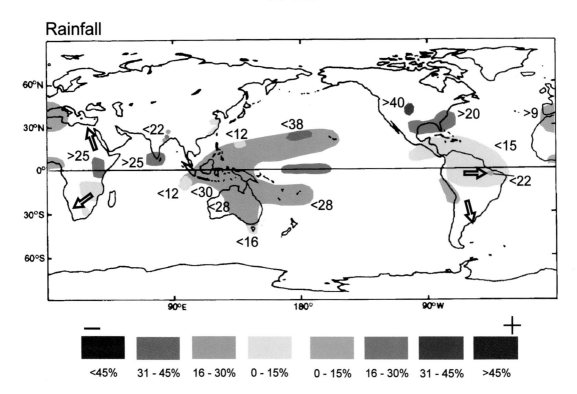

Rainfall

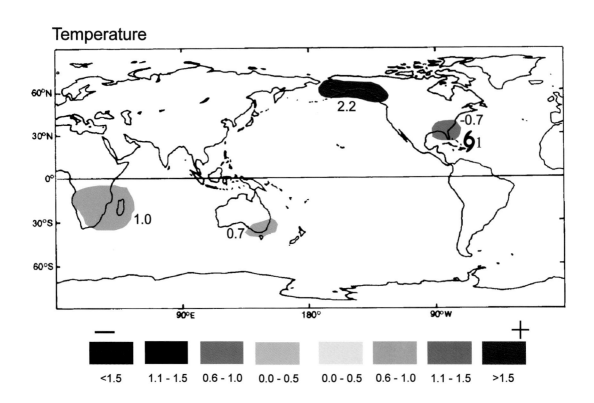

Temperature

1930

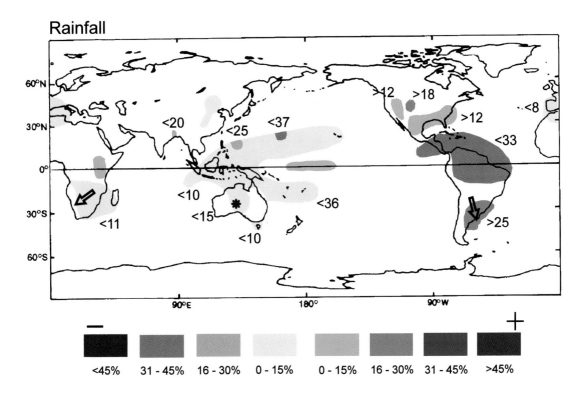

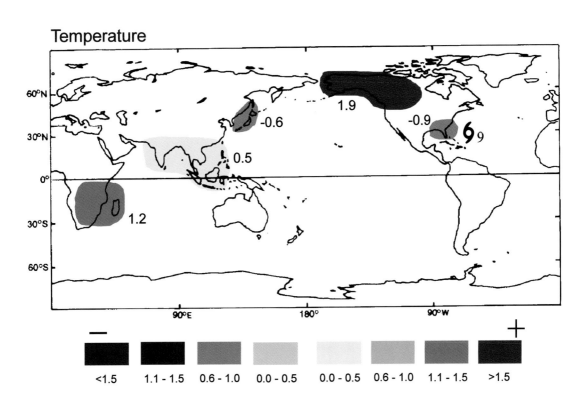

1940

Rainfall

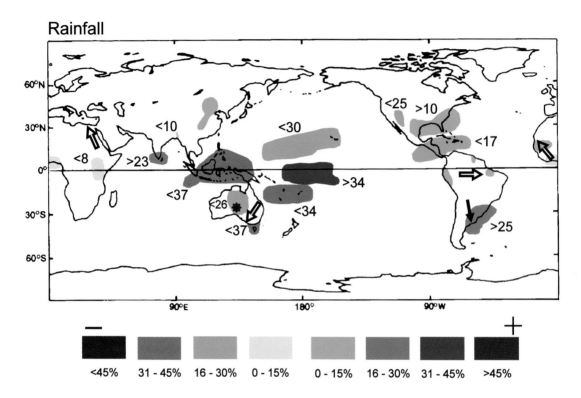

<45%	31 - 45%	16 - 30%	0 - 15%	0 - 15%	16 - 30%	31 - 45%	>45%

Temperature

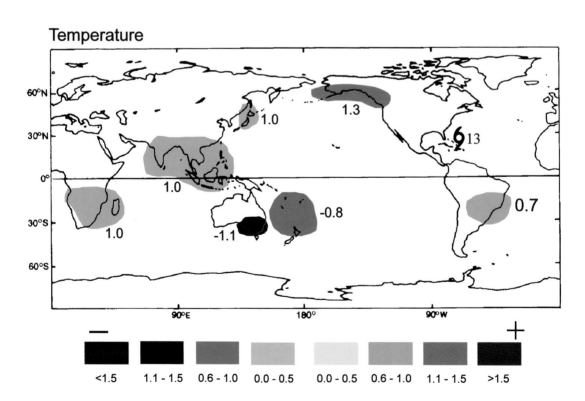

<1.5	1.1 - 1.5	0.6 - 1.0	0.0 - 0.5	0.0 - 0.5	0.6 - 1.0	1.1 - 1.5	>1.5

1941

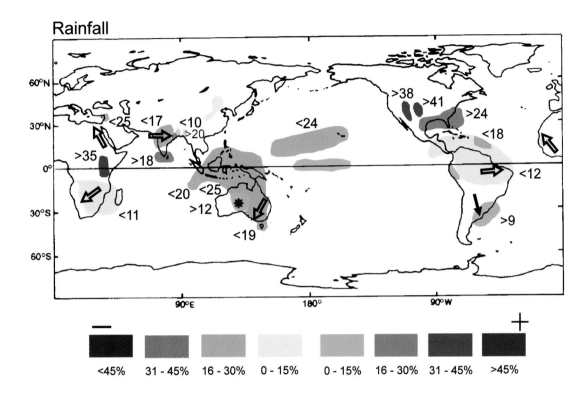

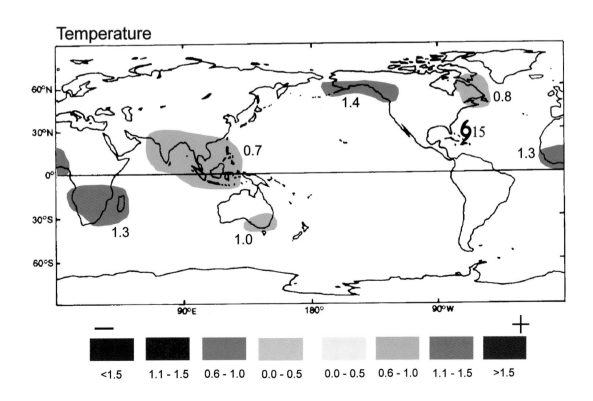

1957

Rainfall

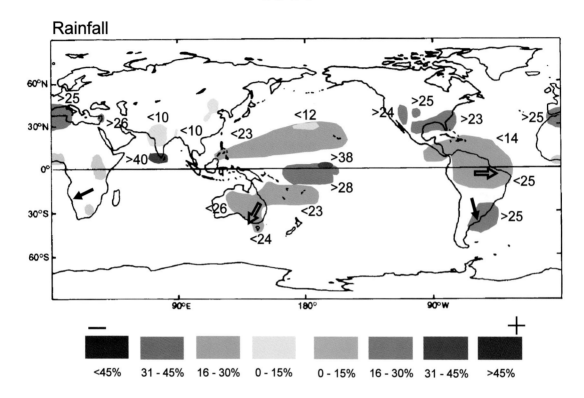

| <45% | 31 - 45% | 16 - 30% | 0 - 15% | 0 - 15% | 16 - 30% | 31 - 45% | >45% |

Temperature

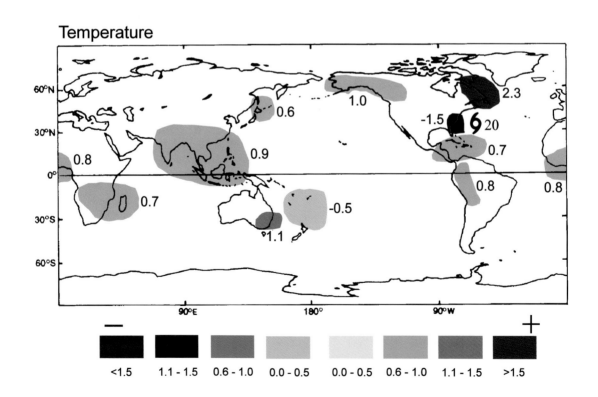

| <1.5 | 1.1 - 1.5 | 0.6 - 1.0 | 0.0 - 0.5 | 0.0 - 0.5 | 0.6 - 1.0 | 1.1 - 1.5 | >1.5 |

1958

Rainfall

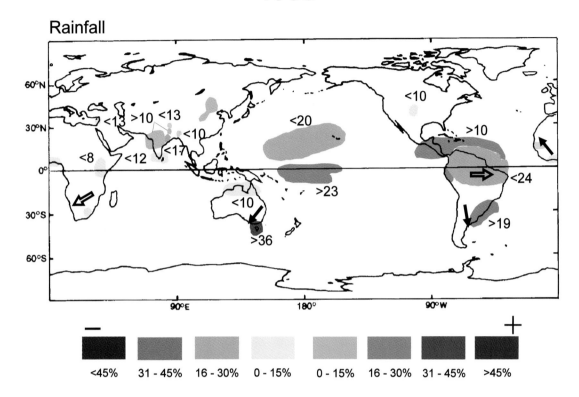

Temperature

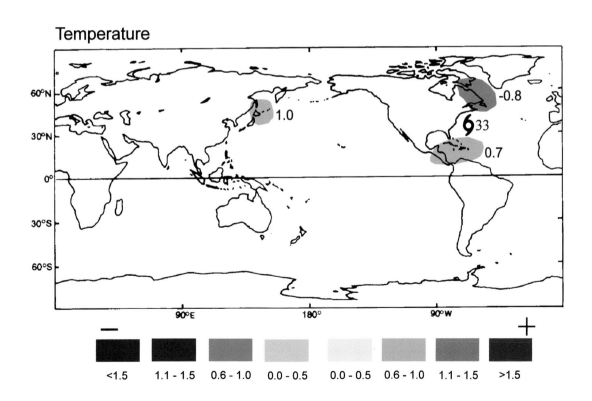

1965

Rainfall

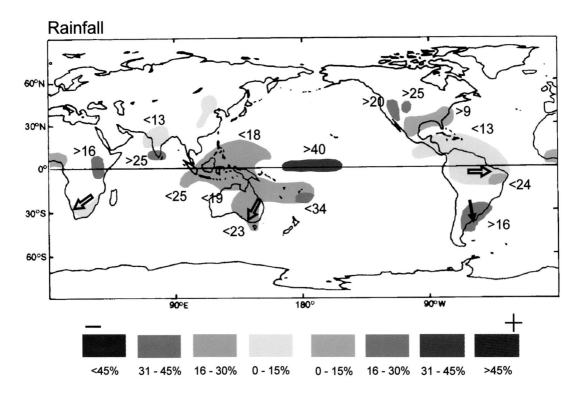

| <45% | 31 - 45% | 16 - 30% | 0 - 15% | 0 - 15% | 16 - 30% | 31 - 45% | >45% |

Temperature

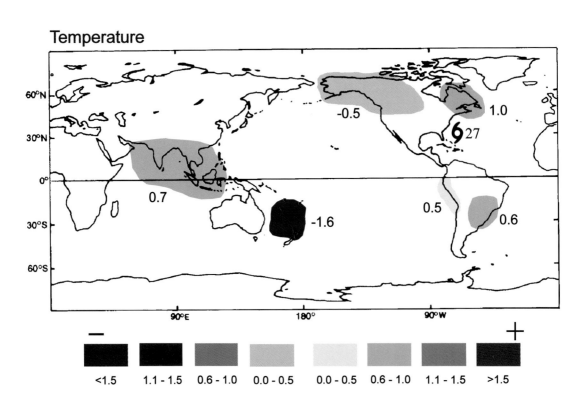

| <1.5 | 1.1 - 1.5 | 0.6 - 1.0 | 0.0 - 0.5 | 0.0 - 0.5 | 0.6 - 1.0 | 1.1 - 1.5 | >1.5 |

1966

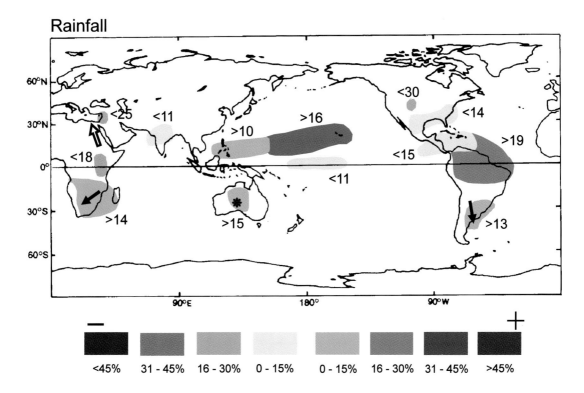

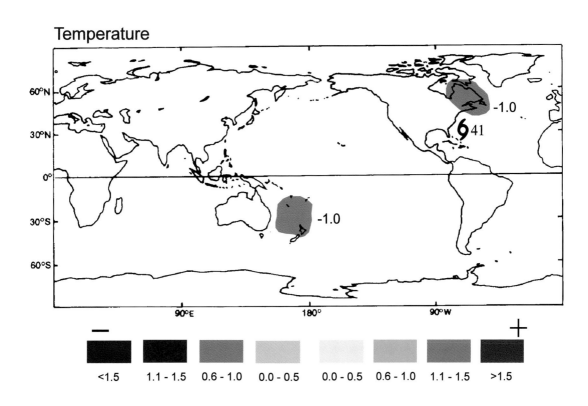

1967

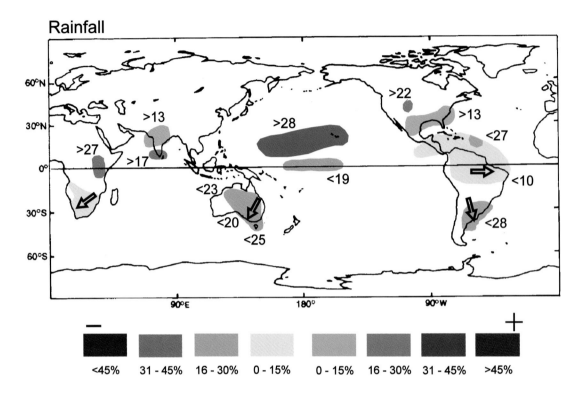

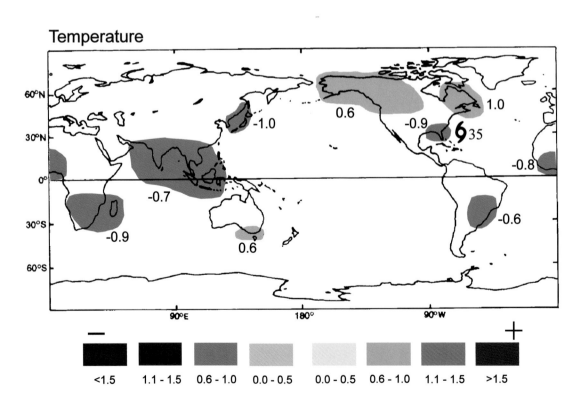

1972

Rainfall

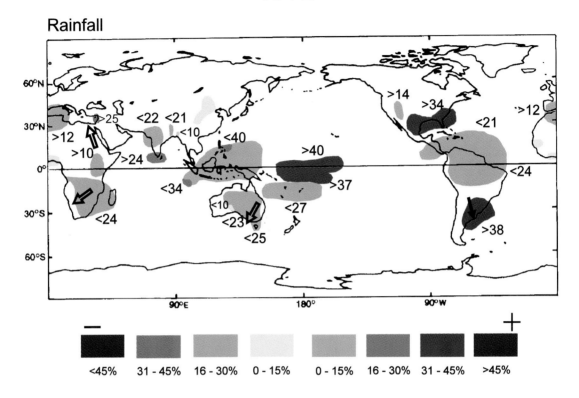

Temperature

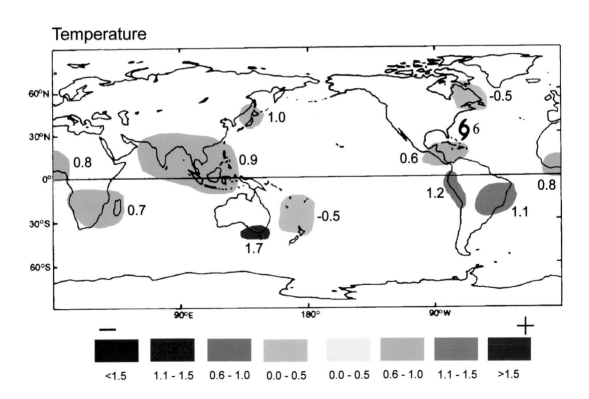

1982

Rainfall

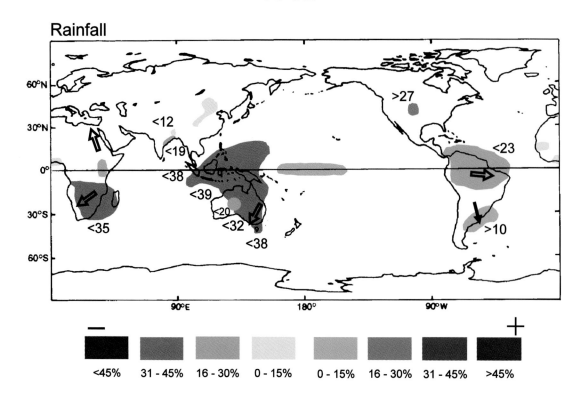

<45%	31 - 45%	16 - 30%	0 - 15%	0 - 15%	16 - 30%	31 - 45%	>45%

Temperature

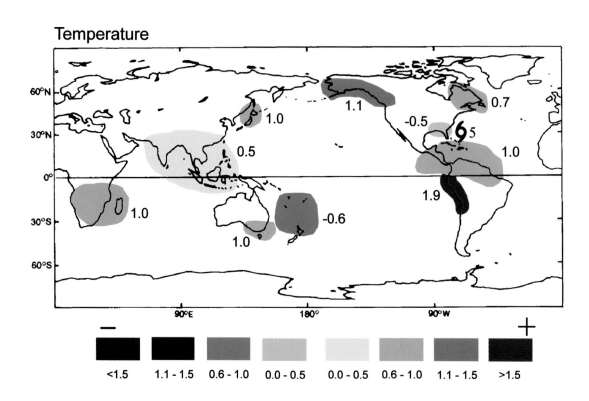

<1.5	1.1 - 1.5	0.6 - 1.0	0.0 - 0.5	0.0 - 0.5	0.6 - 1.0	1.1 - 1.5	>1.5

La Niña Impact Maps

Table 4: La Niña years for which impacts on precipitation, river discharge, air temperature and tropical cyclone frequency are shown			
1889	1910	1950	1971
1898	1916	1954	1973
1903	1917	1955	1975
1908	1924	1956	
1909	1938	1964	

KEY TO IMPACT MAPS

⟨ 24	Hurricane/cyclone frequency
↘	Above-average river flow
↘	Below-average river flow
✻	Lake Eyre flooding

1889

Rainfall

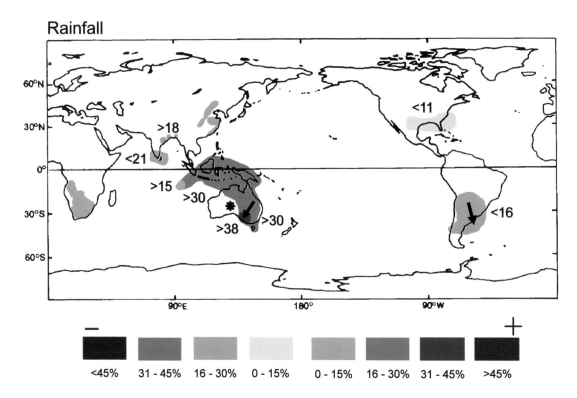

<45%	31 - 45%	16 - 30%	0 - 15%	0 - 15%	16 - 30%	31 - 45%	>45%

Temperature

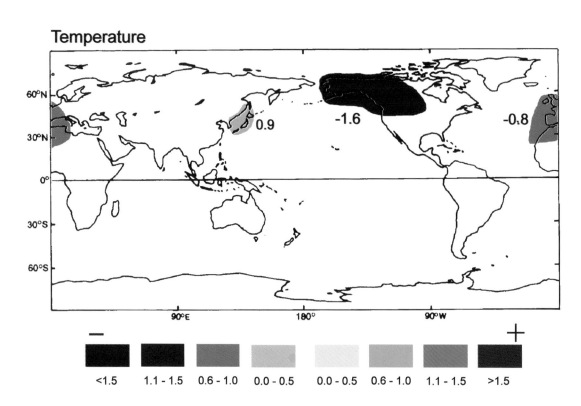

<1.5	1.1 - 1.5	0.6 - 1.0	0.0 - 0.5	0.0 - 0.5	0.6 - 1.0	1.1 - 1.5	>1.5

1898

Rainfall

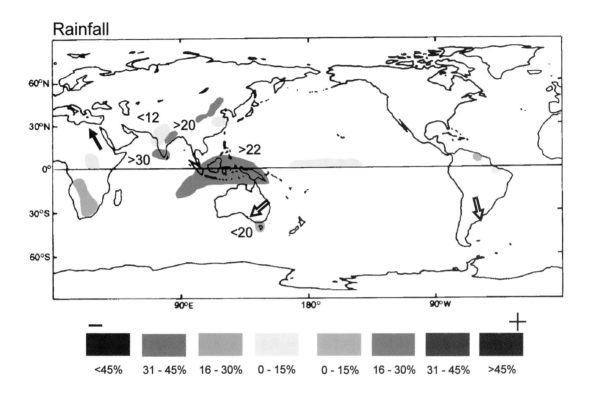

Temperature

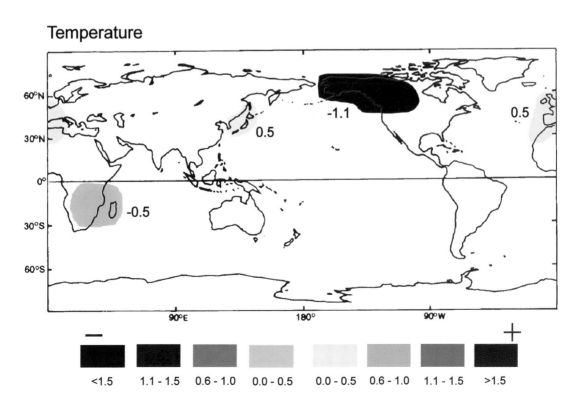

1903

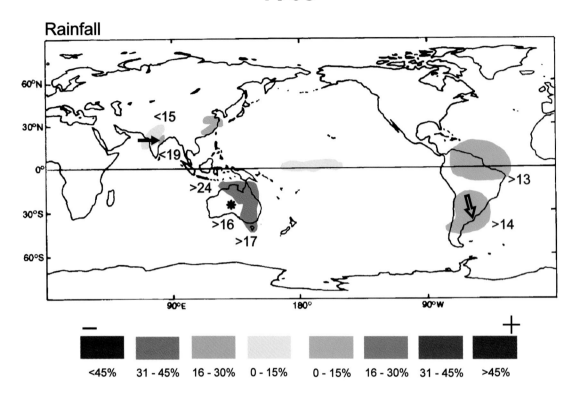

Rainfall

| <45% | 31 - 45% | 16 - 30% | 0 - 15% | 0 - 15% | 16 - 30% | 31 - 45% | >45% |

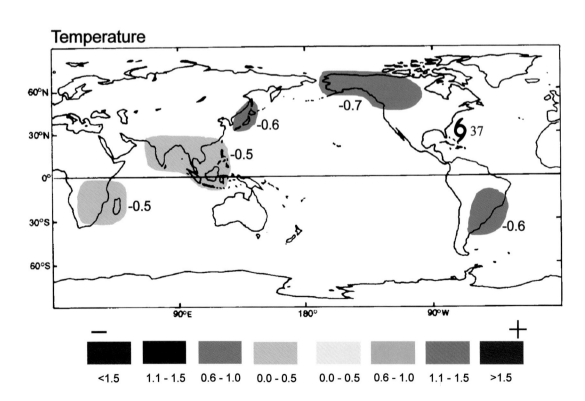

Temperature

| <1.5 | 1.1 - 1.5 | 0.6 - 1.0 | 0.0 - 0.5 | 0.0 - 0.5 | 0.6 - 1.0 | 1.1 - 1.5 | >1.5 |

1908

Rainfall

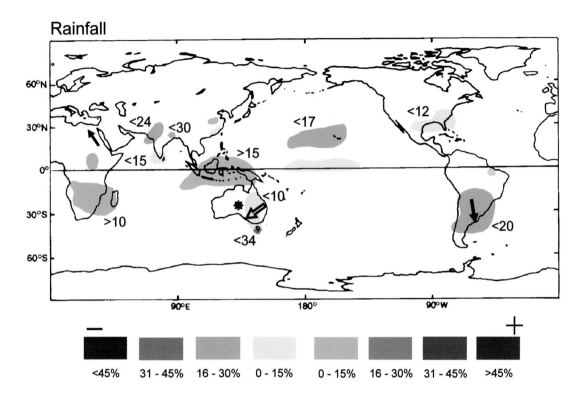

Temperature

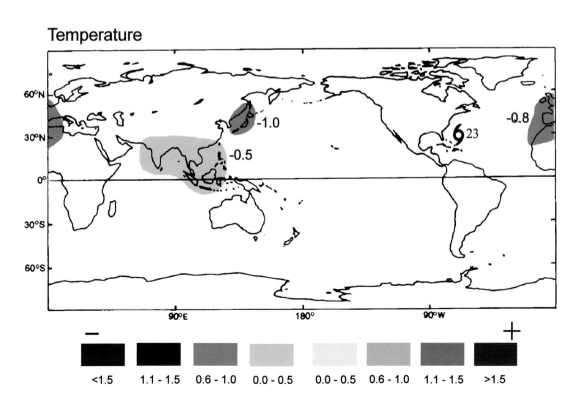

1909

Rainfall

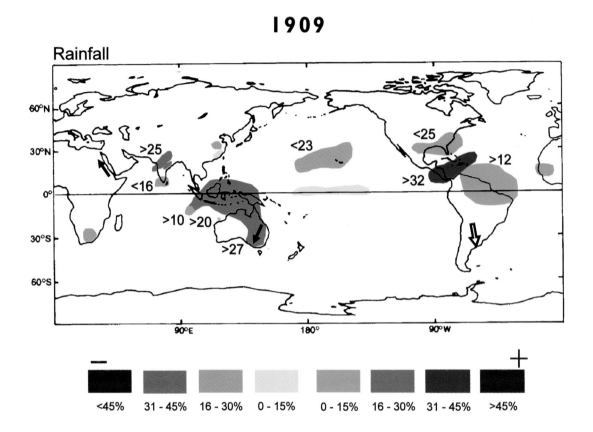

—							+
<45%	31 - 45%	16 - 30%	0 - 15%	0 - 15%	16 - 30%	31 - 45%	>45%

Temperature

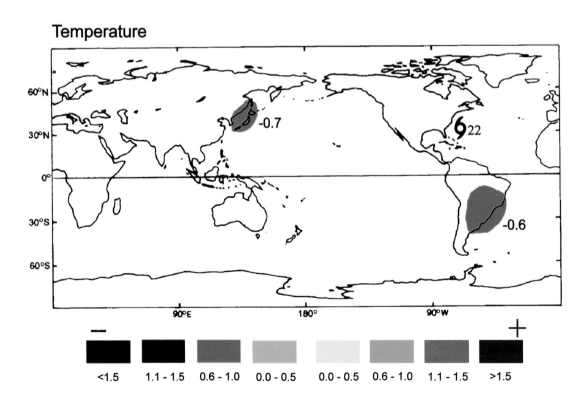

—							+
<1.5	1.1 - 1.5	0.6 - 1.0	0.0 - 0.5	0.0 - 0.5	0.6 - 1.0	1.1 - 1.5	>1.5

1910

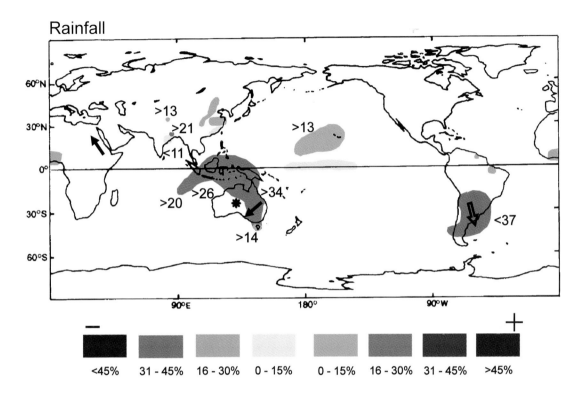

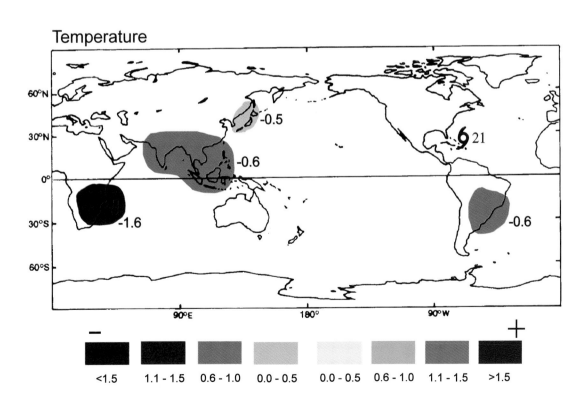

1916

Rainfall

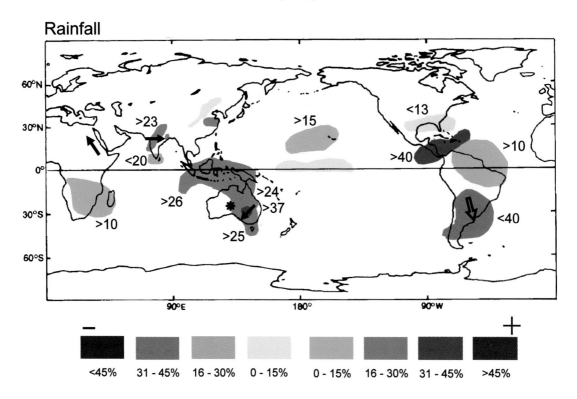

Temperature

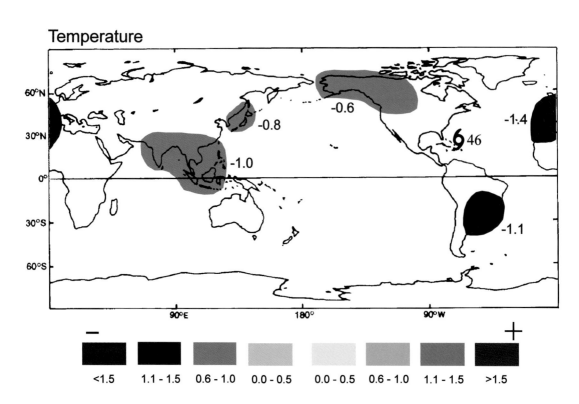

1917

Rainfall

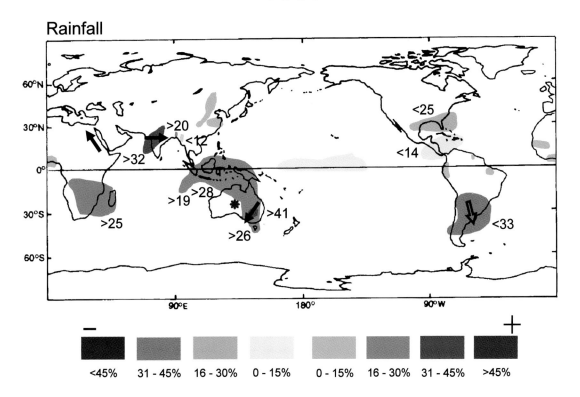

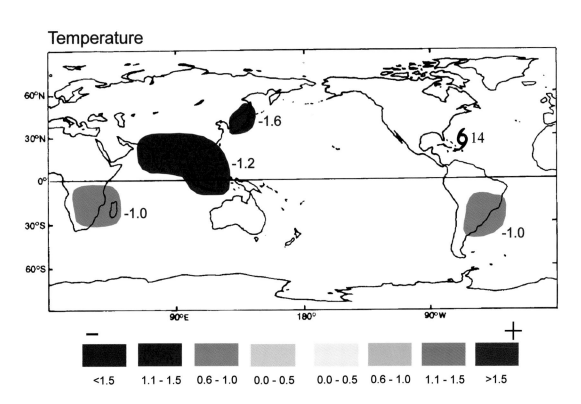

Temperature

1924

Rainfall

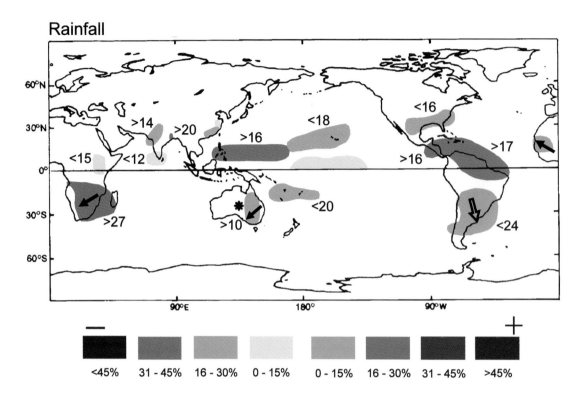

				+
<45%	31 - 45%	16 - 30%	0 - 15%	0 - 15% 16 - 30% 31 - 45% >45%

Temperature

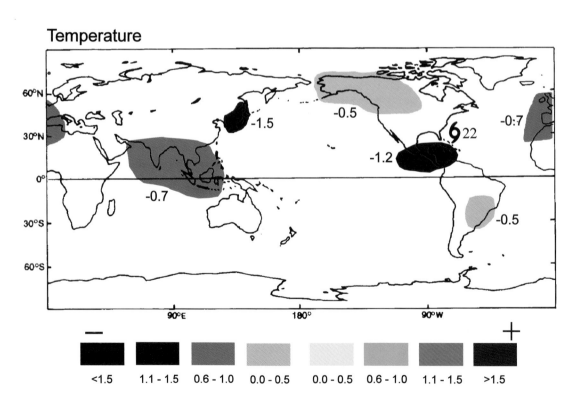

				+
<1.5	1.1 - 1.5	0.6 - 1.0	0.0 - 0.5	0.0 - 0.5 0.6 - 1.0 1.1 - 1.5 >1.5

1938

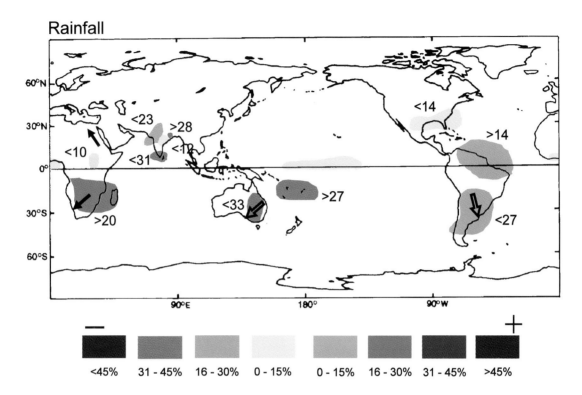

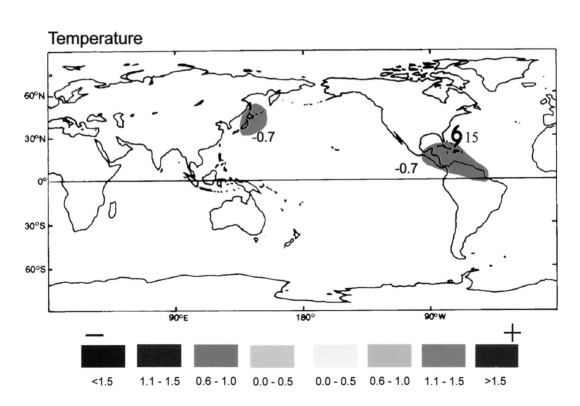

1950

Rainfall

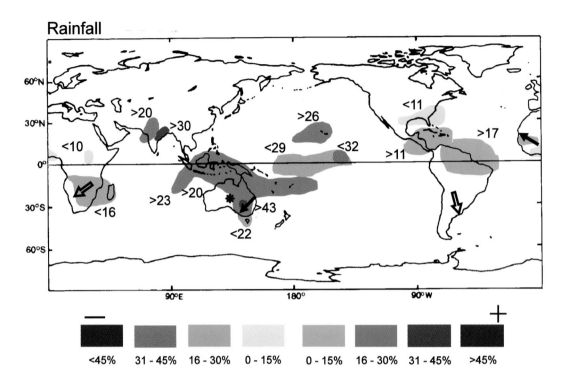

Temperature

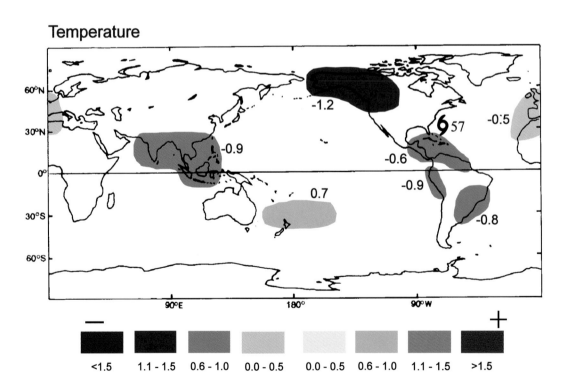

1954

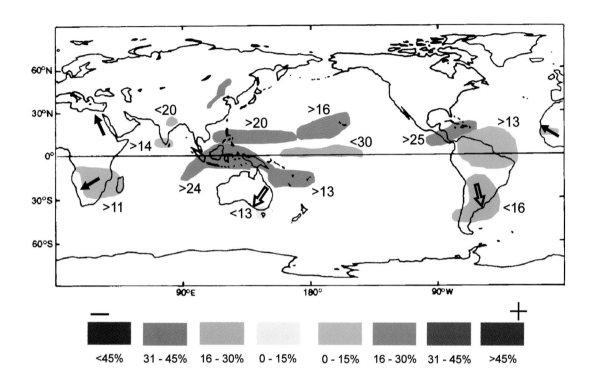

<45%	31 - 45%	16 - 30%	0 - 15%	0 - 15%	16 - 30%	31 - 45%	>45%

Temperature

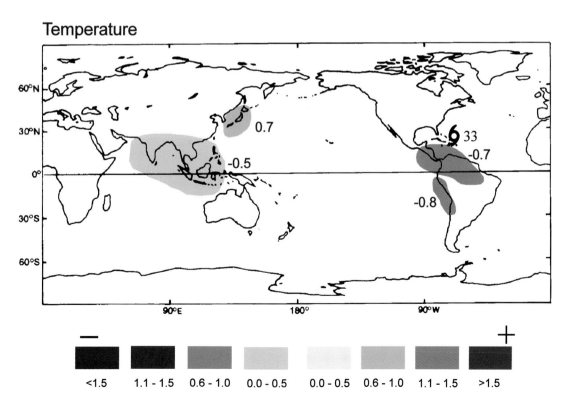

<1.5	1.1 - 1.5	0.6 - 1.0	0.0 - 0.5	0.0 - 0.5	0.6 - 1.0	1.1 - 1.5	>1.5

1955

Rainfall

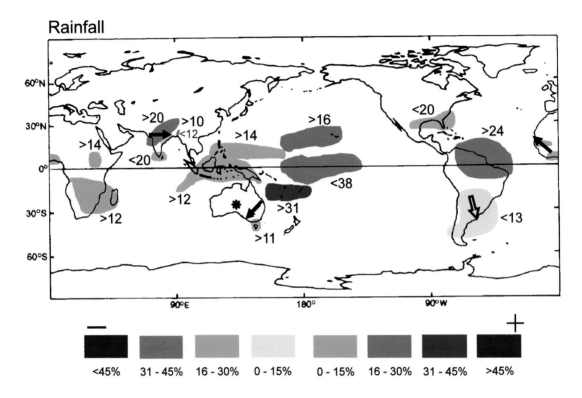

| <45% | 31 - 45% | 16 - 30% | 0 - 15% | 0 - 15% | 16 - 30% | 31 - 45% | >45% |

Temperature

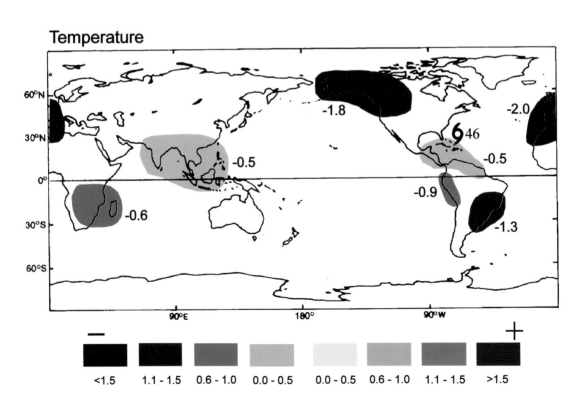

| <1.5 | 1.1 - 1.5 | 0.6 - 1.0 | 0.0 - 0.5 | 0.0 - 0.5 | 0.6 - 1.0 | 1.1 - 1.5 | >1.5 |

1956

Rainfall

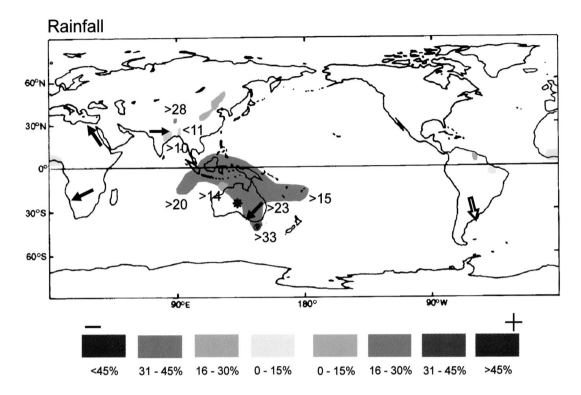

Temperature

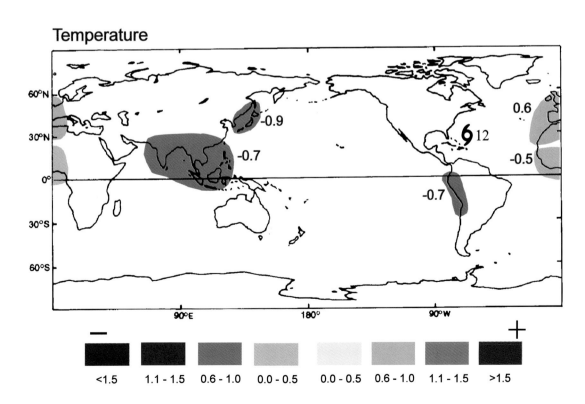

1964

Rainfall

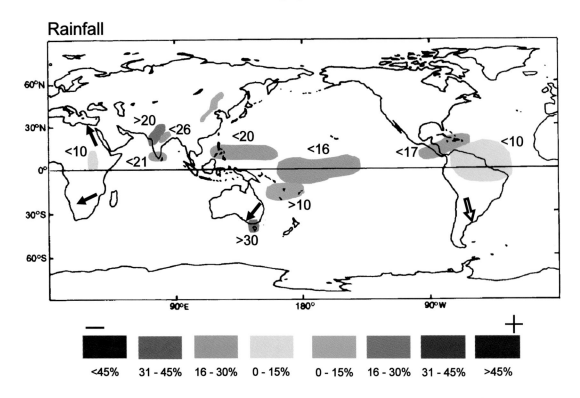

Temperature

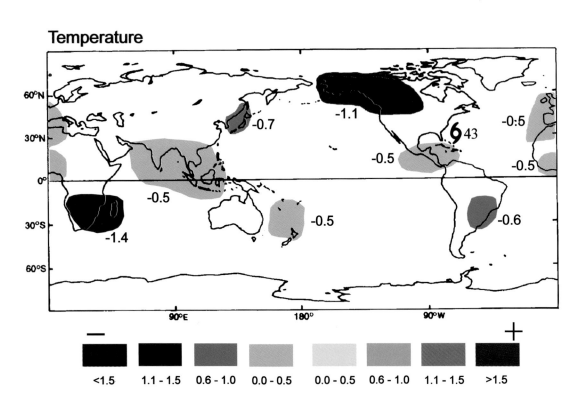

1971

Rainfall

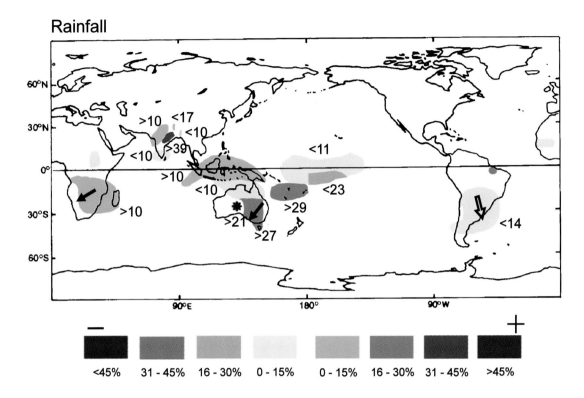

Temperature

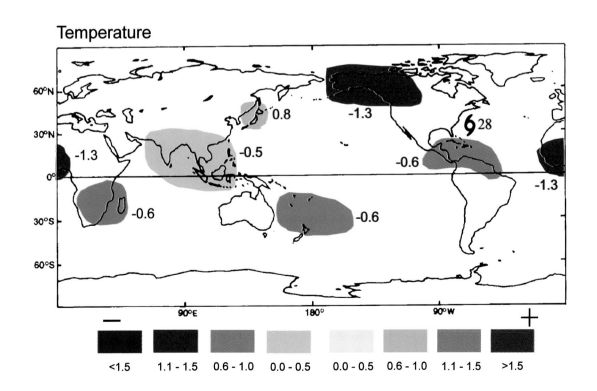

1973

Rainfall

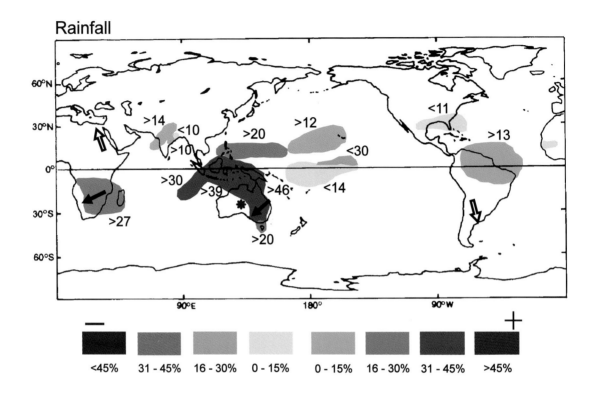

| <45% | 31 - 45% | 16 - 30% | 0 - 15% | 0 - 15% | 16 - 30% | 31 - 45% | >45% |

Temperature

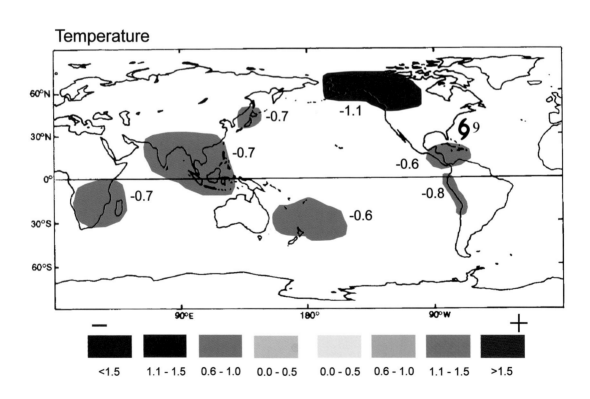

| <1.5 | 1.1 - 1.5 | 0.6 - 1.0 | 0.0 - 0.5 | 0.0 - 0.5 | 0.6 - 1.0 | 1.1 - 1.5 | >1.5 |

1975

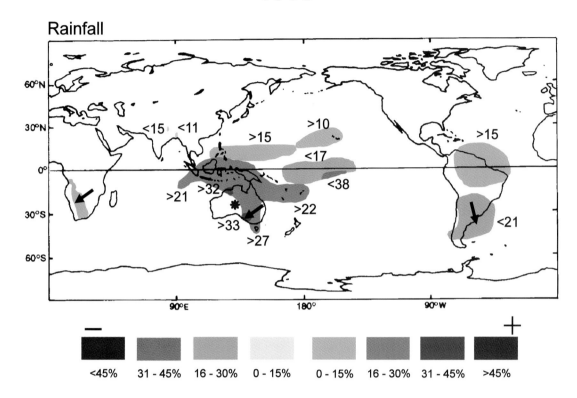

Rainfall

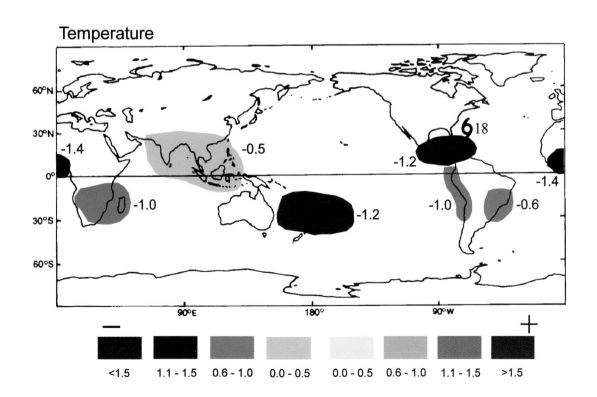

Temperature

9. Filtered Normalised Monthly Anomalies of MSLP and SST Since 1871

Global 32–88 month bandpass filtered, normalised mean sea level pressure (hPa) and sea surface temperature (°C) anomalies are shown every third month (Jan, Apr, Jul, Oct) for each year in the book. The CD-ROM shows the 18–35 and the 32–88 month bandpass filtered, normalised mean sea level pressure (hPa) and sea surface temperature (°C) anomalies for each month. All anomalies are with respect to the 1961–1990 mean.

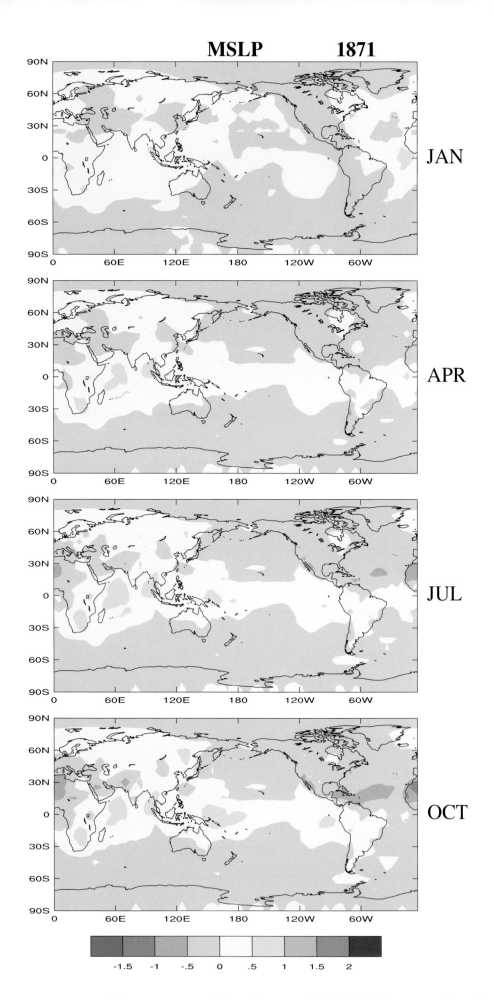

MSLP   **1871**

JAN

APR

JUL

OCT

-1.5   -1   -.5   0   .5   1   1.5   2

SST   **1871**

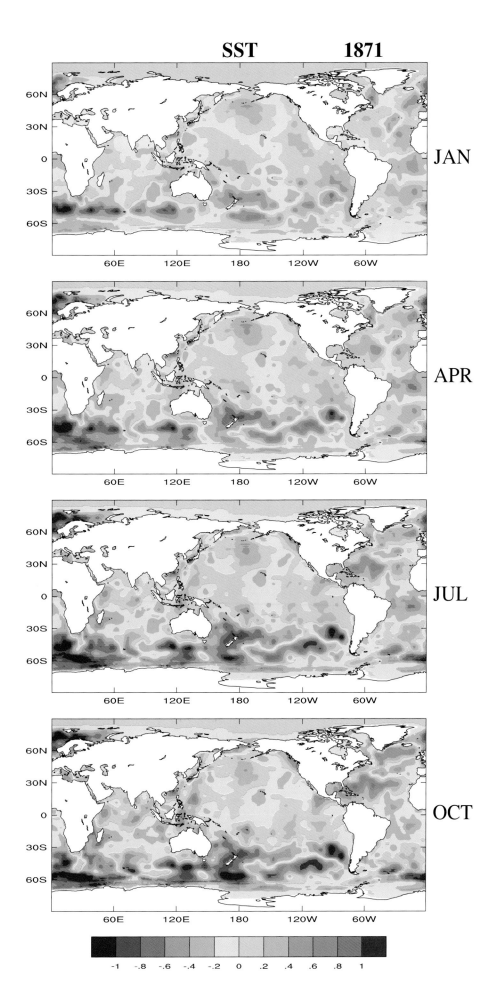

JAN

APR

JUL

OCT

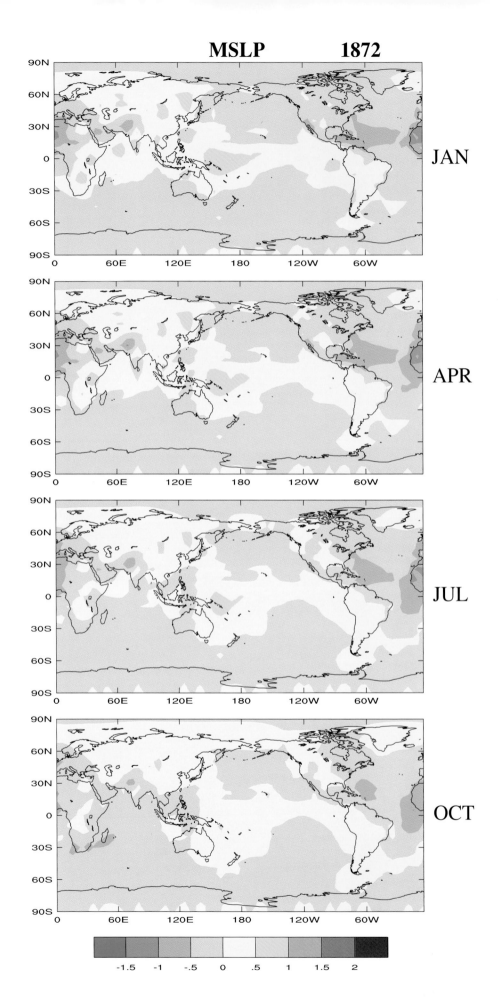

MSLP **1872**

JAN

APR

JUL

OCT

-1.5 -1 -.5 0 .5 1 1.5 2

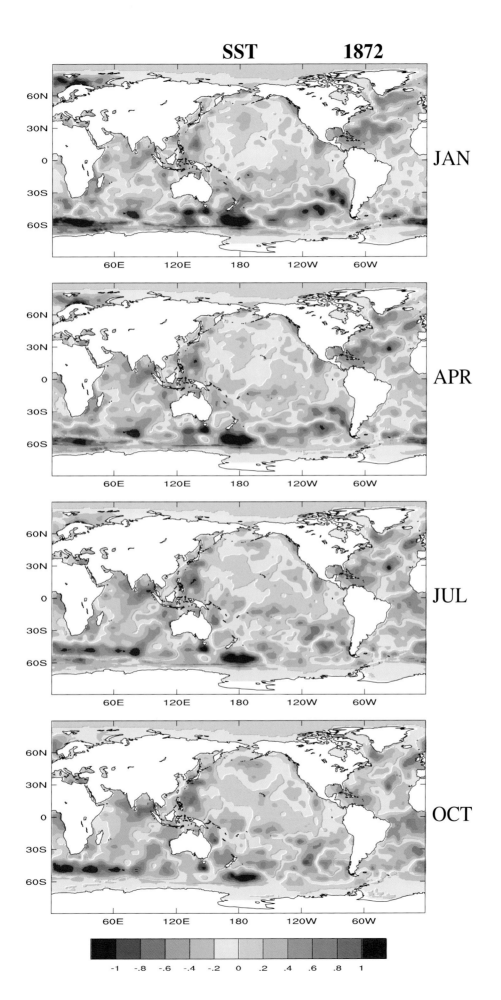

SST 1872

JAN

APR

JUL

OCT

-1 -.8 -.6 -.4 -.2 0 .2 .4 .6 .8 1

MSLP   **1873**

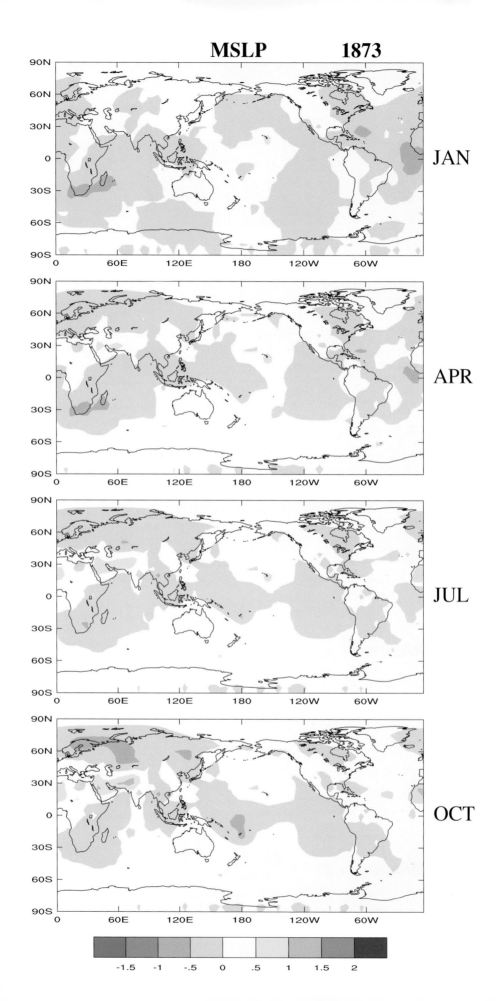

JAN

APR

JUL

OCT

-1.5   -1   -.5   0   .5   1   1.5   2

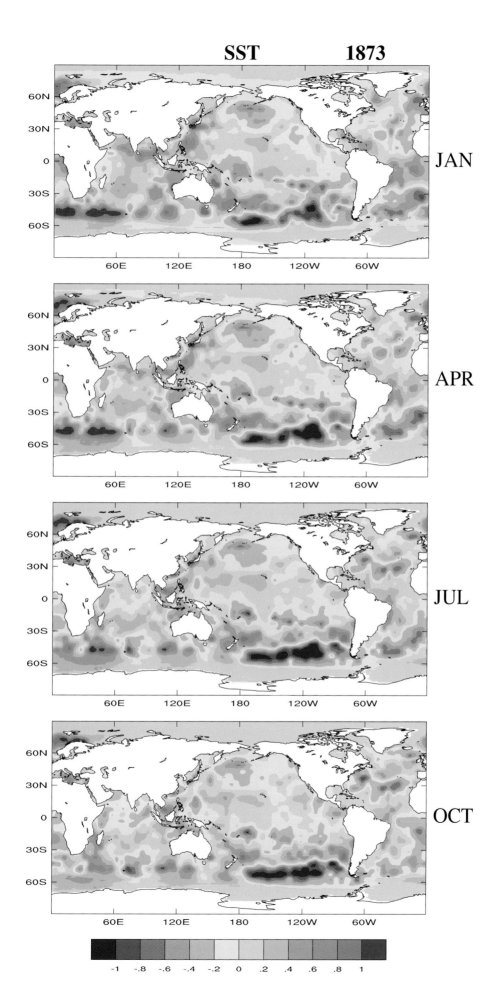

SST   **1873**

JAN

APR

JUL

OCT

-1   -.8   -.6   -.4   -.2   0   .2   .4   .6   .8   1

MSLP 1874

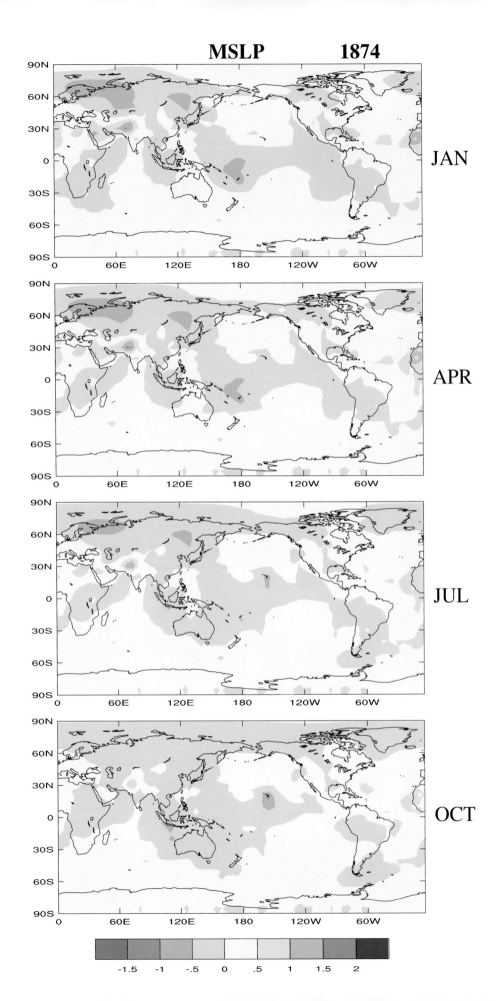

JAN

APR

JUL

OCT

SST   **1874**

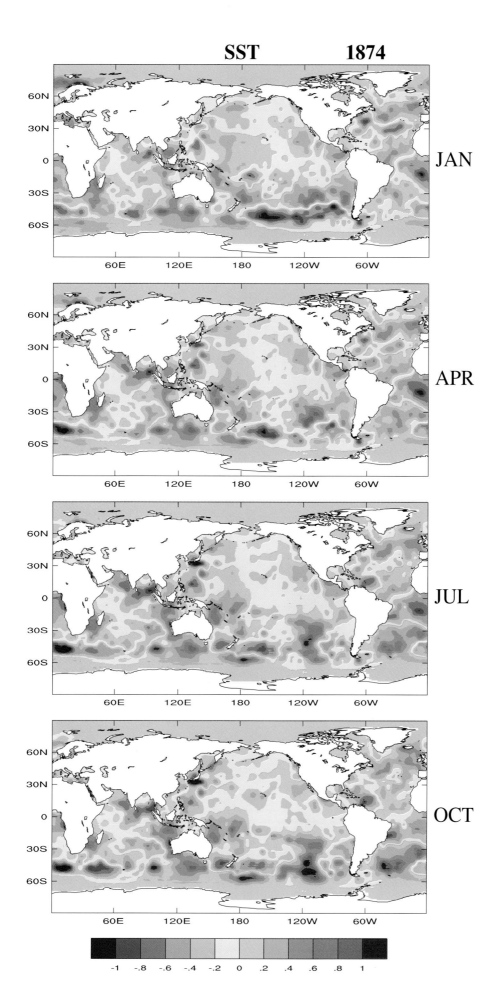

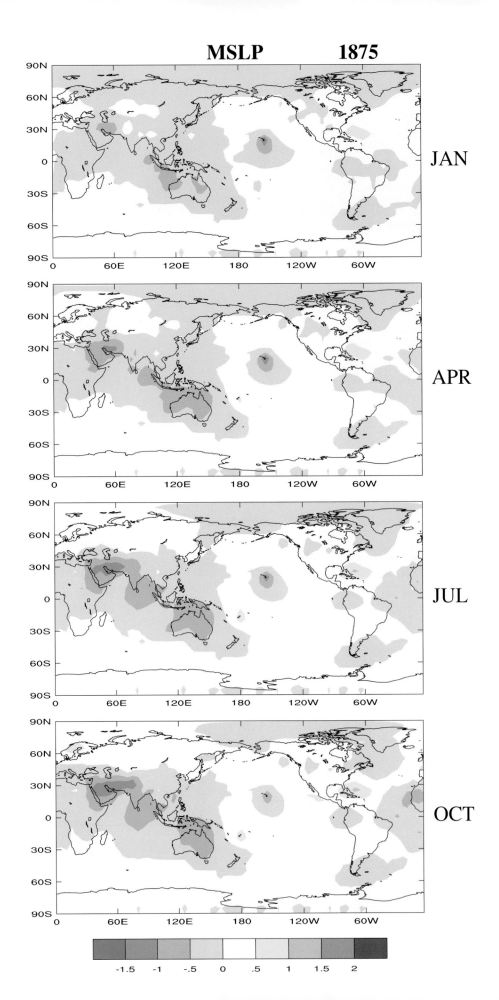

SST   **1875**

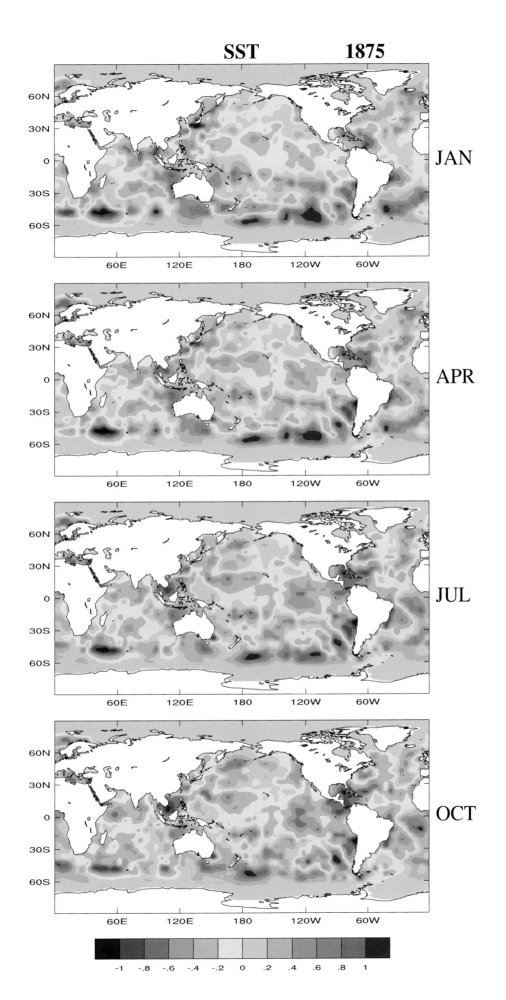

JAN

APR

JUL

OCT

-1   -.8   -.6   -.4   -.2   0   .2   .4   .6   .8   1

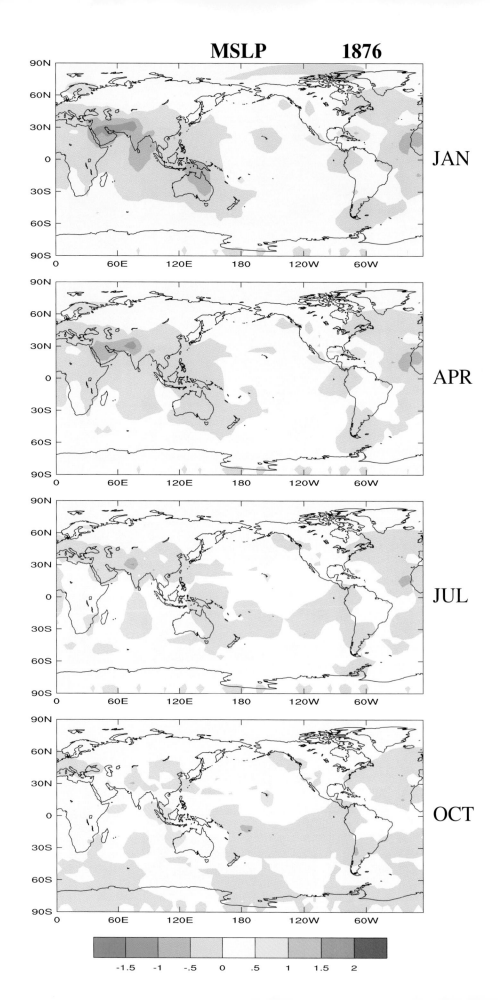

MSLP   1876

JAN

APR

JUL

OCT

-1.5   -1   -.5   0   .5   1   1.5   2

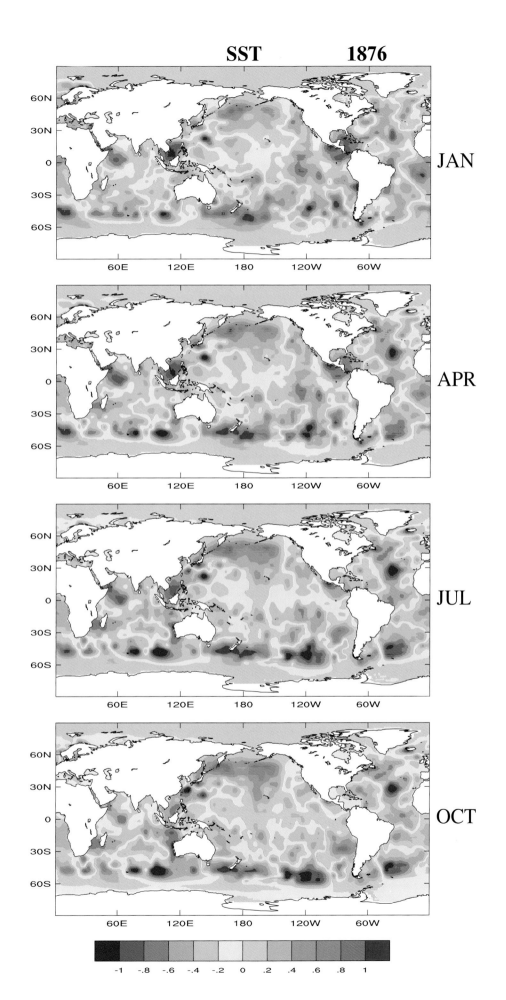

SST   1876

JAN

APR

JUL

OCT

-1 -.8 -.6 -.4 -.2   0   .2   .4   .6   .8   1

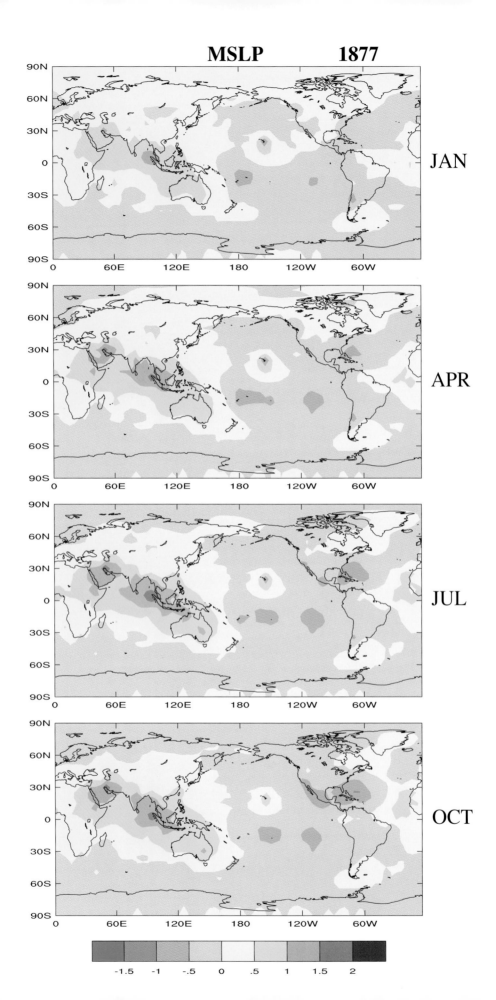

MSLP   **1877**

JAN

APR

JUL

OCT

-1.5   -1   -.5   0   .5   1   1.5   2

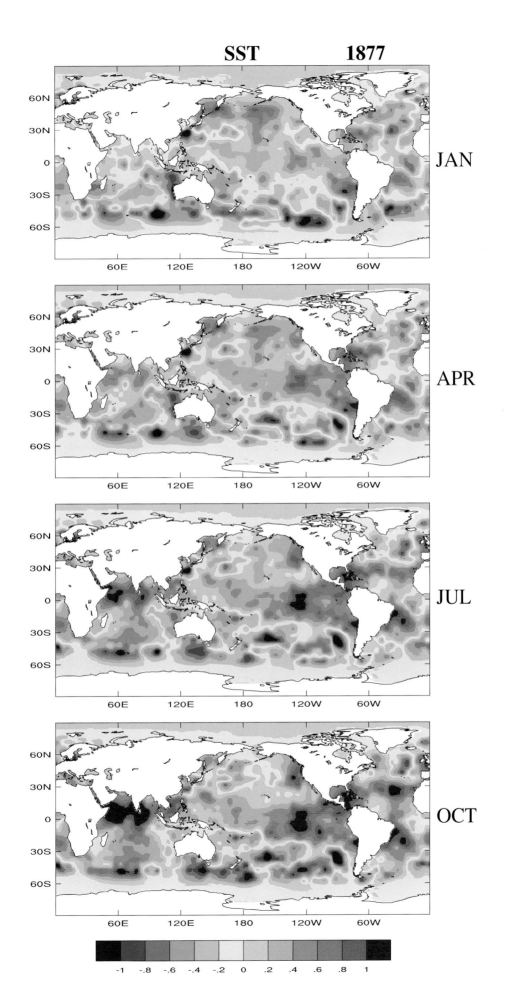

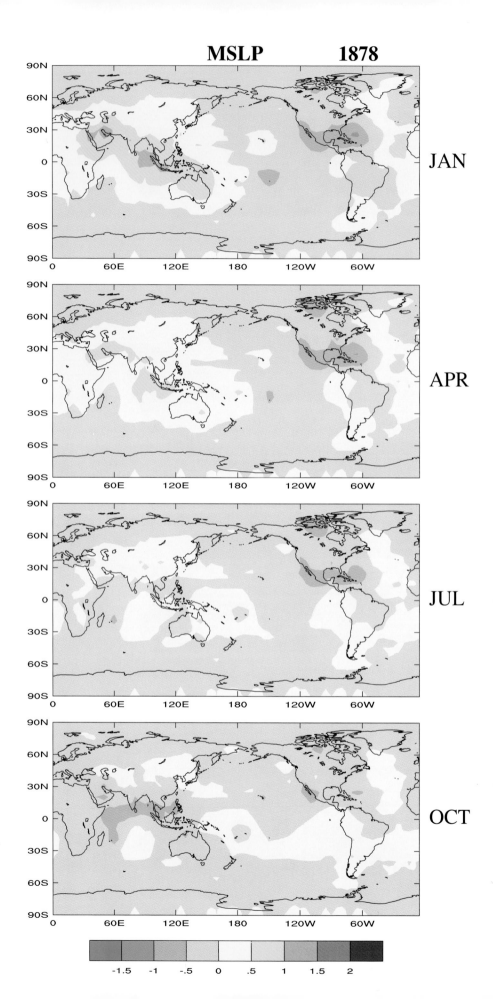

MSLP   **1878**

JAN

APR

JUL

OCT

-1.5   -1   -.5   0   .5   1   1.5   2

SST   1878

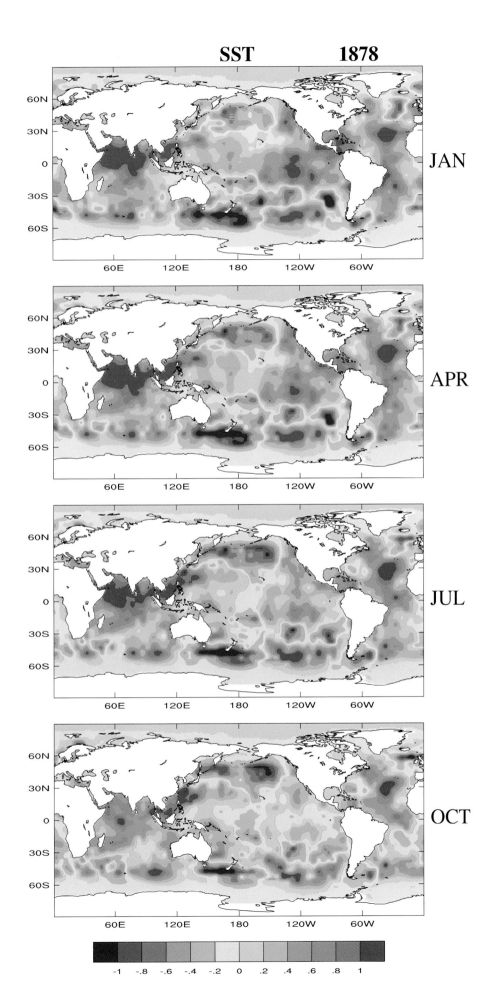

JAN

APR

JUL

OCT

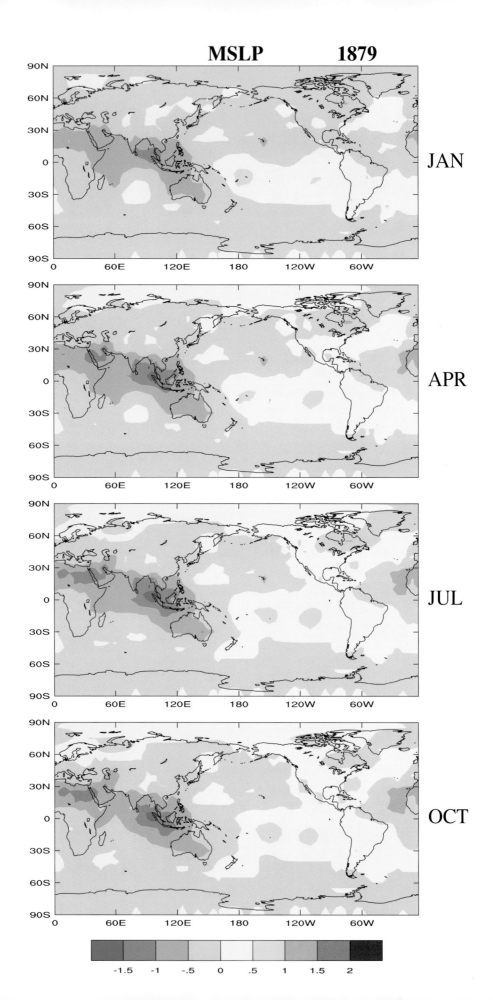

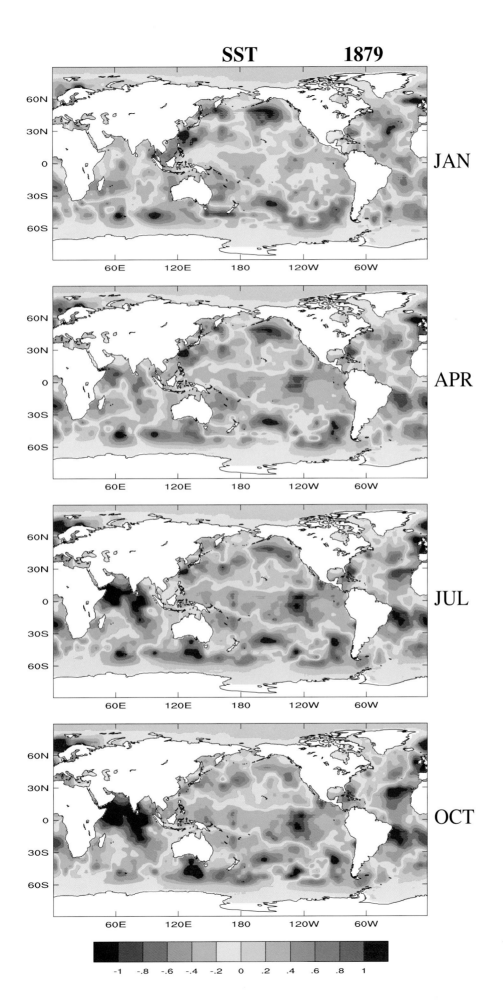

SST   **1879**

JAN

APR

JUL

OCT

-1   -.8   -.6   -.4   -.2   0   .2   .4   .6   .8   1

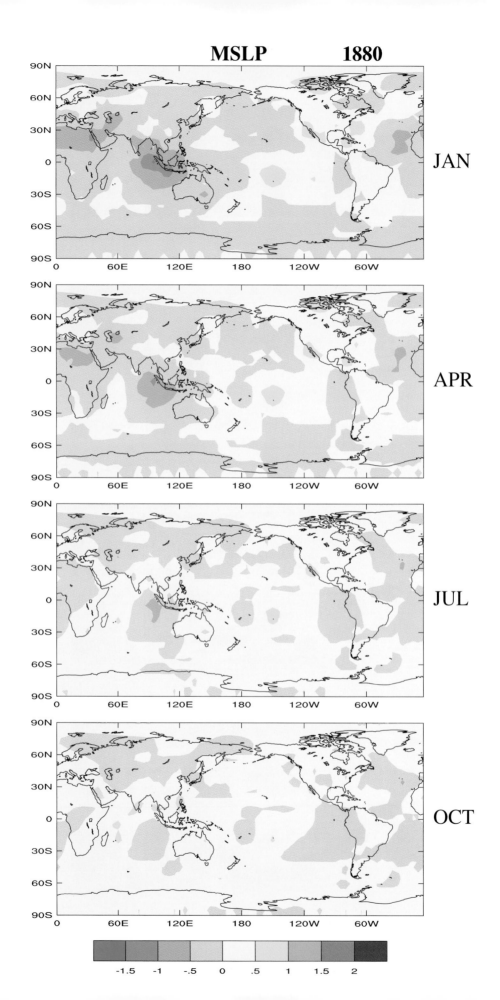

MSLP   1880

JAN

APR

JUL

OCT

-1.5   -1   -.5   0   .5   1   1.5   2

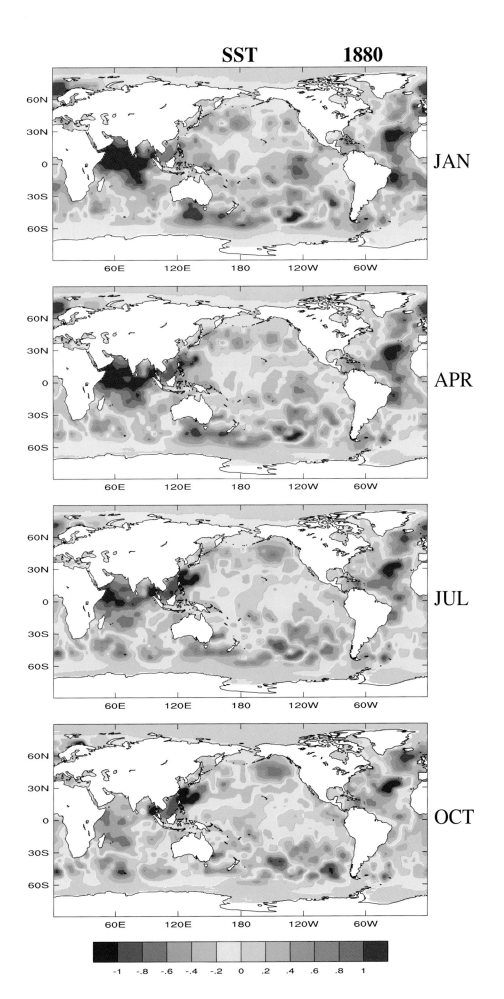

SST   1880

JAN

APR

JUL

OCT

-1 -.8 -.6 -.4 -.2 0 .2 .4 .6 .8 1

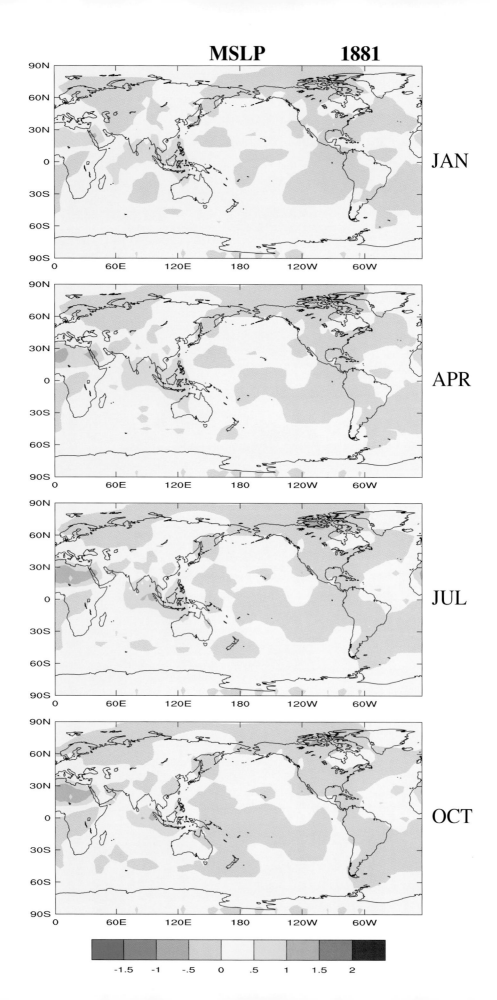

MSLP   **1881**

JAN

APR

JUL

OCT

SST   **1881**

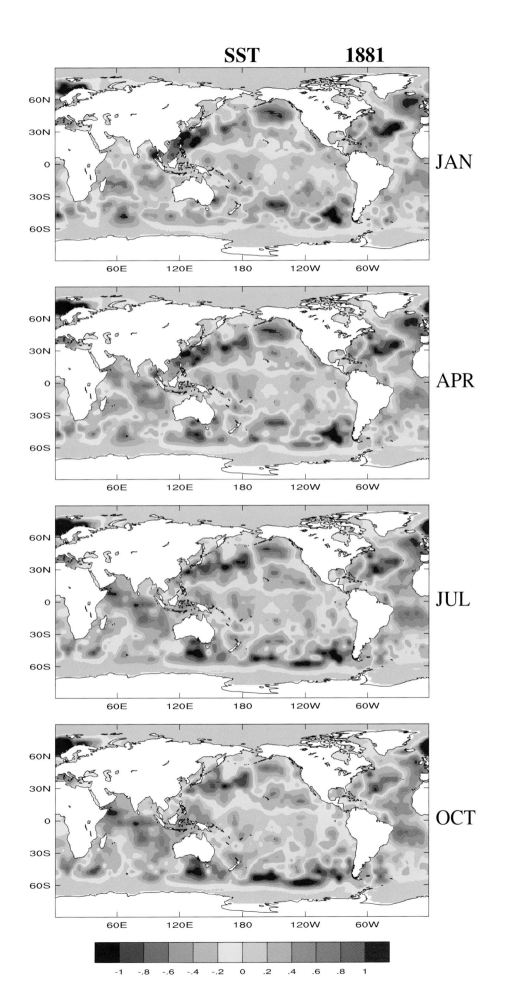

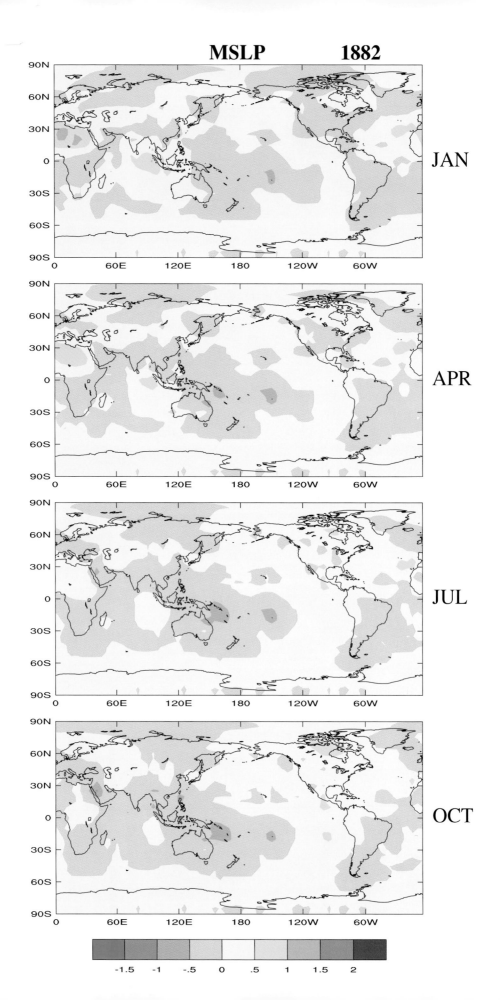

MSLP   **1882**

JAN

APR

JUL

OCT

-1.5   -1   -.5   0   .5   1   1.5   2

SST   **1882**

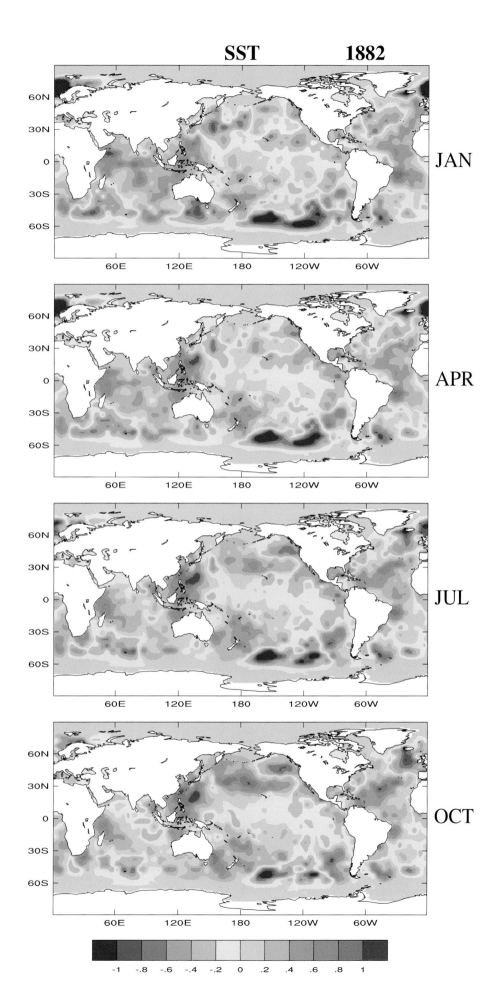

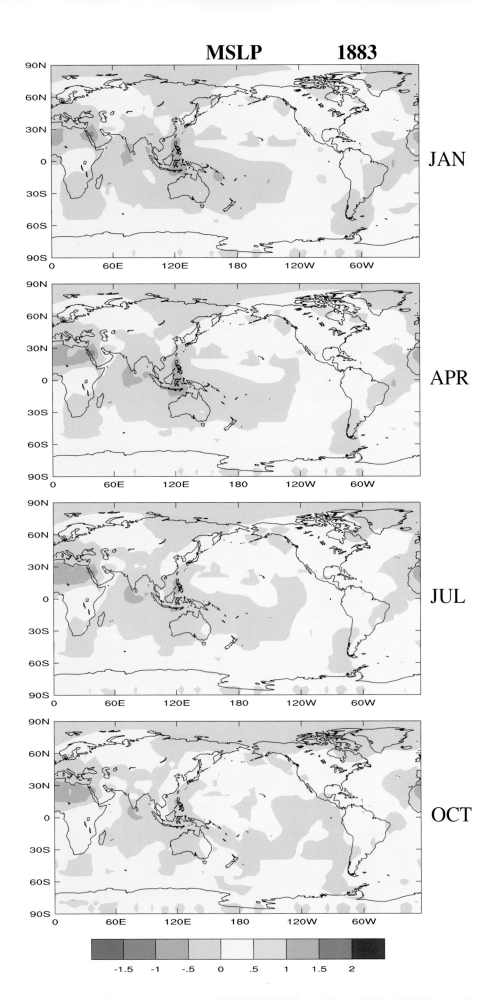

MSLP   **1883**

JAN

APR

JUL

OCT

-1.5   -1   -.5   0   .5   1   1.5   2

SST   **1883**

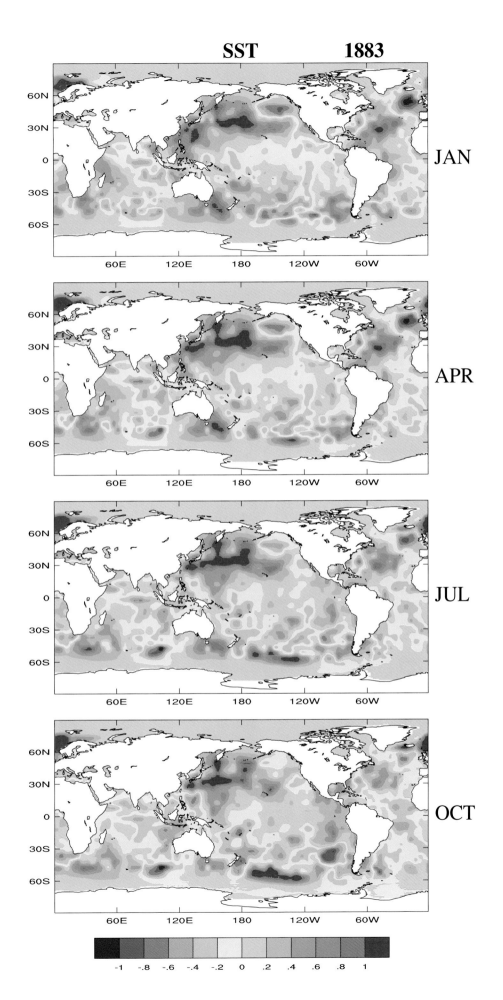

JAN

APR

JUL

OCT

-1   -.8   -.6   -.4   -.2   0   .2   .4   .6   .8   1

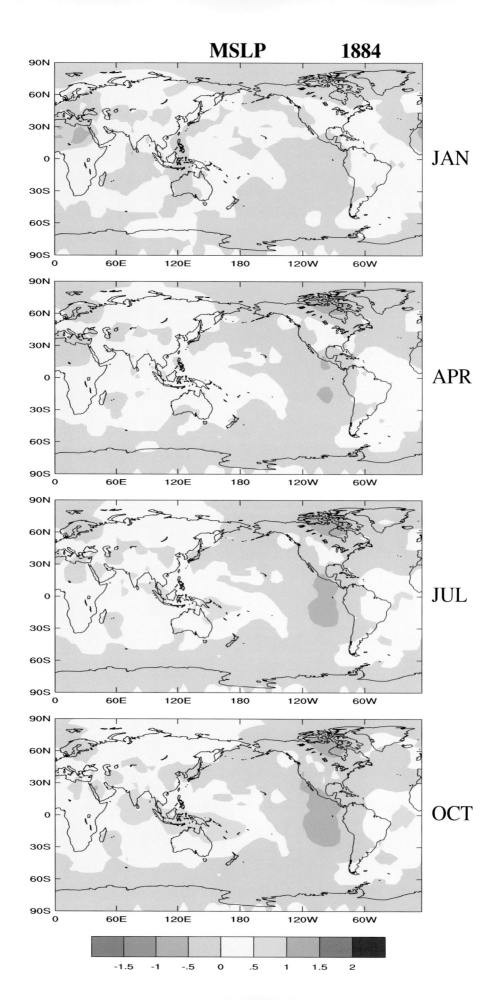

MSLP 1884

JAN

APR

JUL

OCT

-1.5 -1 -.5 0 .5 1 1.5 2

SST 1884

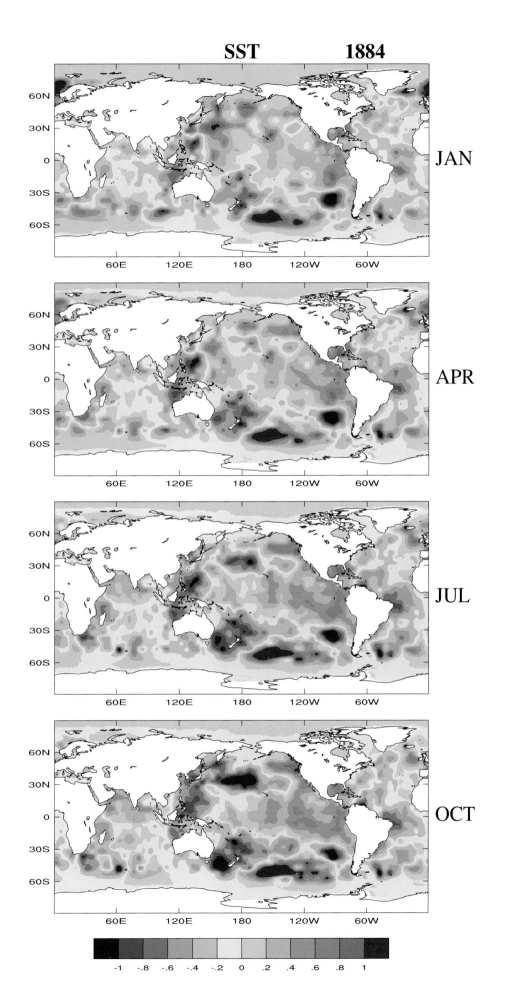

JAN

APR

JUL

OCT

-1 -.8 -.6 -.4 -.2 0 .2 .4 .6 .8 1

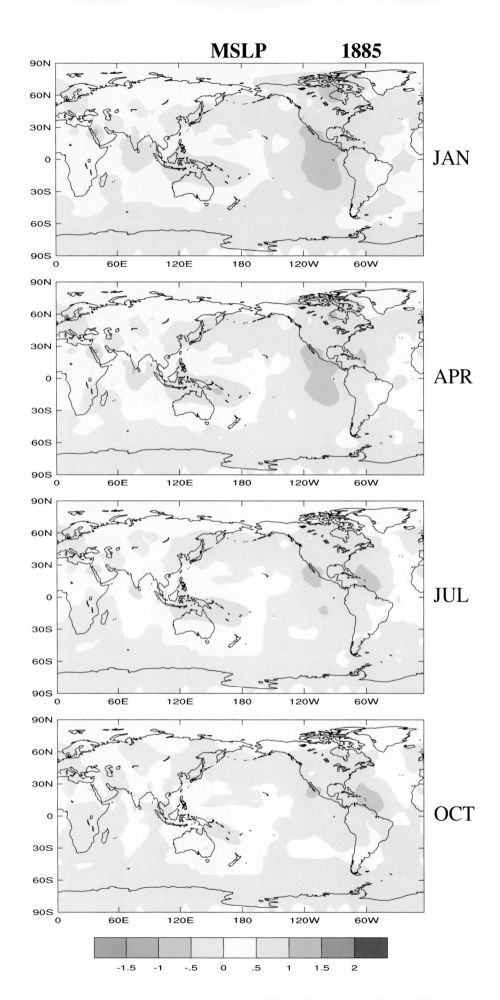

MSLP   1885

JAN

APR

JUL

OCT

-1.5   -1   -.5   0   .5   1   1.5   2

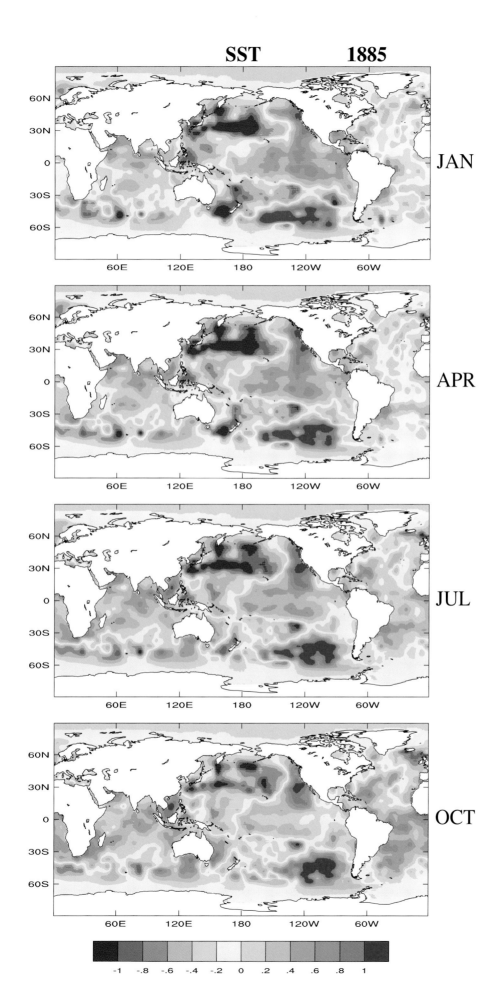

SST   1885

JAN

APR

JUL

OCT

-1 -.8 -.6 -.4 -.2 0 .2 .4 .6 .8 1

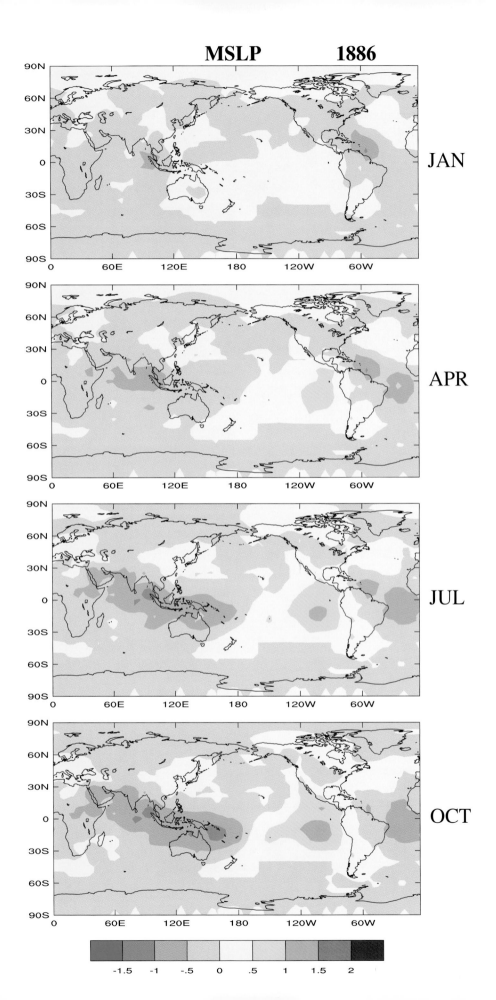

MSLP 1886

JAN

APR

JUL

OCT

-1.5 -1 -.5 0 .5 1 1.5 2

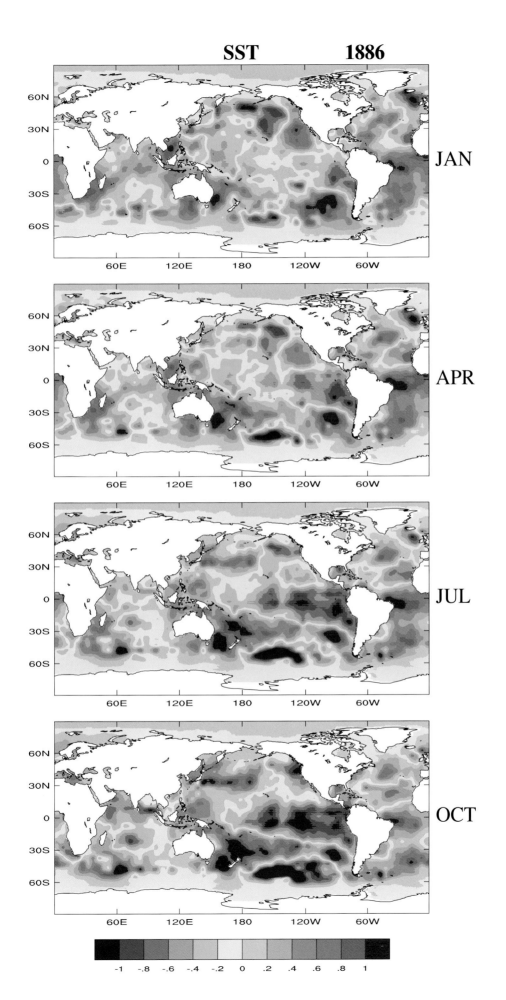

SST   1886

JAN

APR

JUL

OCT

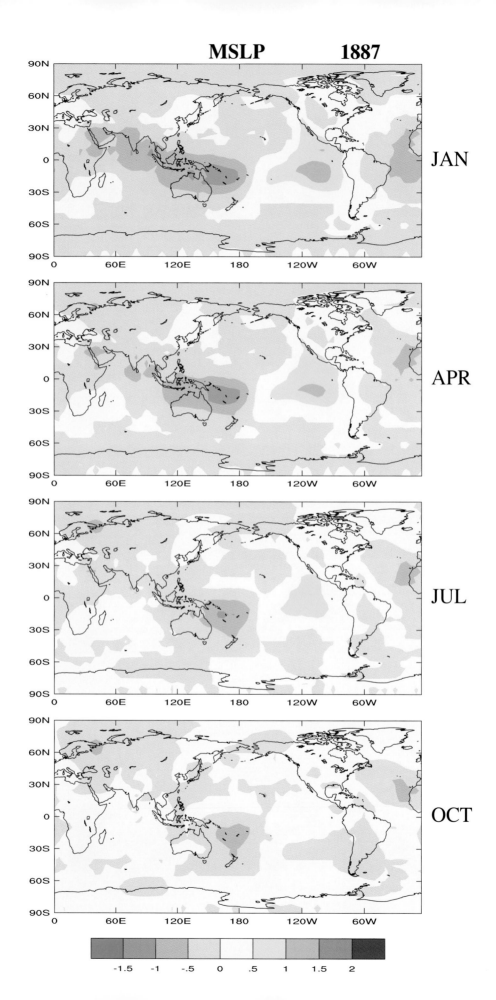

MSLP   **1887**

JAN

APR

JUL

OCT

-1.5   -1   -.5   0   .5   1   1.5   2

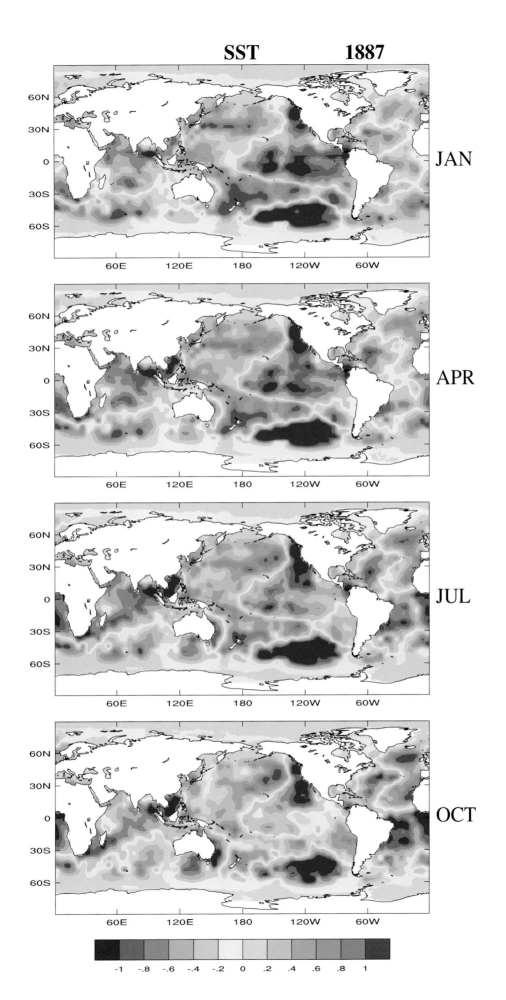

SST   1887

JAN

APR

JUL

OCT

-1   -.8   -.6   -.4   -.2   0   .2   .4   .6   .8   1

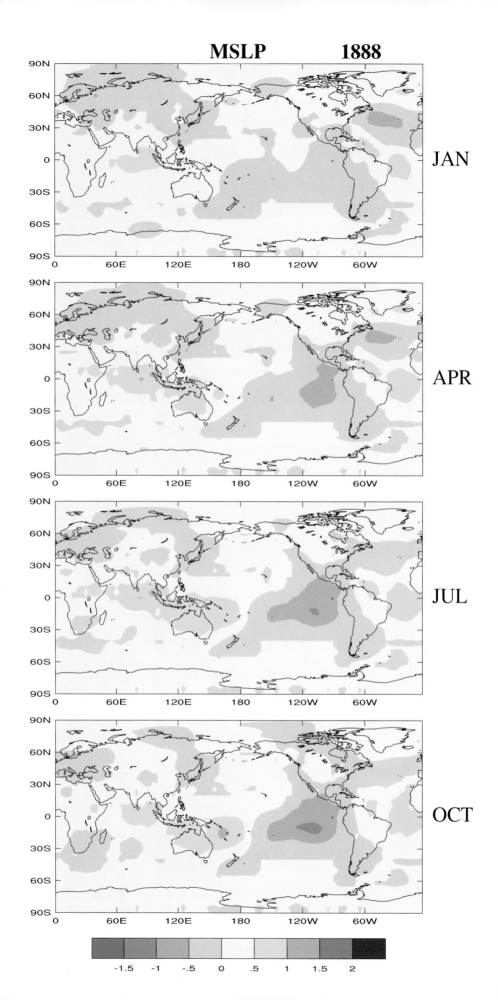

MSLP   1888

JAN

APR

JUL

OCT

-1.5   -1   -.5   0   .5   1   1.5   2

SST   **1888**

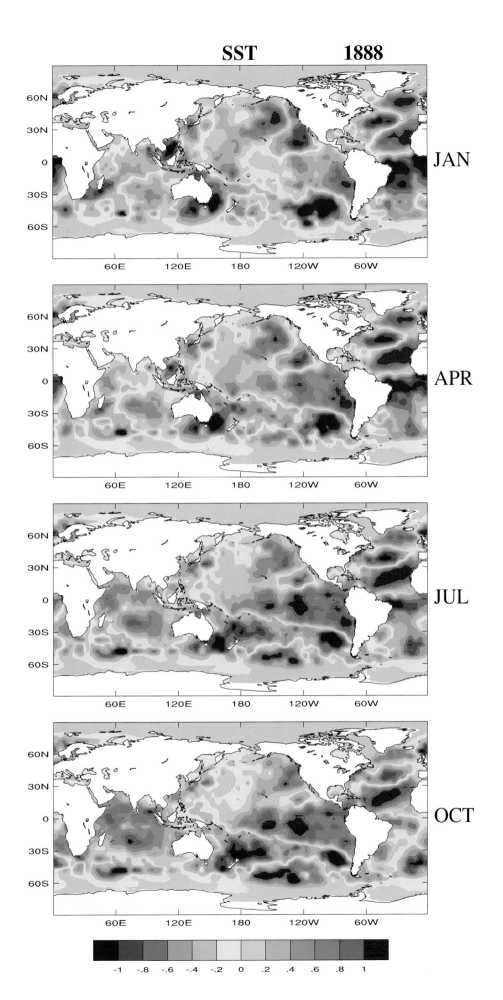

JAN

APR

JUL

OCT

-1   -.8   -.6   -.4   -.2   0   .2   .4   .6   .8   1

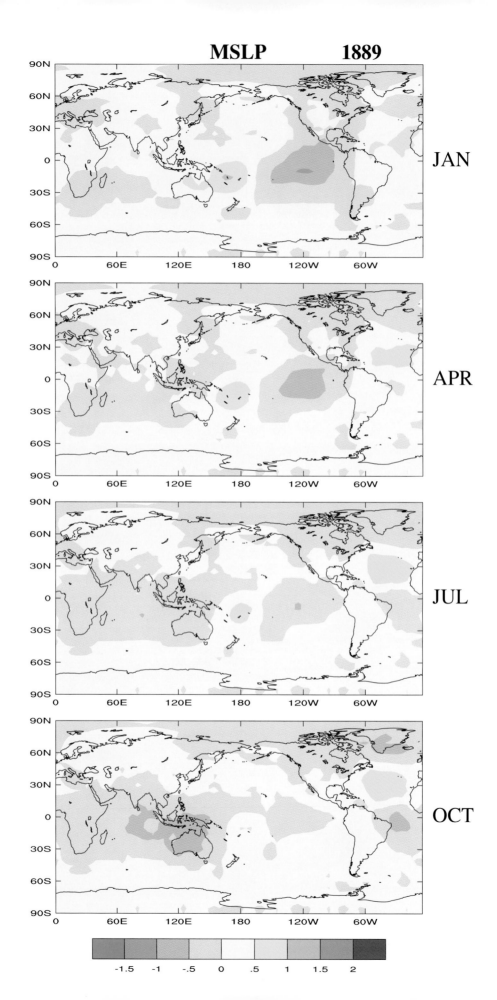

SST　　**1889**

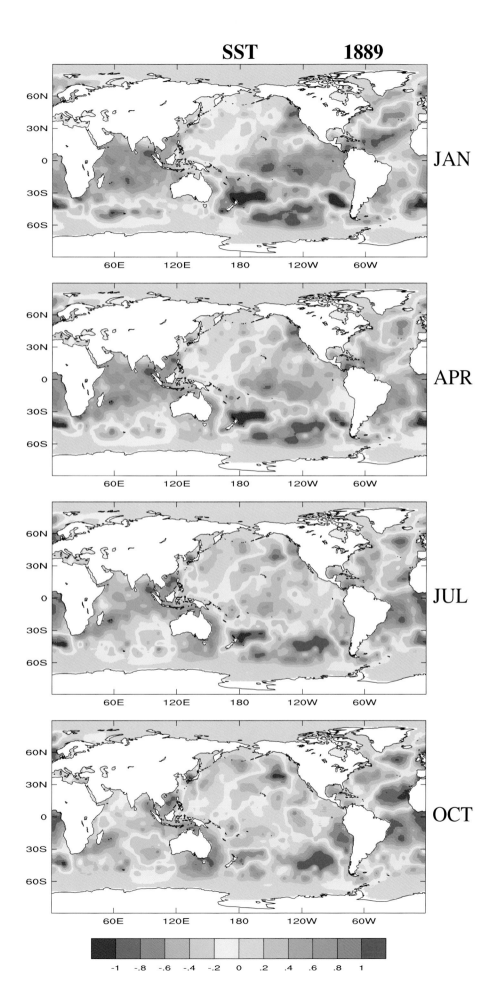

JAN

APR

JUL

OCT

-1　-.8　-.6　-.4　-.2　0　.2　.4　.6　.8　1

MSLP   **1890**

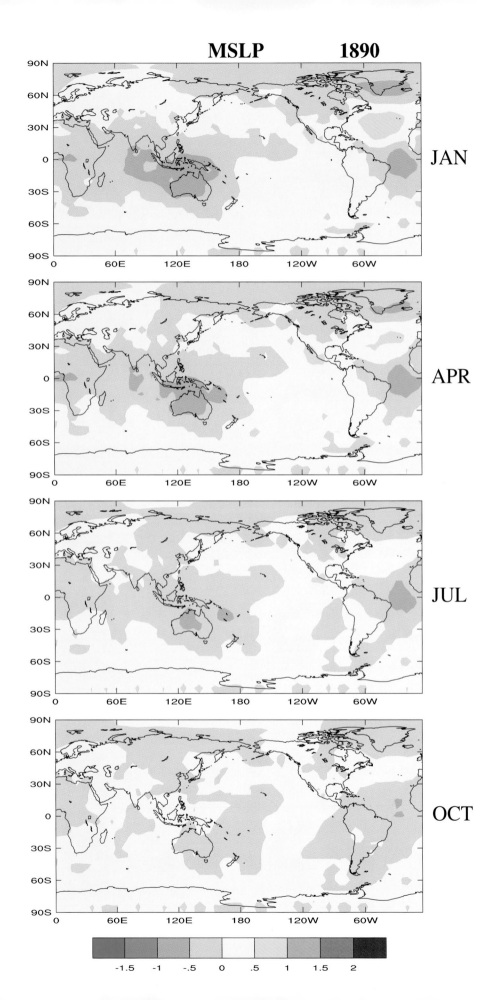

SST   1890

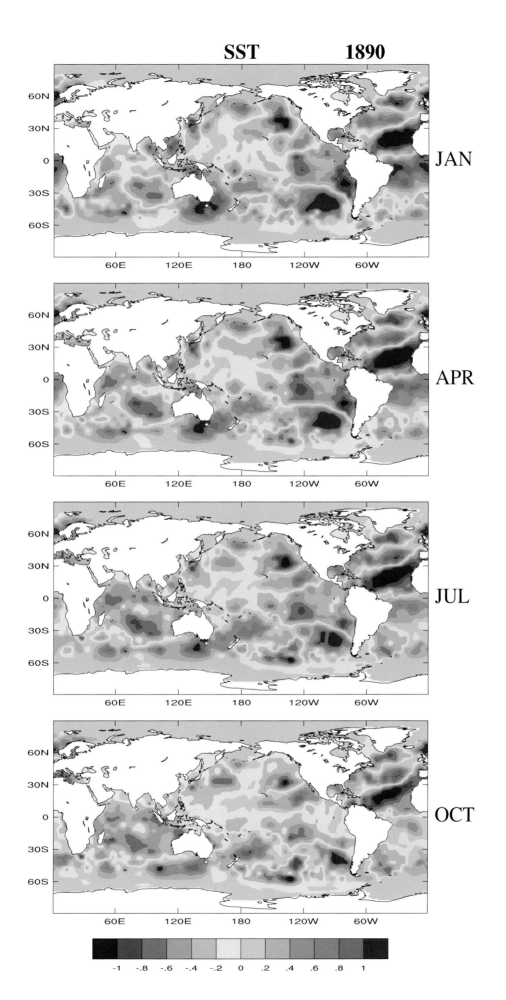

MSLP   **1891**

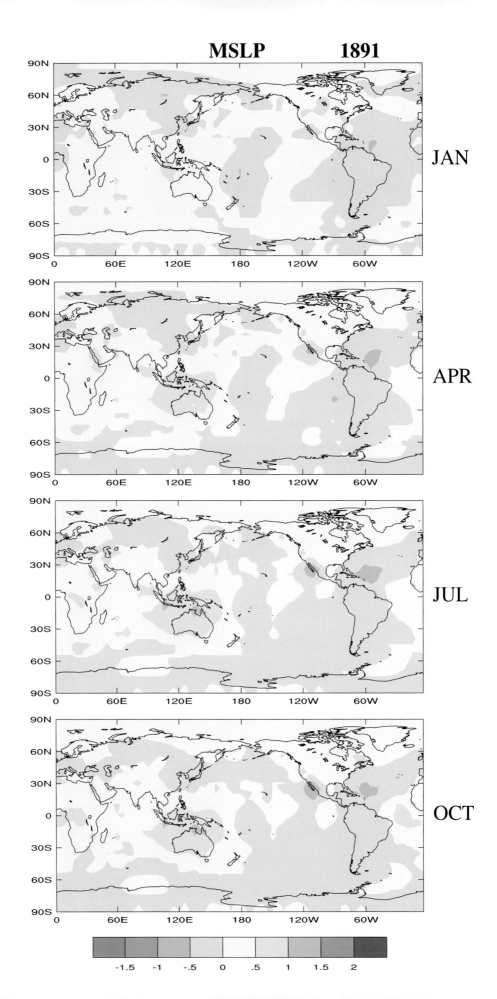

JAN

APR

JUL

OCT

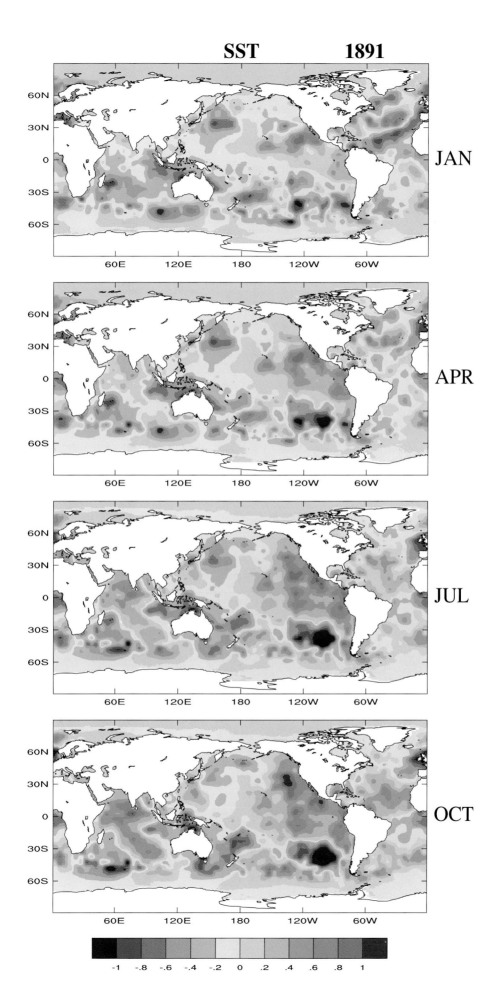

SST   1891

JAN

APR

JUL

OCT

-1 -.8 -.6 -.4 -.2   0   .2   .4   .6   .8   1

199

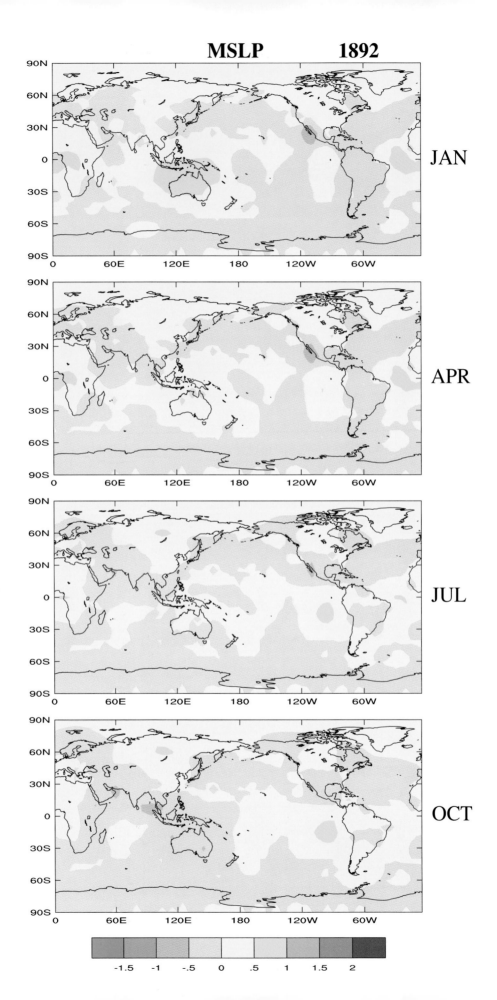

MSLP   **1892**

JAN

APR

JUL

OCT

-1.5   -1   -.5   0   .5   1   1.5   2

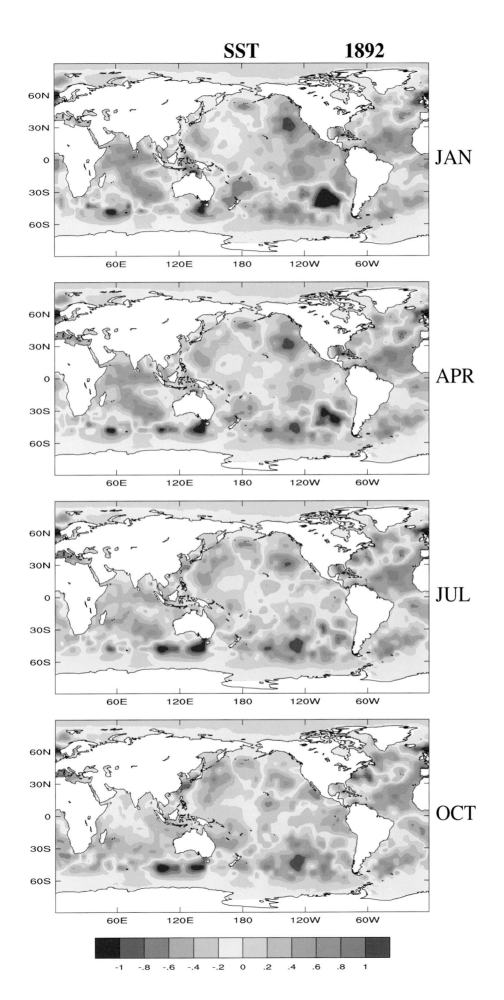

SST   1892

JAN

APR

JUL

OCT

-1 -.8 -.6 -.4 -.2 0 .2 .4 .6 .8 1

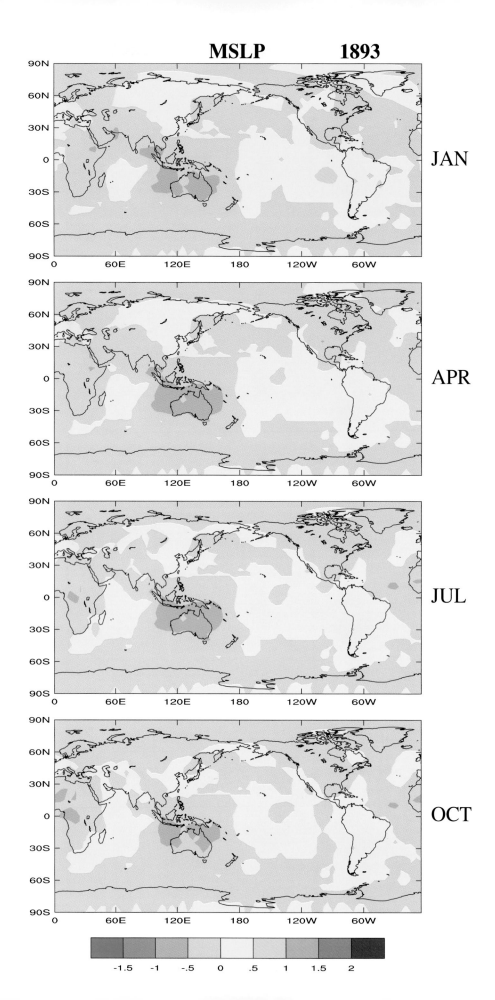

SST 1893

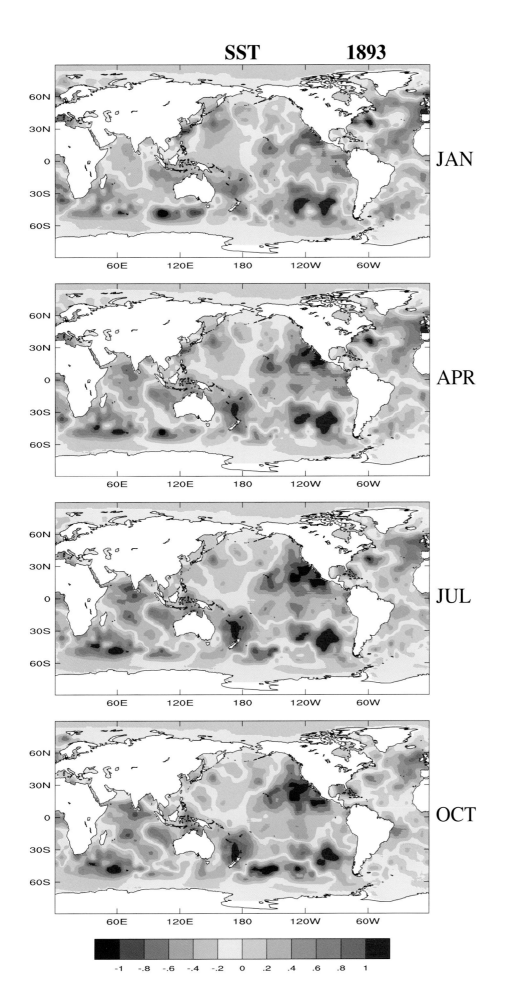

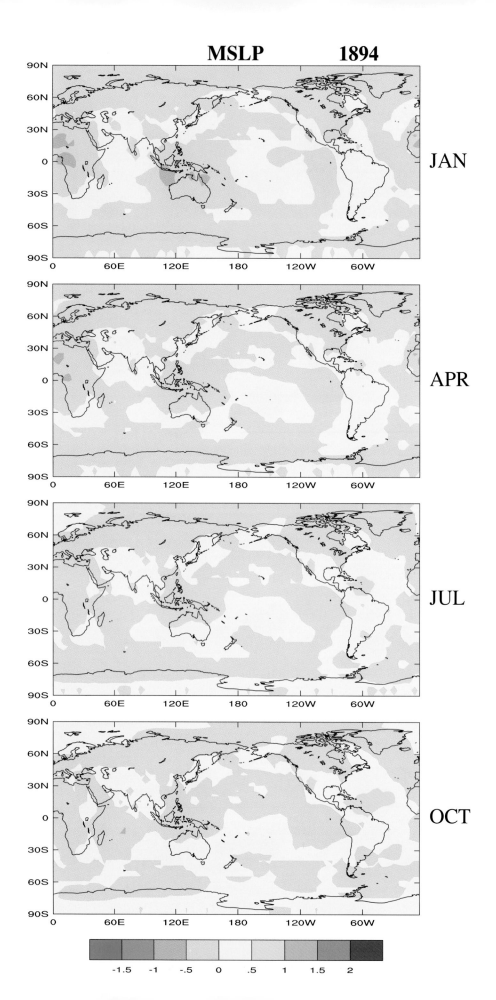

MSLP 1894

JAN

APR

JUL

OCT

-1.5   -1   -.5   0   .5   1   1.5   2

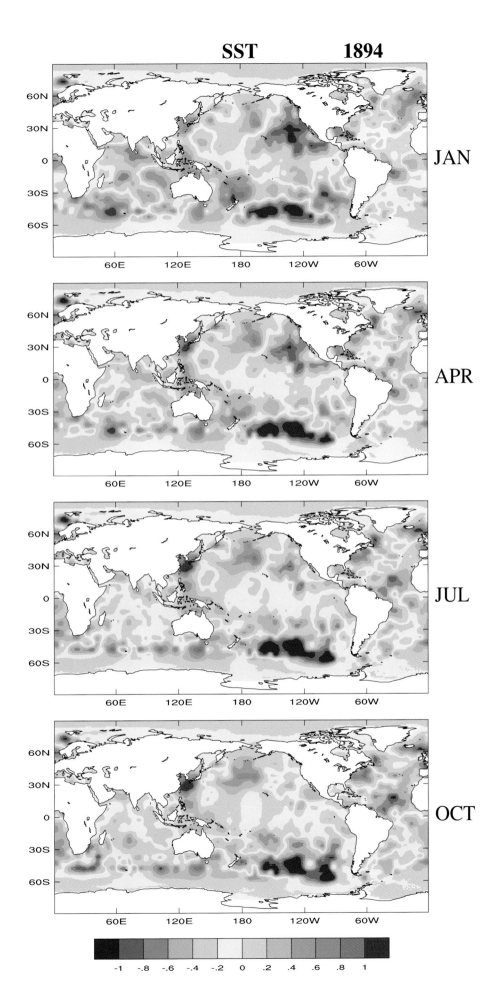

SST   1894

JAN

APR

JUL

OCT

-1 -.8 -.6 -.4 -.2 0 .2 .4 .6 .8 1

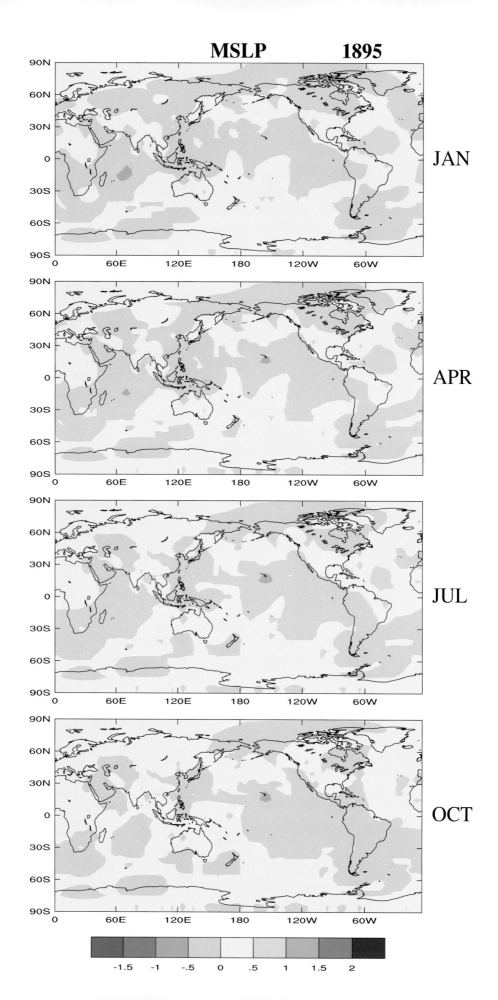

MSLP   **1895**

JAN

APR

JUL

OCT

-1.5   -1   -.5   0   .5   1   1.5   2

SST   1895

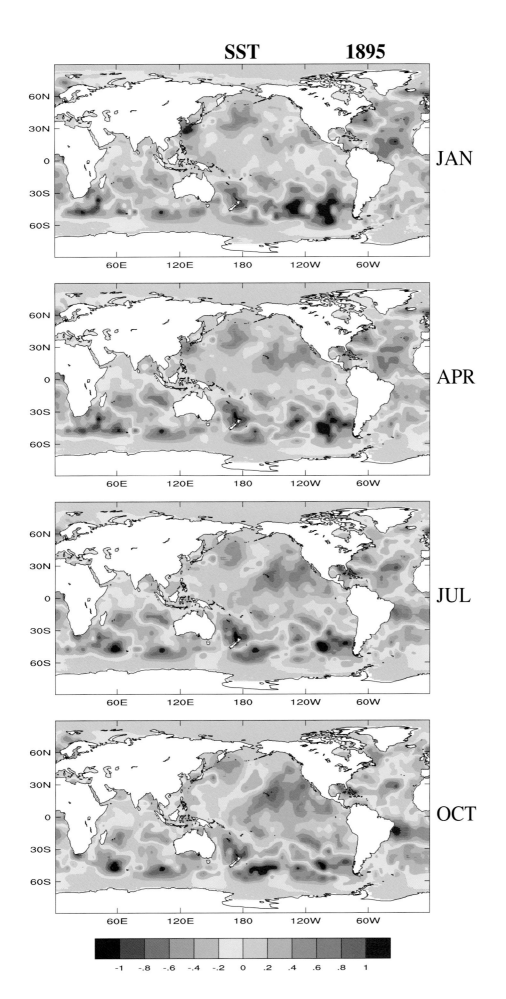

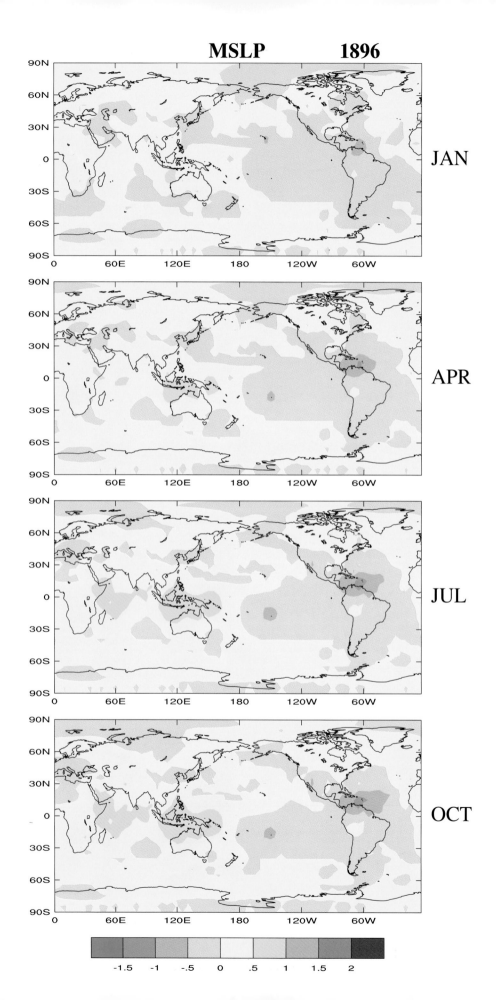

MSLP 1896

JAN

APR

JUL

OCT

-1.5 -1 -.5 0 .5 1 1.5 2

208

SST   1896

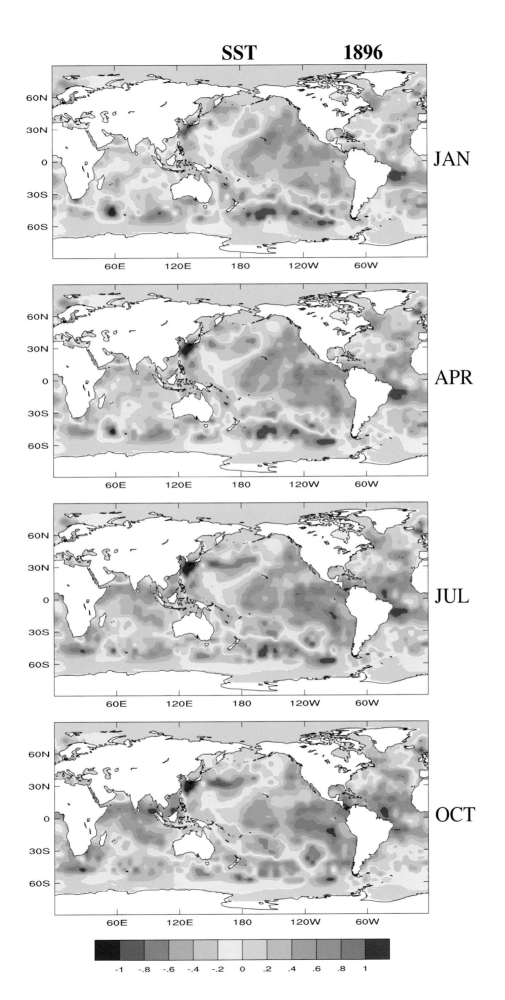

JAN

APR

JUL

OCT

| -1 | -.8 | -.6 | -.4 | -.2 | 0 | .2 | .4 | .6 | .8 | 1 |

MSLP   **1897**

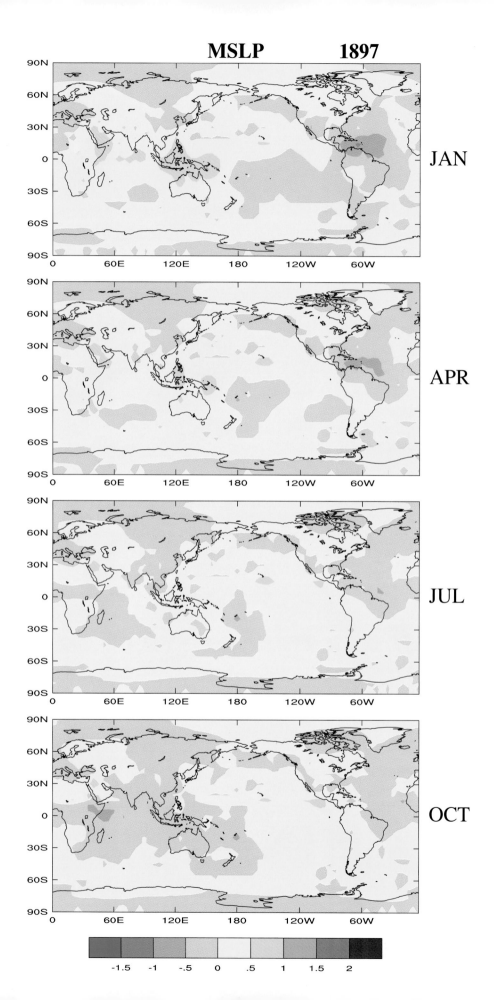

JAN

APR

JUL

OCT

-1.5   -1   -.5   0   .5   1   1.5   2

SST **1897**

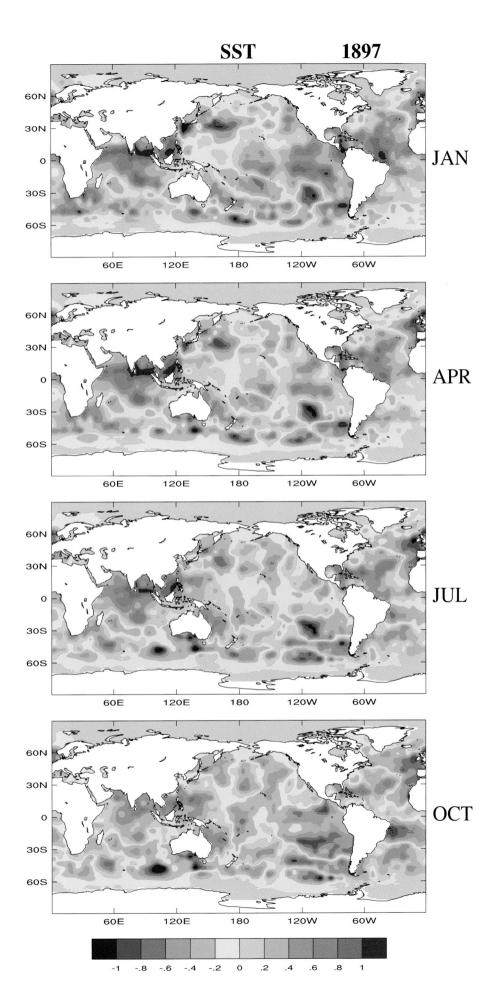

JAN

APR

JUL

OCT

-1 -.8 -.6 -.4 -.2 0 .2 .4 .6 .8 1

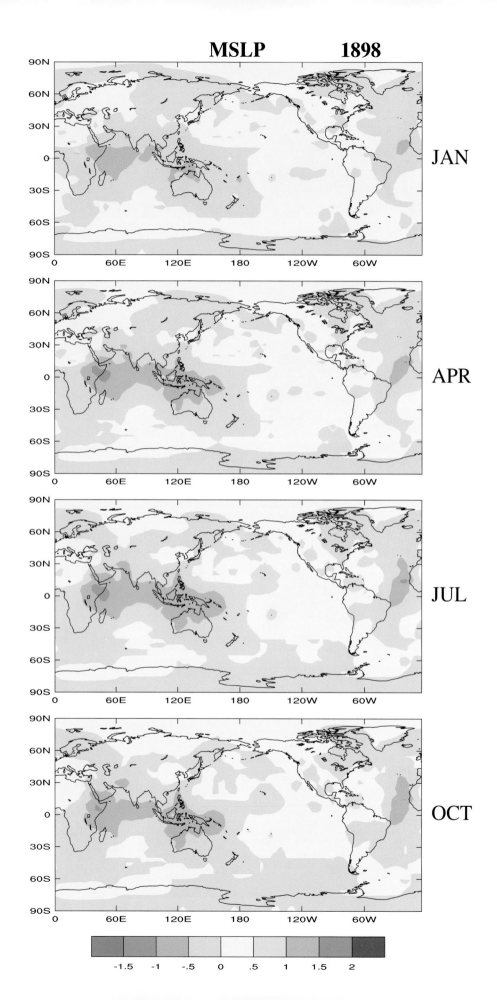

SST   1898

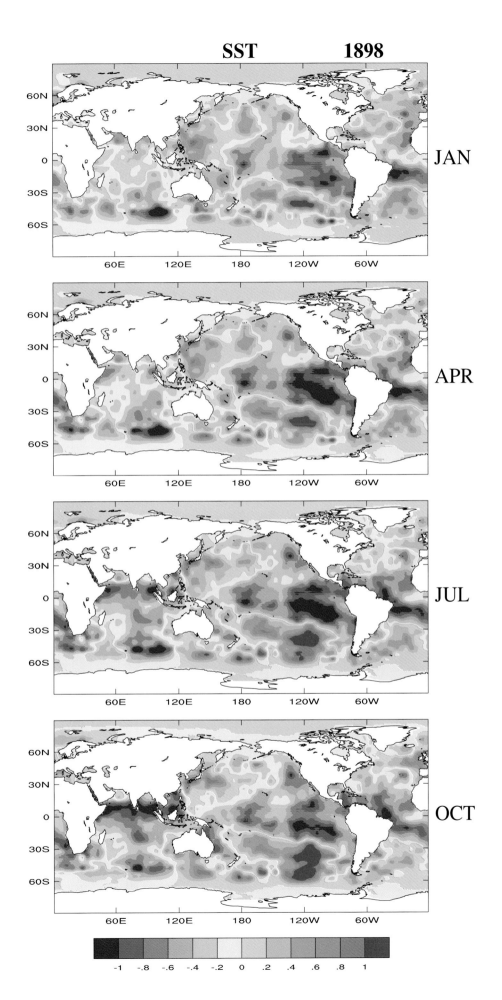

JAN

APR

JUL

OCT

213

MSLP   **1899**

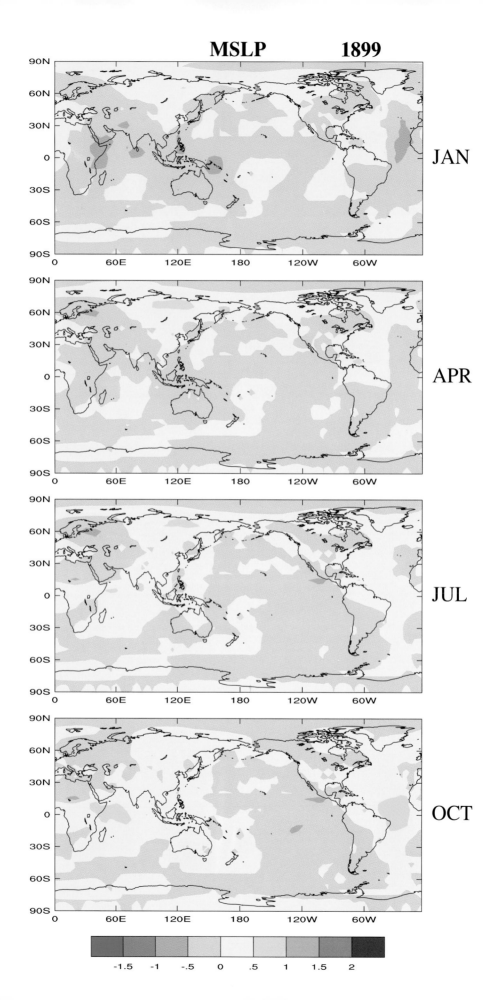

JAN

APR

JUL

OCT

-1.5   -1   -.5   0   .5   1   1.5   2

SST   **1899**

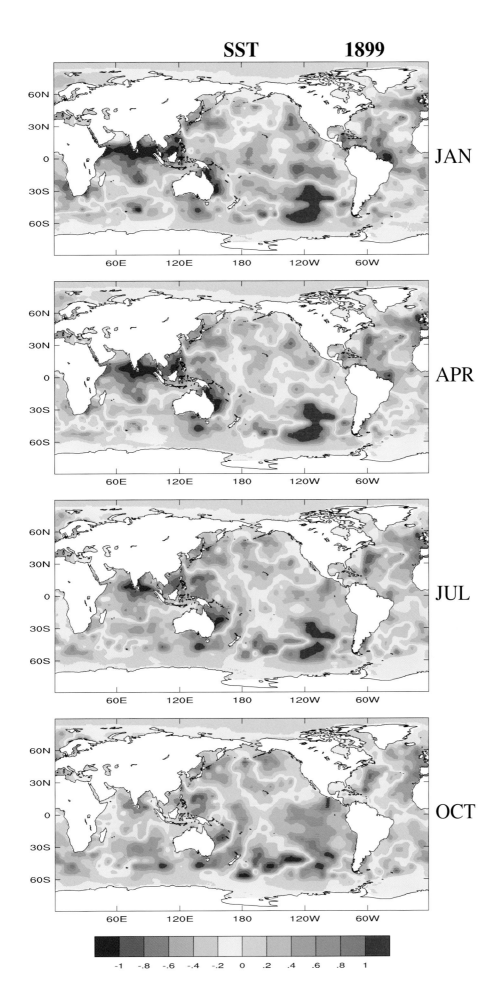

JAN

APR

JUL

OCT

-1   -.8   -.6   -.4   -.2   0   .2   .4   .6   .8   1

215

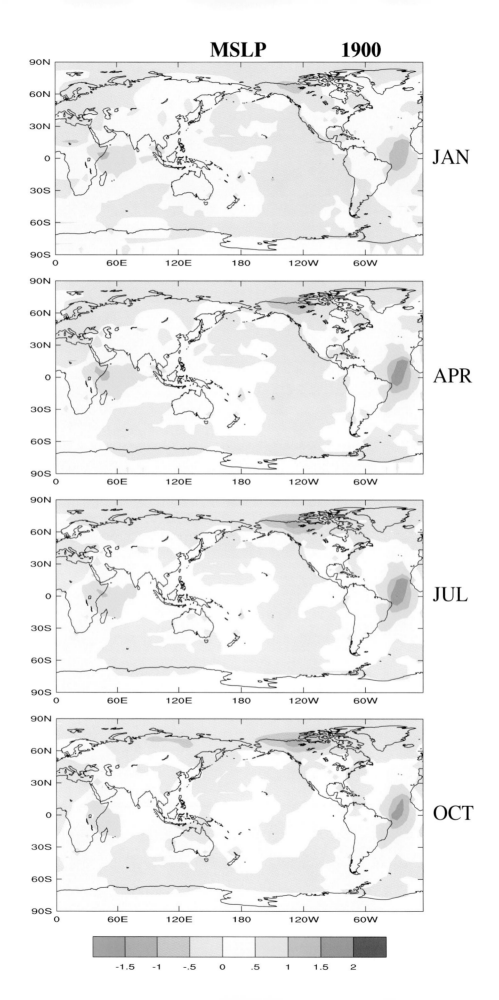

MSLP   **1900**

JAN

APR

JUL

OCT

-1.5   -1   -.5   0   .5   1   1.5   2

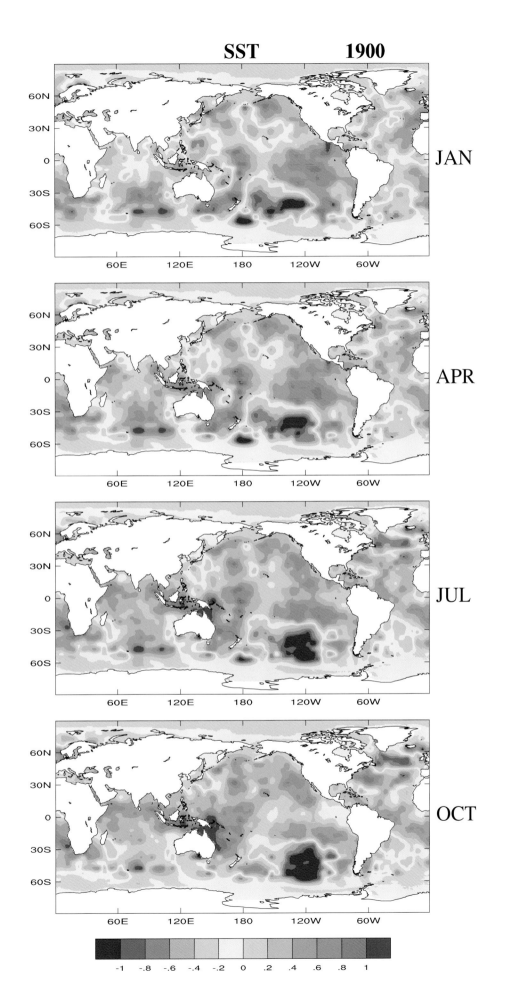

217

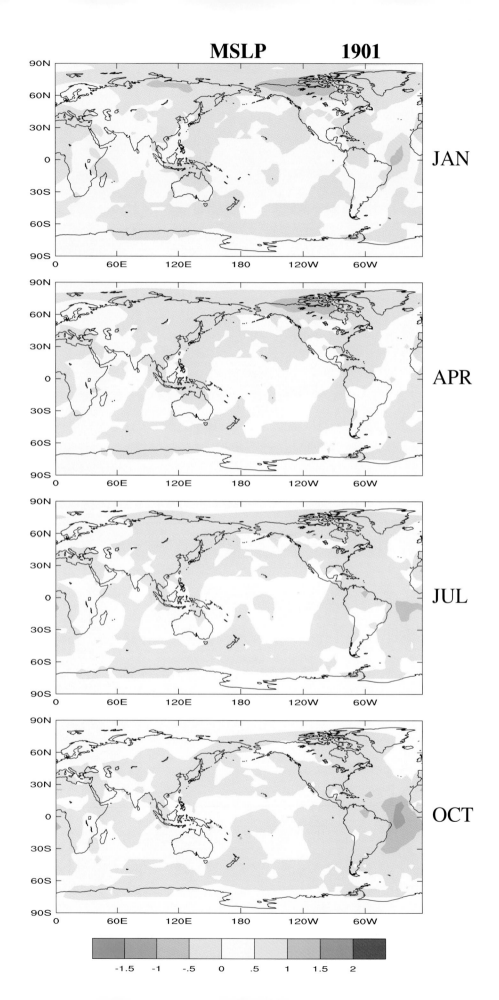

MSLP   **1901**

JAN

APR

JUL

OCT

-1.5   -1   -.5   0   .5   1   1.5   2

SST   1901

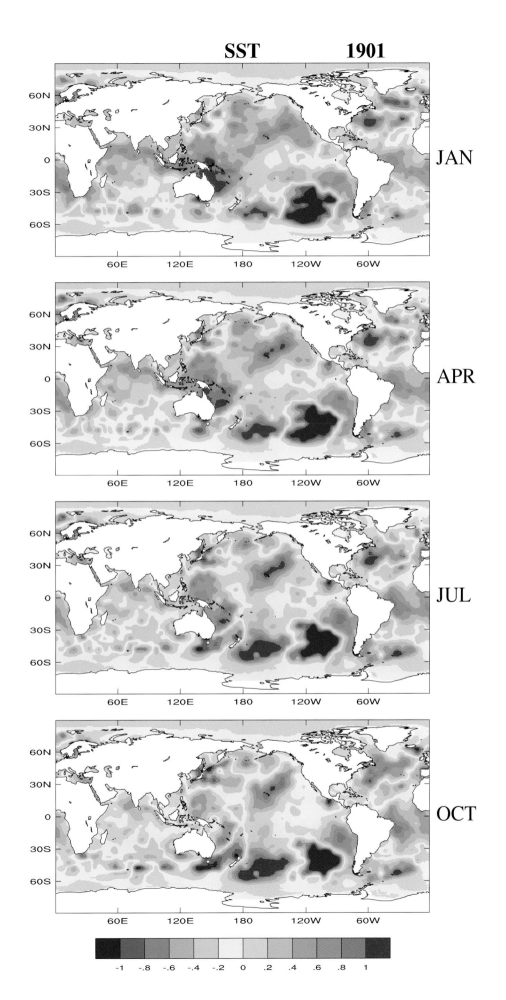

MSLP 1902

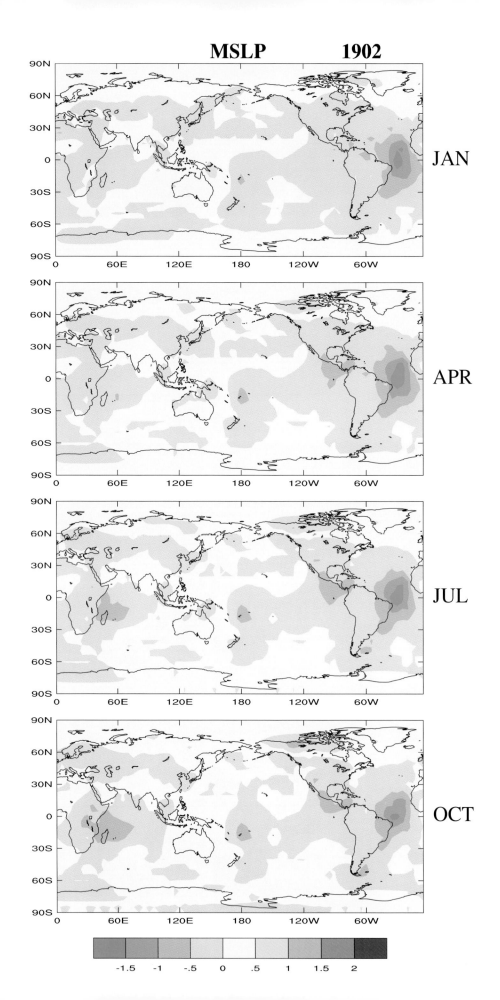

JAN

APR

JUL

OCT

-1.5 -1 -.5 0 .5 1 1.5 2

SST   **1902**

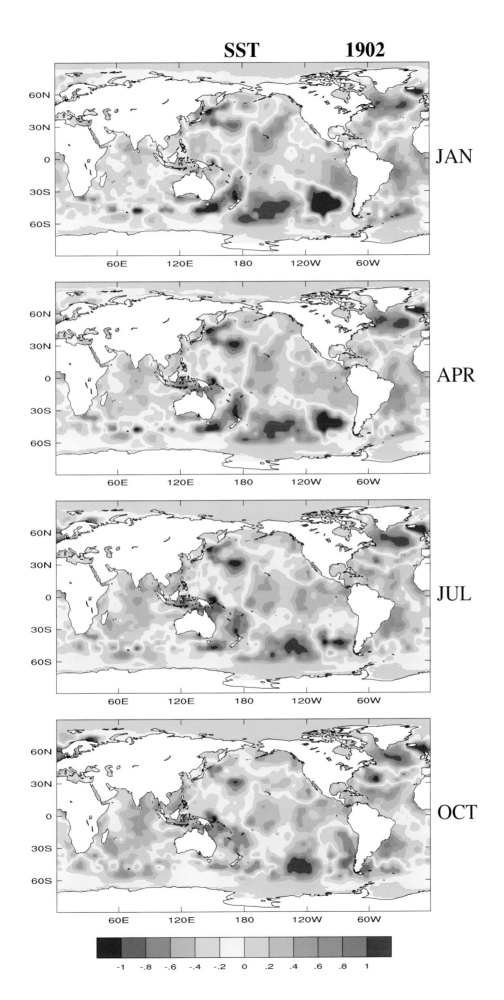

MSLP   **1903**

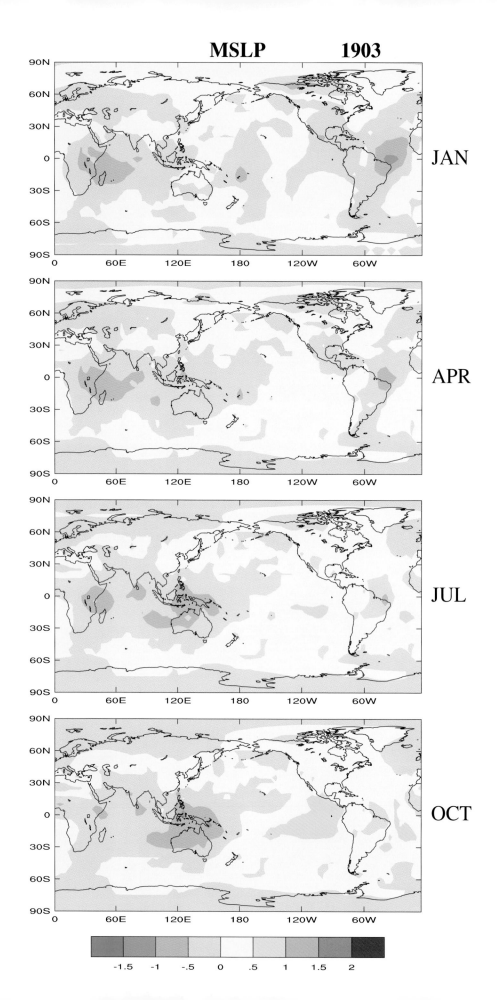

JAN

APR

JUL

OCT

-1.5   -1   -.5   0   .5   1   1.5   2

SST   **1903**

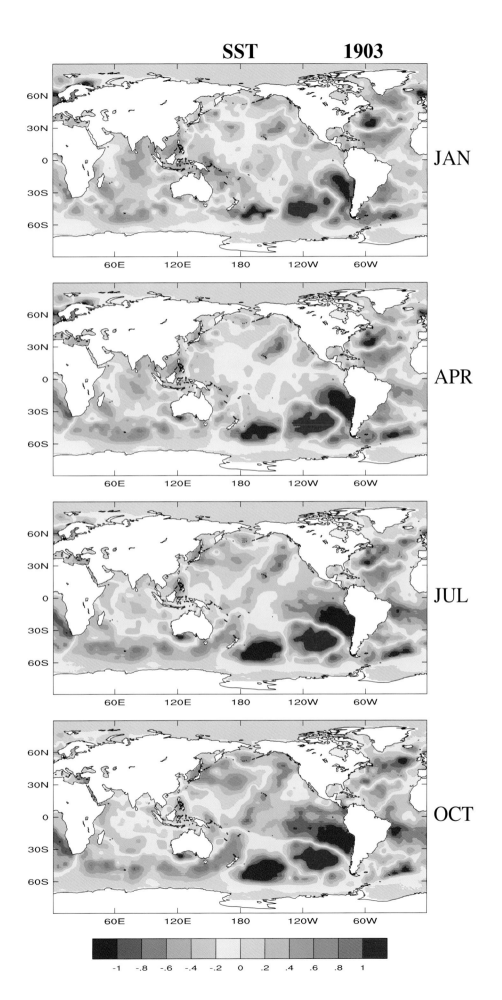

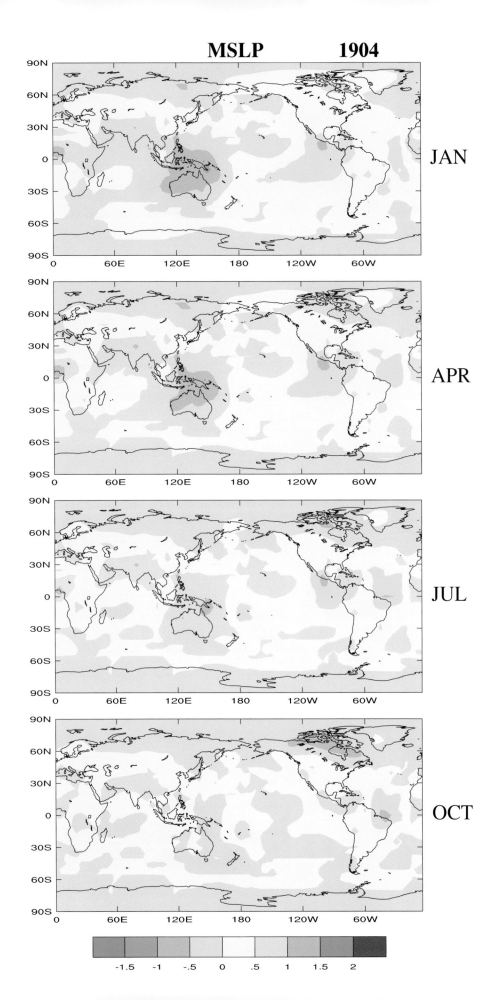

MSLP   **1904**

JAN

APR

JUL

OCT

-1.5   -1   -.5   0   .5   1   1.5   2

SST   **1904**

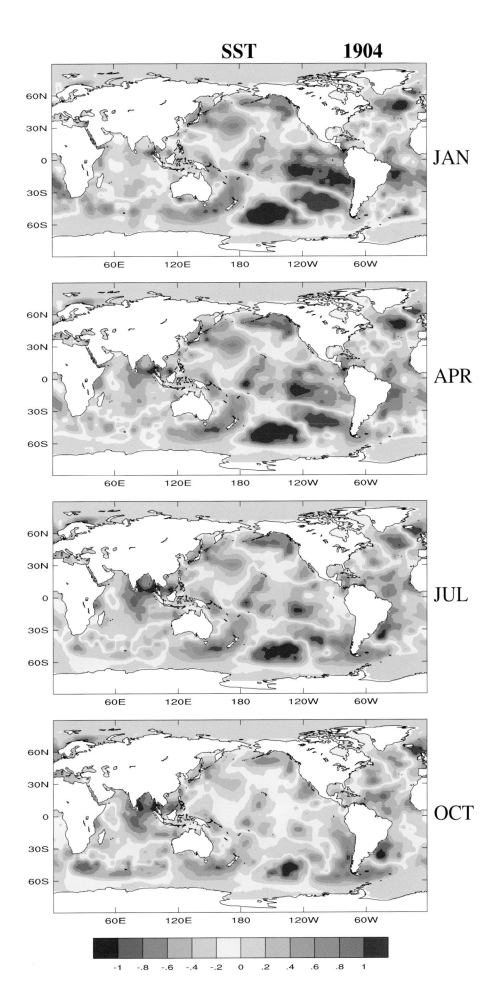

JAN

APR

JUL

OCT

-1 -.8 -.6 -.4 -.2 0 .2 .4 .6 .8 1

MSLP   1905

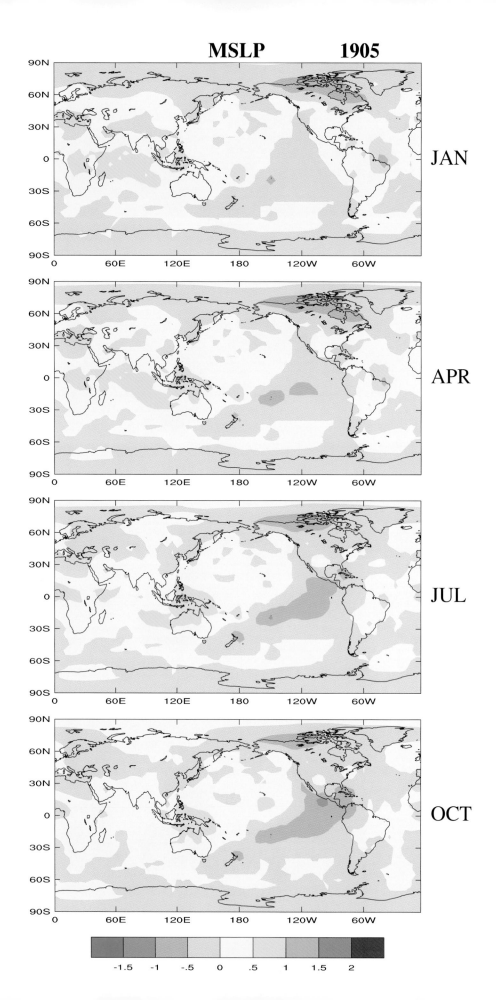

JAN

APR

JUL

OCT

-1.5   -1   -.5   0   .5   1   1.5   2

SST   **1905**

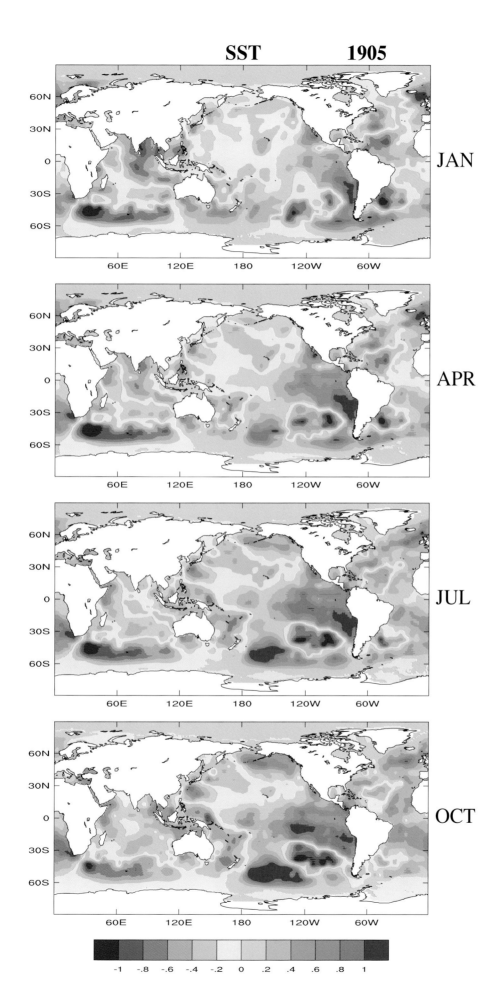

227

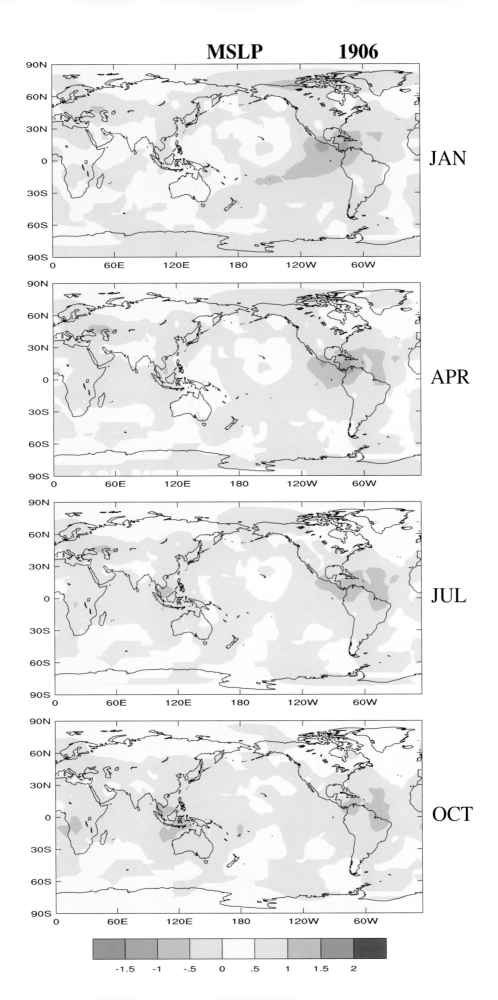

MSLP　　　　**1906**

JAN

APR

JUL

OCT

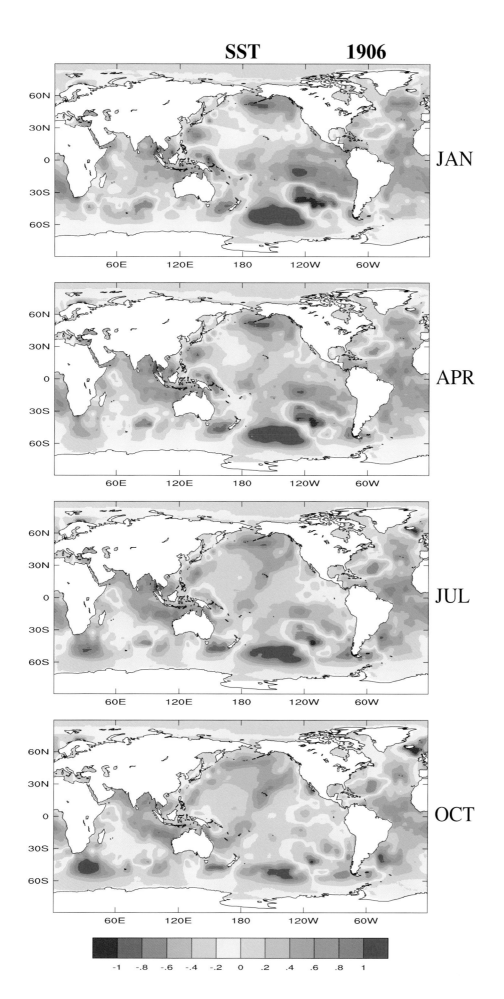

SST　　　1906

JAN

APR

JUL

OCT

-1 -.8 -.6 -.4 -.2 0 .2 .4 .6 .8 1

229

MSLP 1907

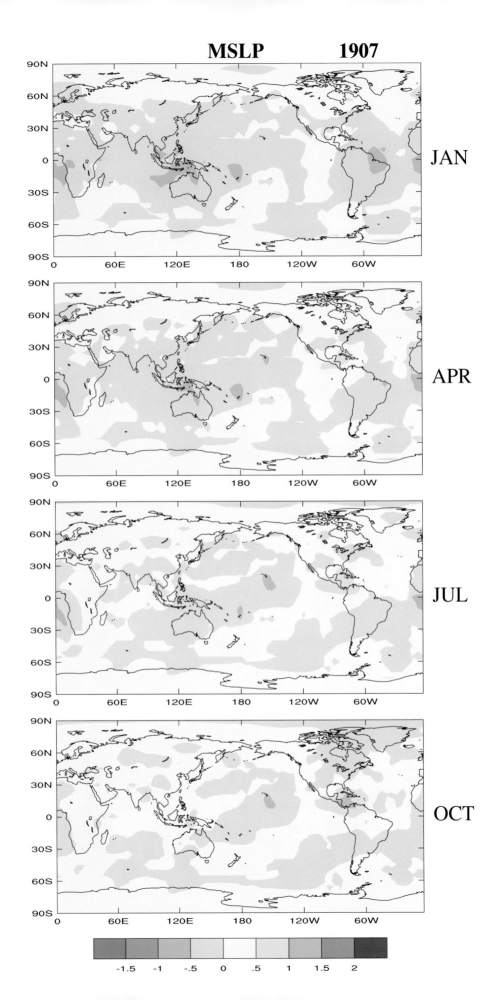

JAN

APR

JUL

OCT

SST 1907

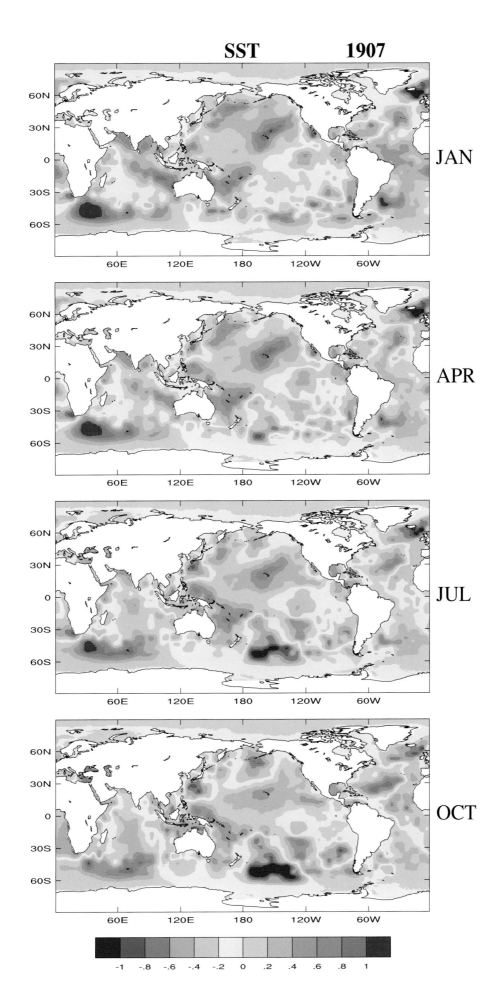

JAN

APR

JUL

OCT

231

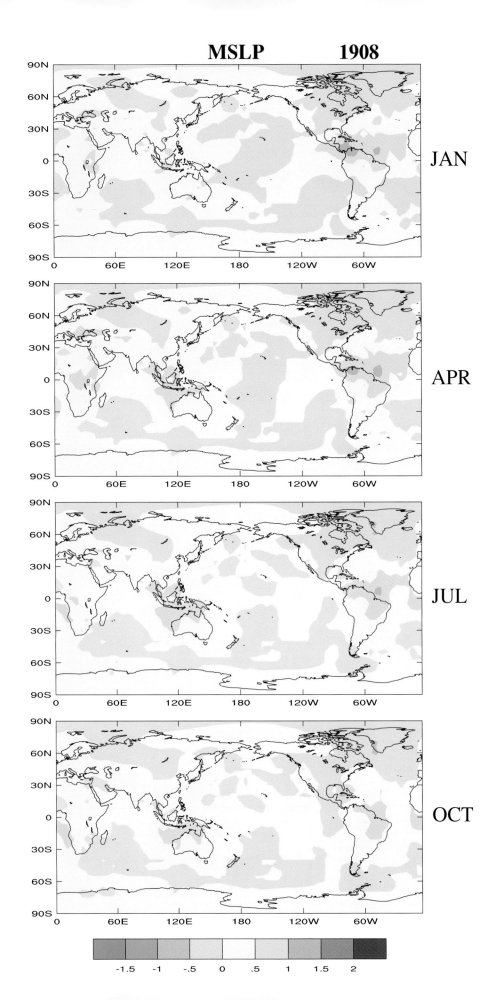

MSLP   1908

JAN

APR

JUL

OCT

-1.5   -1   -.5   0   .5   1   1.5   2

232

SST   1908

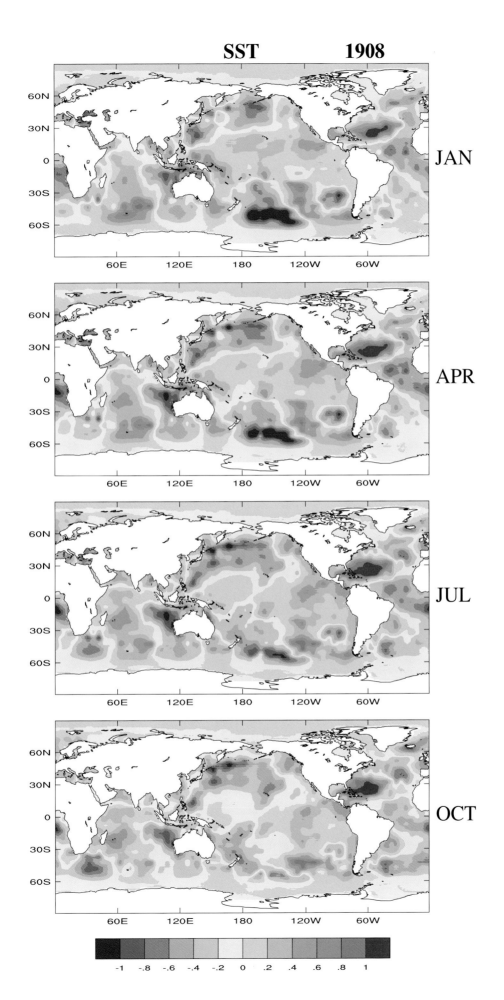

JAN

APR

JUL

OCT

-1 -.8 -.6 -.4 -.2 0 .2 .4 .6 .8 1

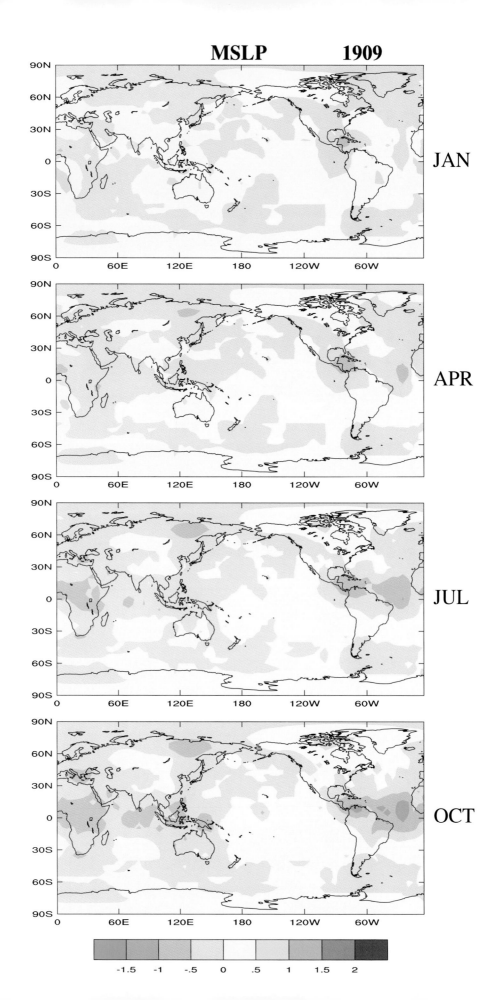

MSLP   **1909**

JAN

APR

JUL

OCT

-1.5   -1   -.5   0   .5   1   1.5   2

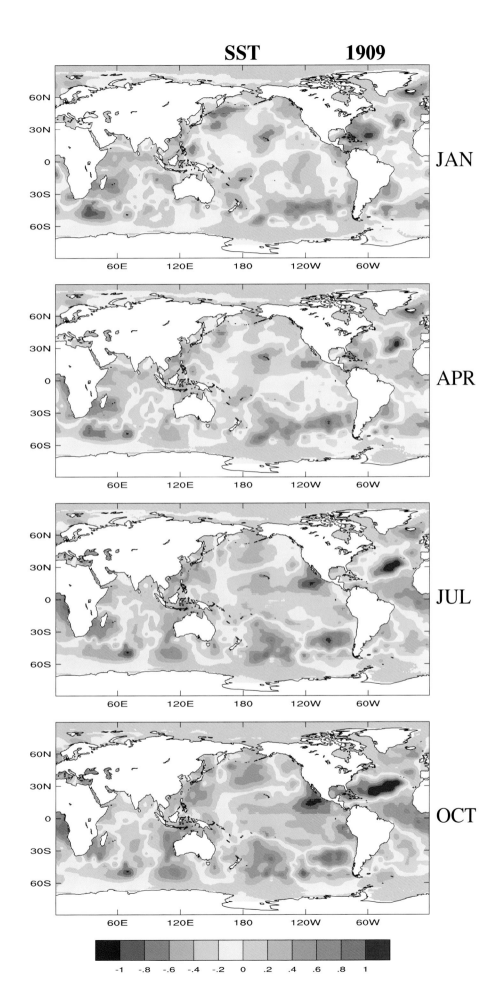

SST   **1909**

JAN

APR

JUL

OCT

-1 -.8 -.6 -.4 -.2 0 .2 .4 .6 .8 1

235

MSLP   1910

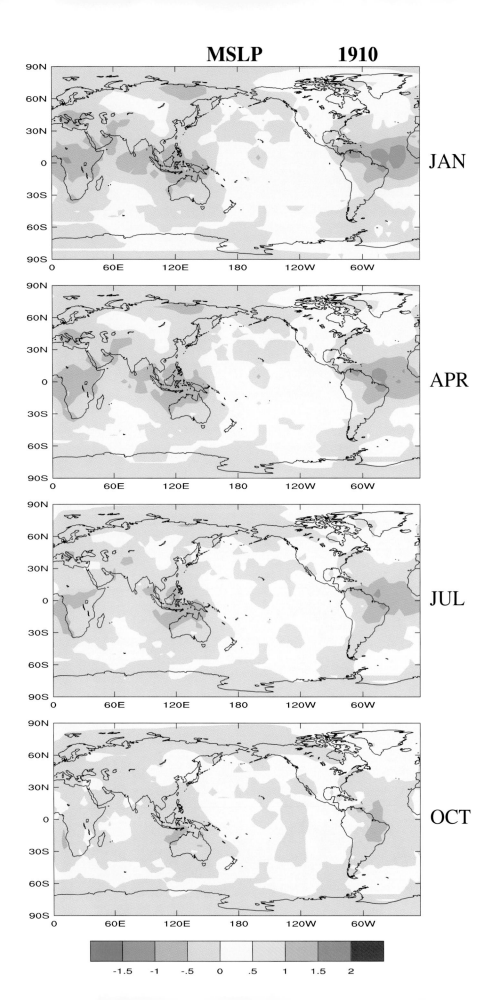

JAN

APR

JUL

OCT

-1.5   -1   -.5   0   .5   1   1.5   2

SST   1910

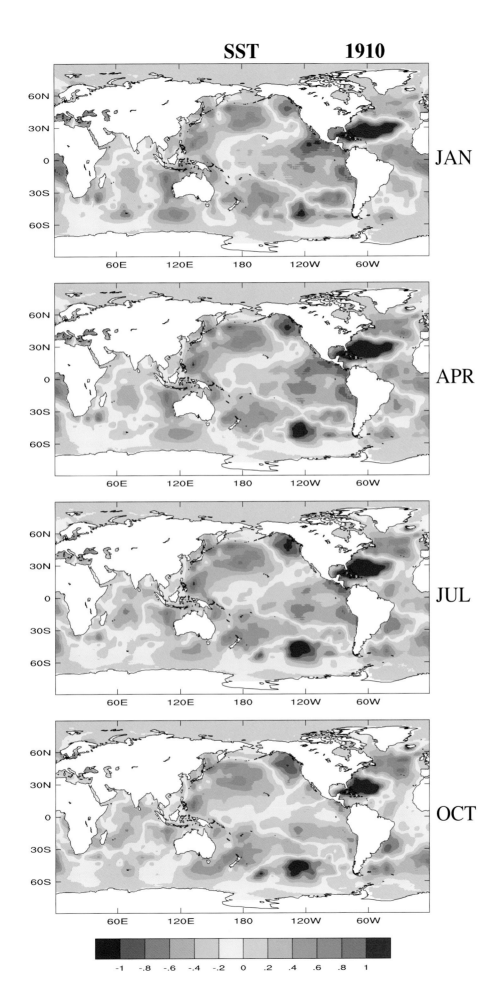

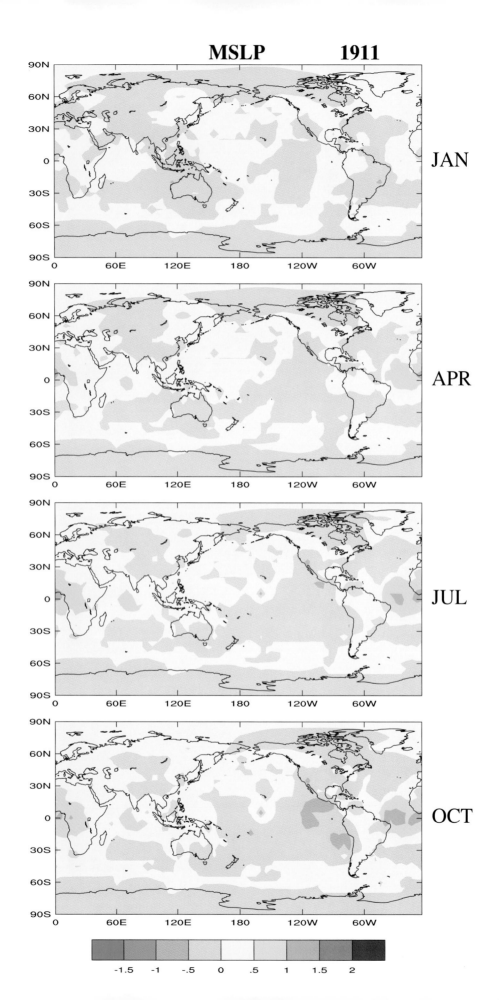

MSLP   **1911**

JAN

APR

JUL

OCT

-1.5   -1   -.5   0   .5   1   1.5   2

SST **1911**

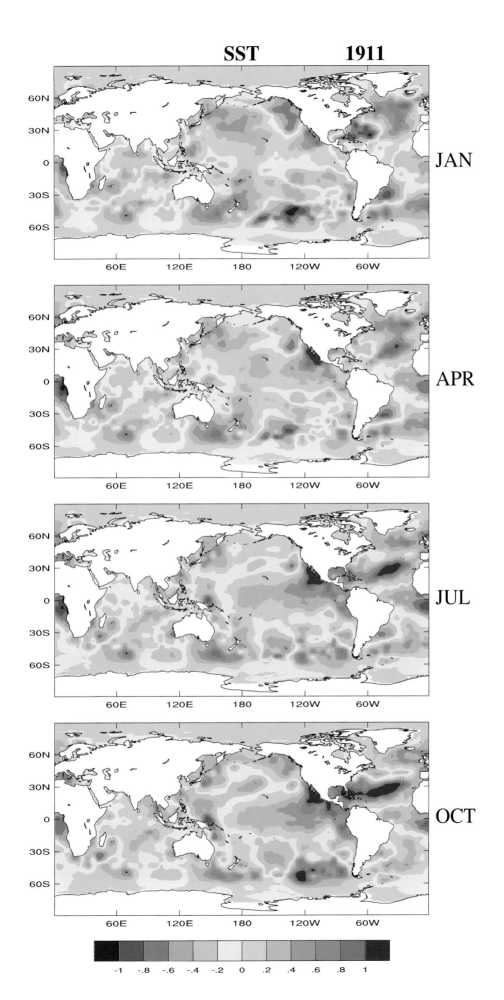

JAN

APR

JUL

OCT

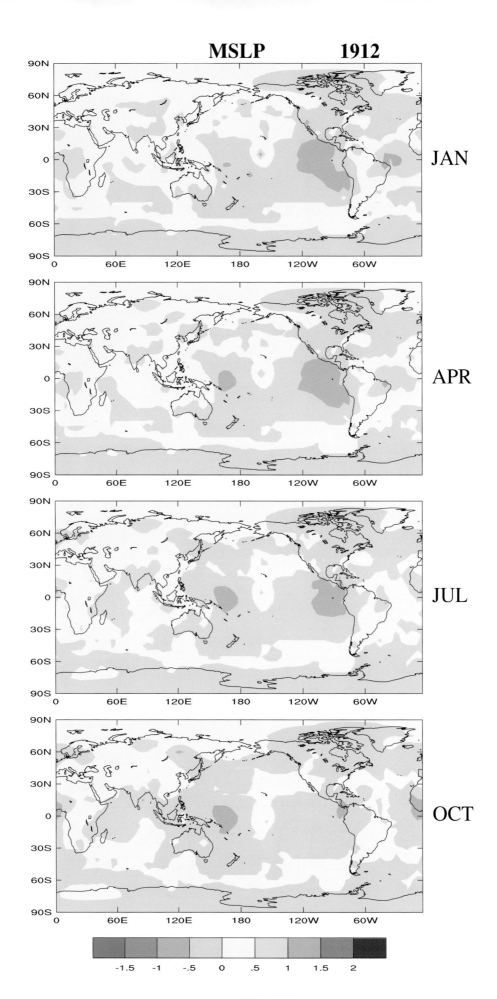

MSLP   1912

JAN

APR

JUL

OCT

-1.5   -1   -.5   0   .5   1   1.5   2

SST **1912**

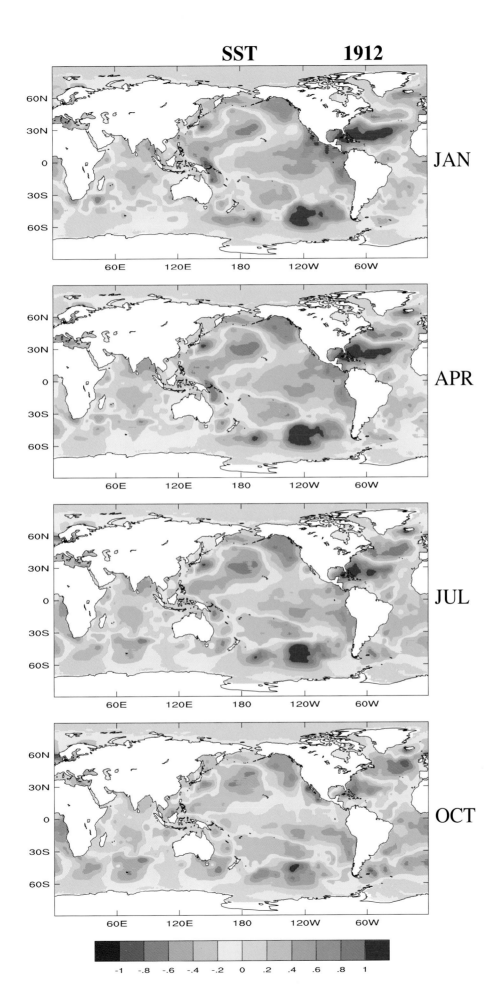

JAN

APR

JUL

OCT

-1 -.8 -.6 -.4 -.2 0 .2 .4 .6 .8 1

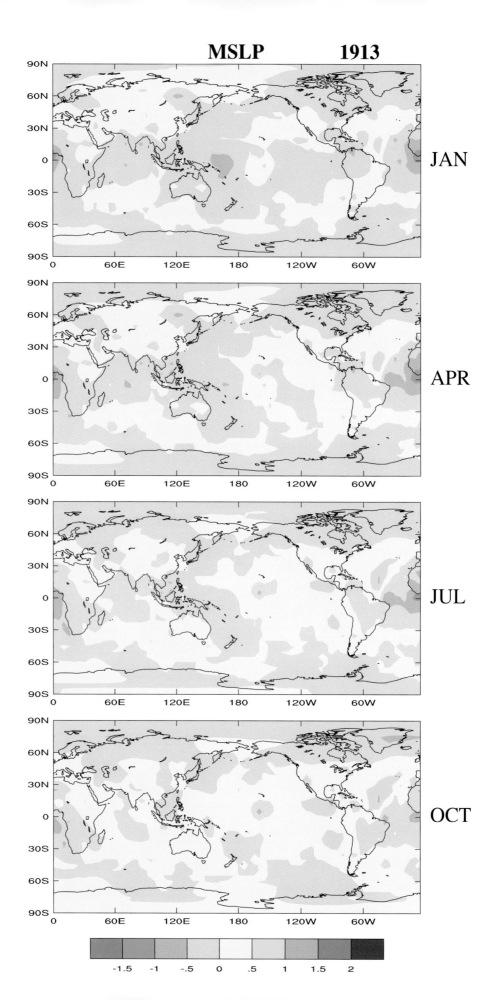

MSLP   1913

JAN

APR

JUL

OCT

-1.5   -1   -.5   0   .5   1   1.5   2

242

SST   1913

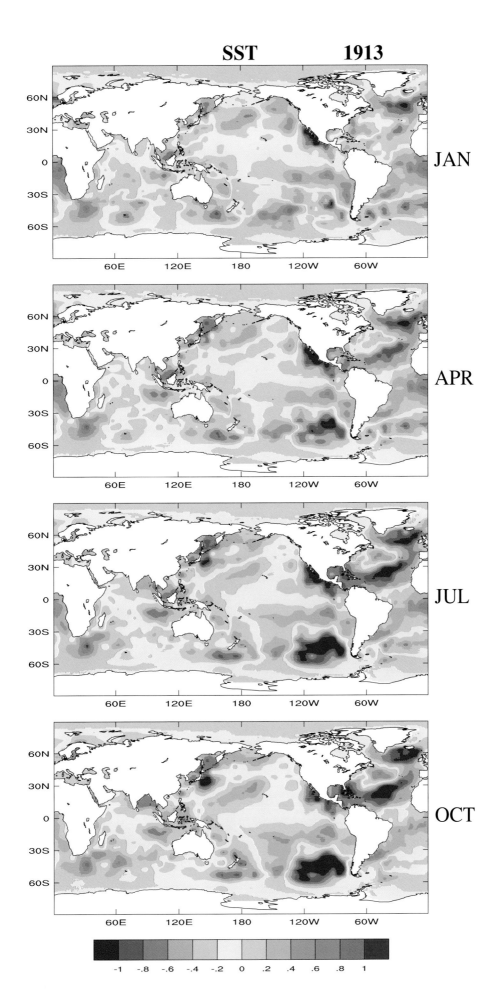

JAN

APR

JUL

OCT

-1   -.8   -.6   -.4   -.2   0   .2   .4   .6   .8   1

MSLP   **1914**

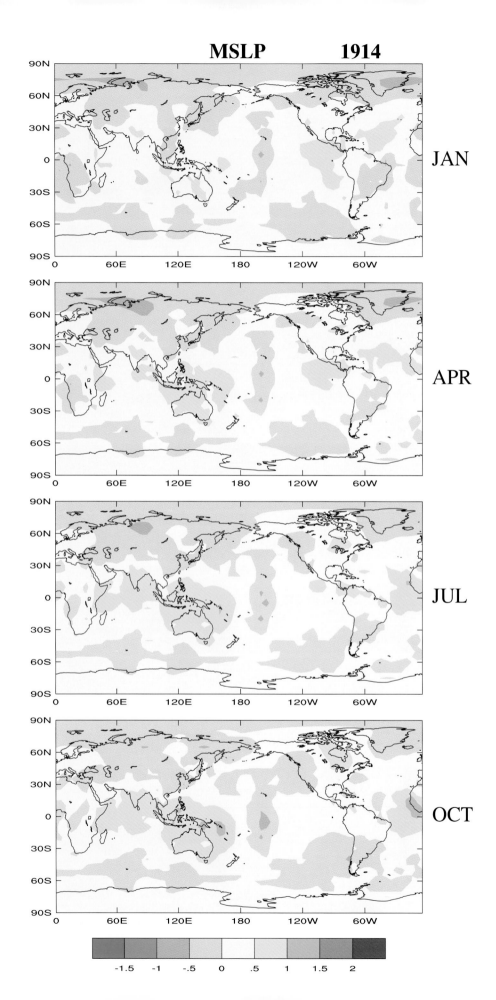

JAN

APR

JUL

OCT

-1.5   -1   -.5   0   .5   1   1.5   2

SST   1914

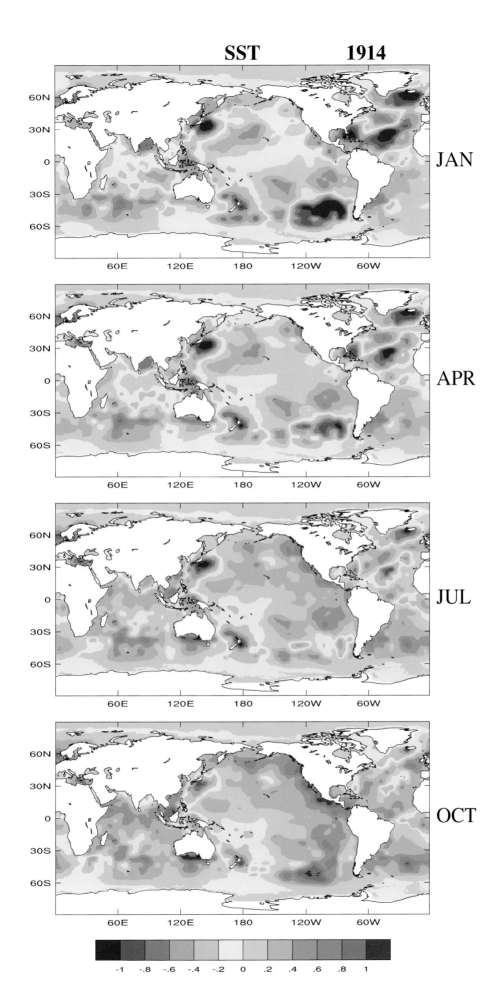

MSLP   **1915**

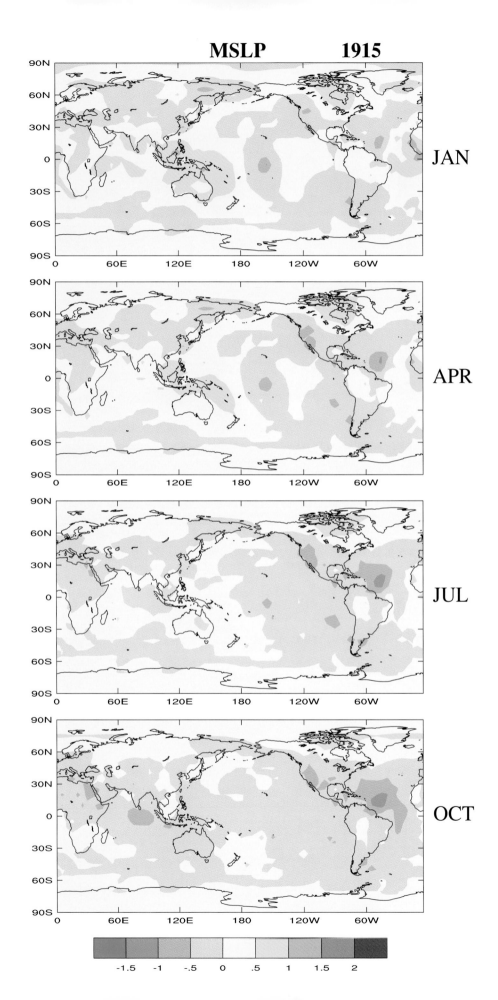

JAN

APR

JUL

OCT

SST   **1915**

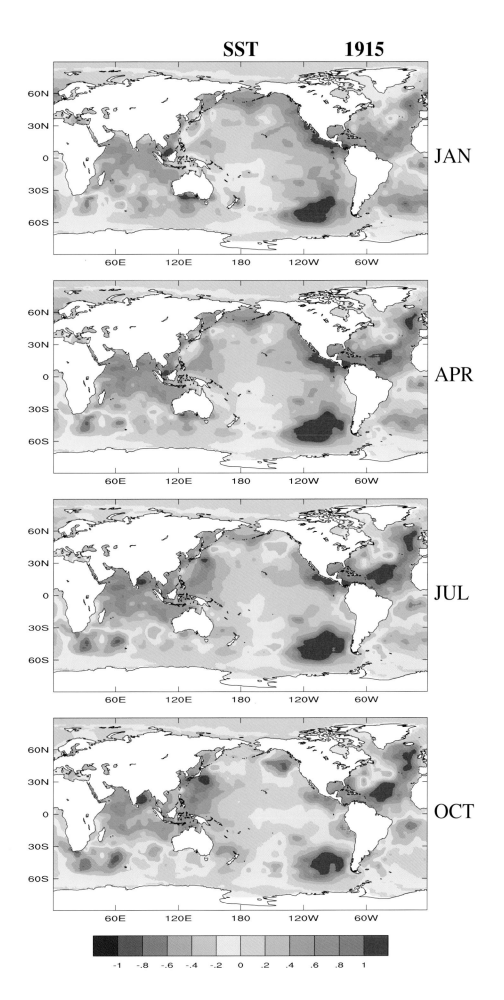

JAN

APR

JUL

OCT

-1   -.8   -.6   -.4   -.2   0   .2   .4   .6   .8   1

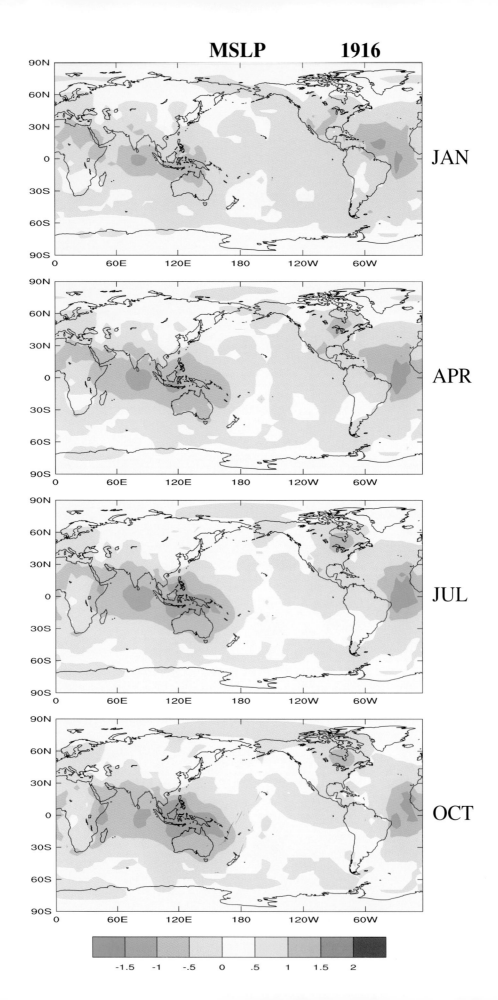

MSLP   1916

JAN

APR

JUL

OCT

-1.5   -1   -.5   0   .5   1   1.5   2

SST   1916

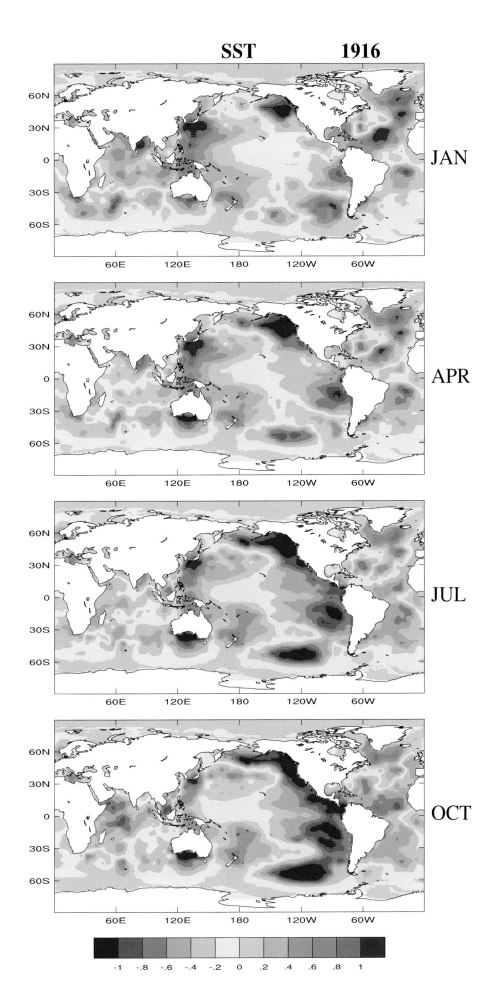

JAN

APR

JUL

OCT

-1 -.8 -.6 -.4 -.2 0 .2 .4 .6 .8 1

249

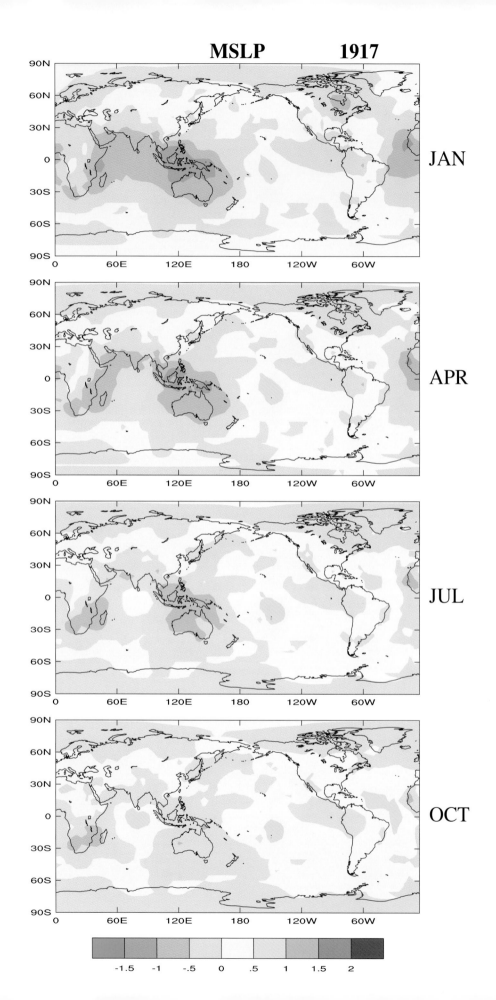

MSLP   **1917**

JAN

APR

JUL

OCT

-1.5   -1   -.5   0   .5   1   1.5   2

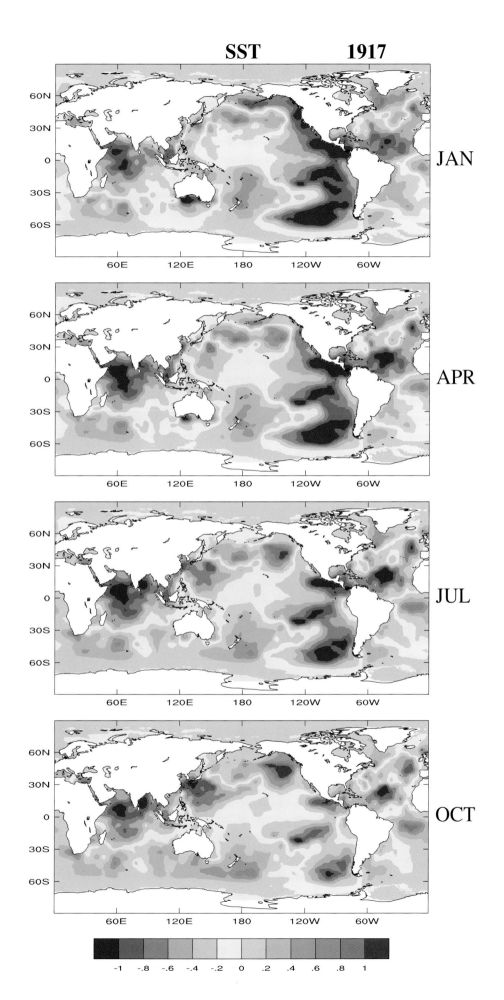

SST   1917

JAN

APR

JUL

OCT

-1   -.8   -.6   -.4   -.2   0   .2   .4   .6   .8   1

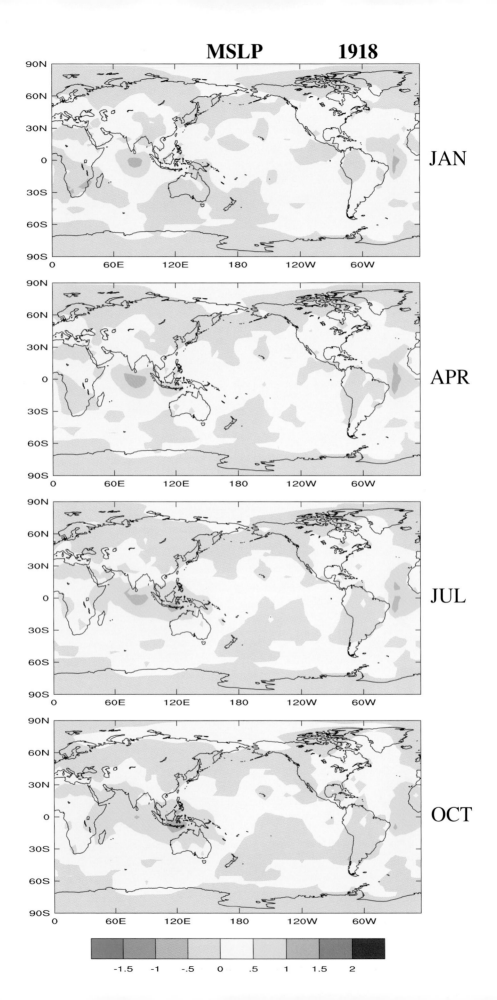

MSLP   1918

JAN

APR

JUL

OCT

-1.5   -1   -.5   0   .5   1   1.5   2

SST   1918

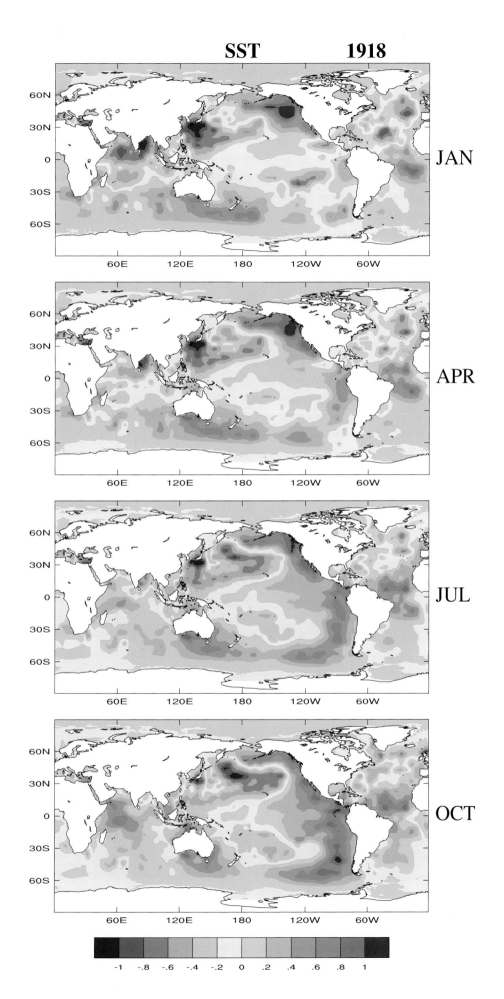

JAN

APR

JUL

OCT

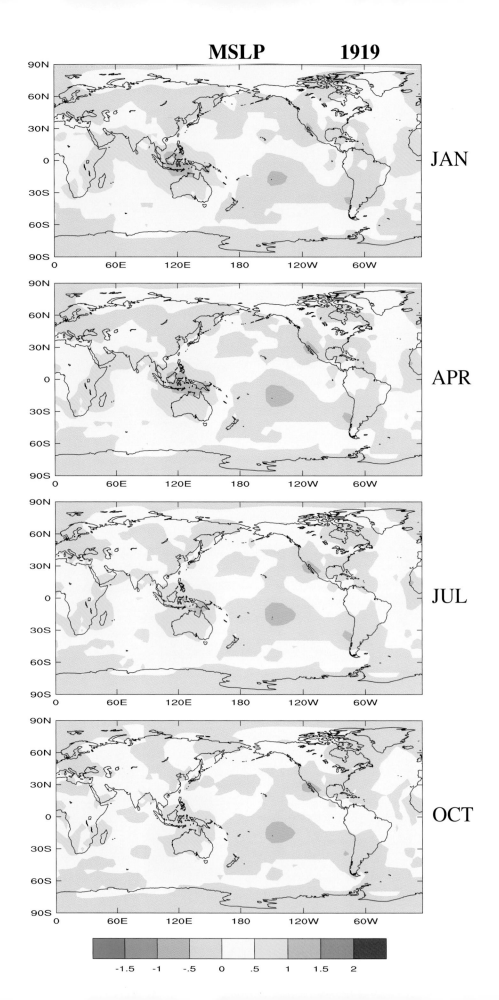

MSLP   **1919**

JAN

APR

JUL

OCT

-1.5   -1   -.5   0   .5   1   1.5   2

SST   **1919**

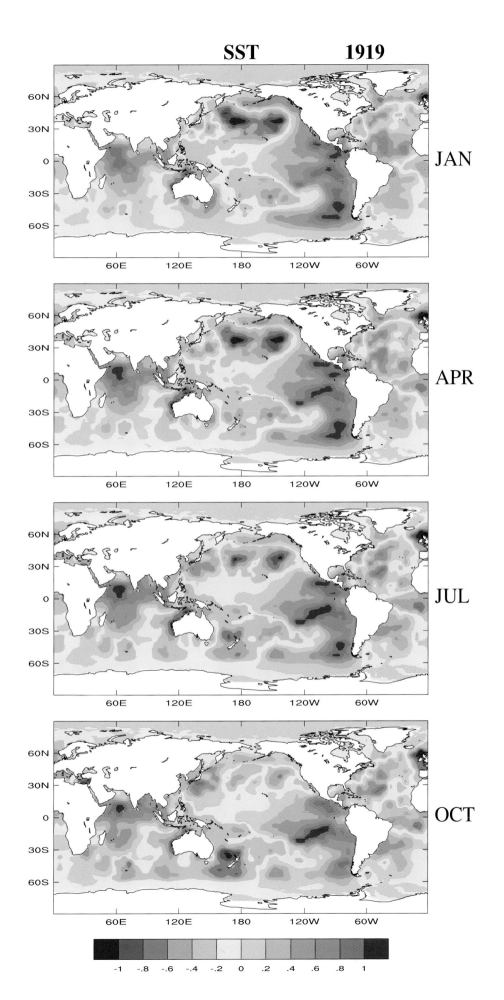

JAN

APR

JUL

OCT

-1   -.8   -.6   -.4   -.2   0   .2   .4   .6   .8   1

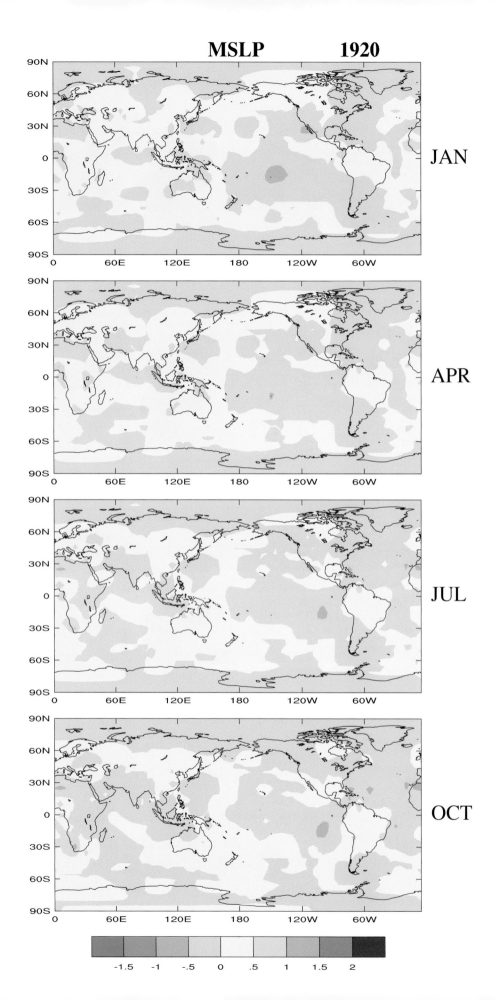

SST   **1920**

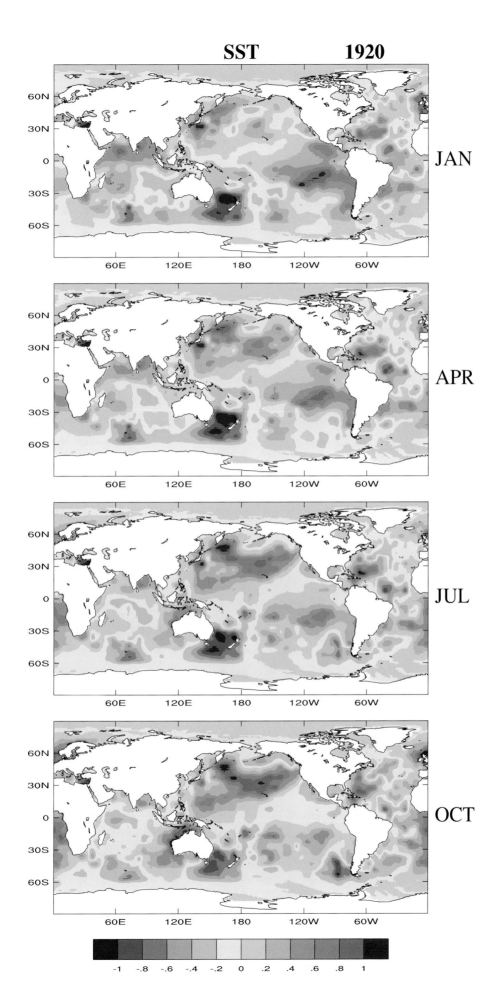

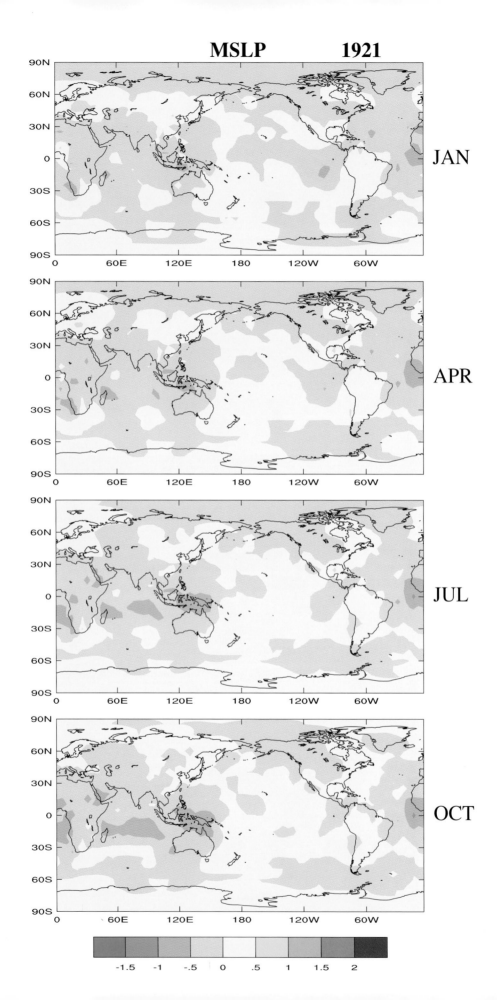

MSLP   1921

JAN

APR

JUL

OCT

SST   **1921**

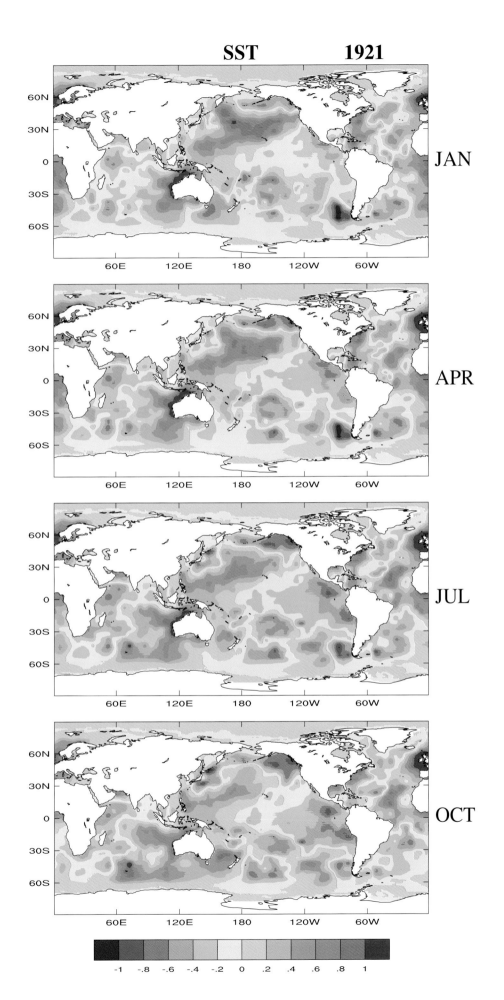

JAN

APR

JUL

OCT

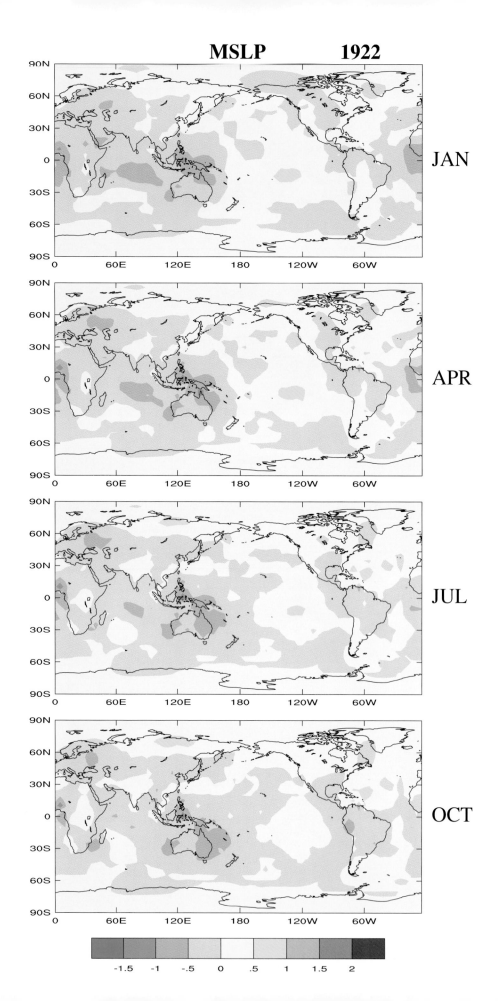

MSLP   **1922**

JAN

APR

JUL

OCT

-1.5   -1   -.5   0   .5   1   1.5   2

SST   1922

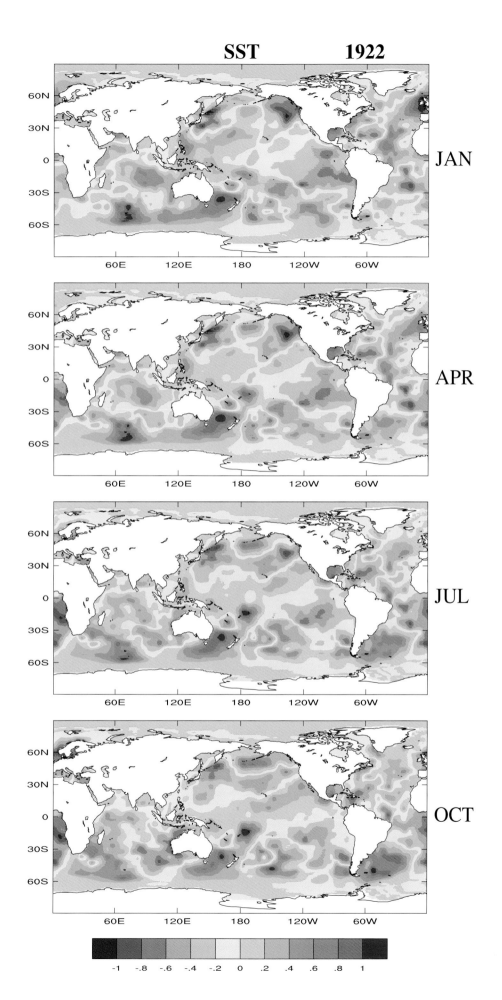

JAN

APR

JUL

OCT

MSLP 1923

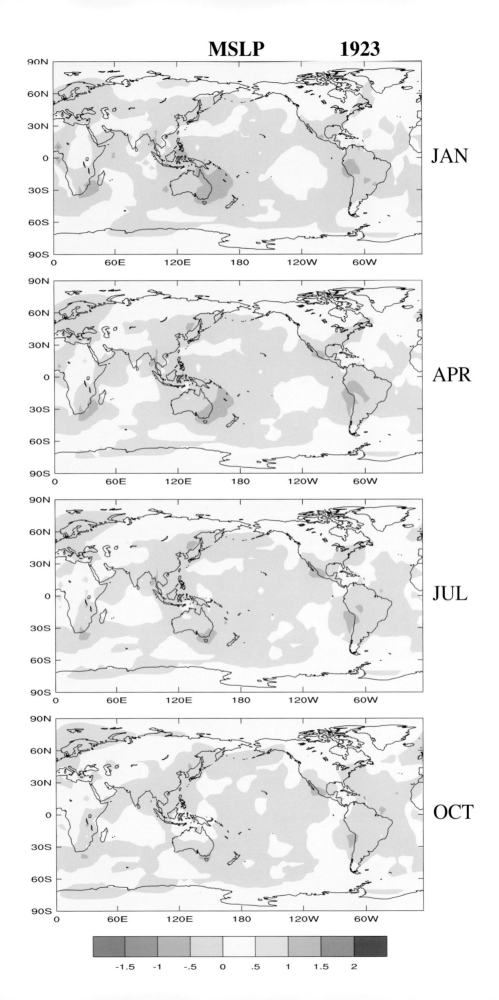

JAN

APR

JUL

OCT

SST   1923

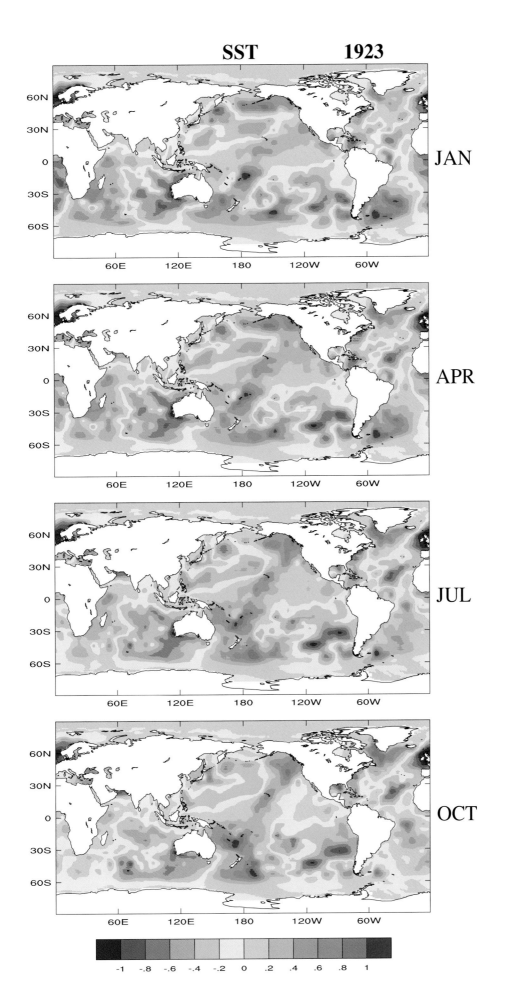

JAN

APR

JUL

OCT

-1   -.8   -.6   -.4   -.2   0   .2   .4   .6   .8   1

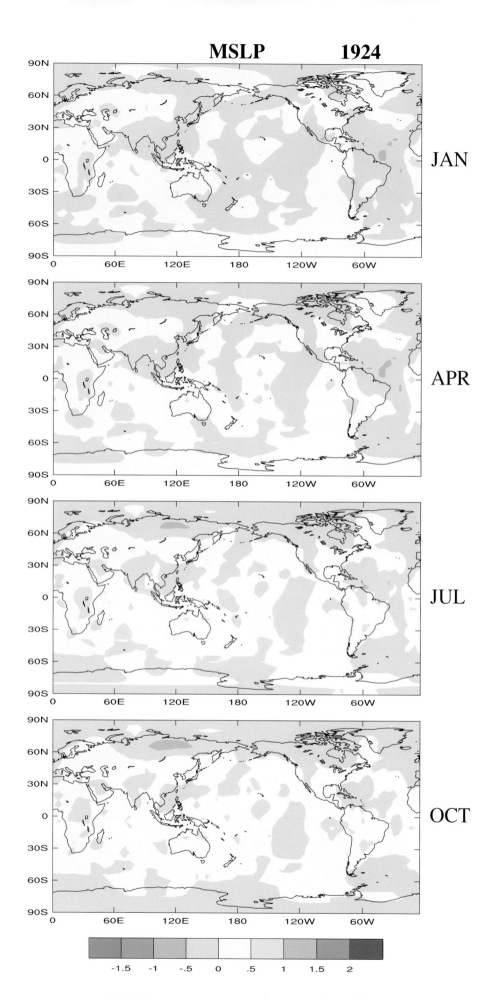

MSLP   **1924**

JAN

APR

JUL

OCT

-1.5   -1   -.5   0   .5   1   1.5   2

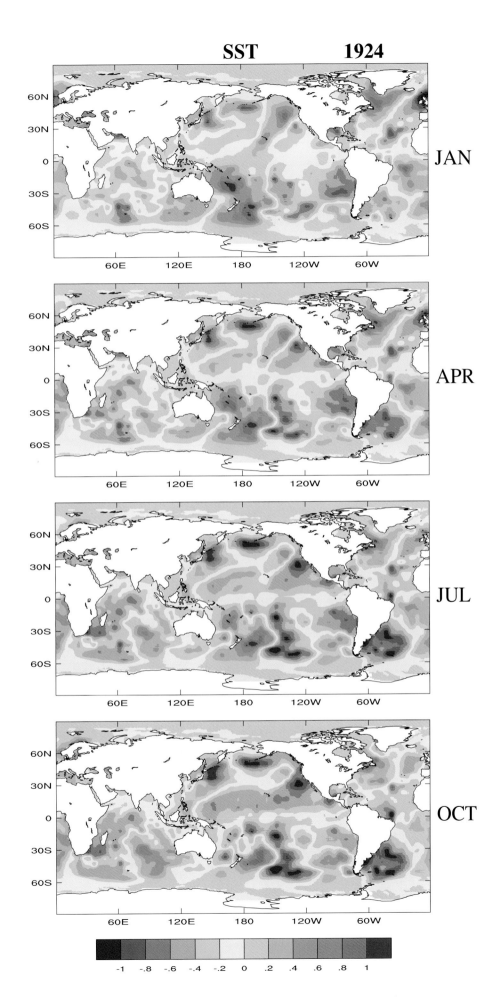

SST   1924

JAN

APR

JUL

OCT

-1   -.8   -.6   -.4   -.2   0   .2   .4   .6   .8   1

265

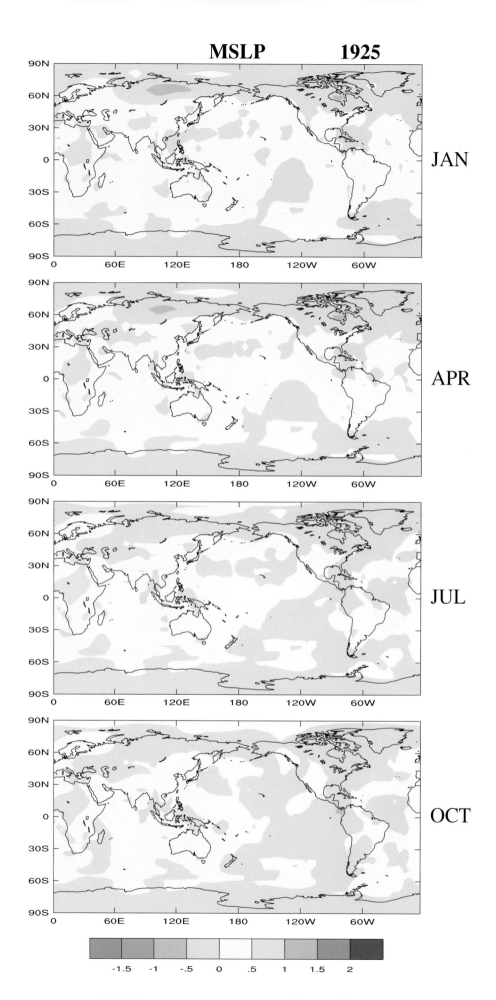

MSLP   1925

JAN

APR

JUL

OCT

| | -1.5 | -1 | -.5 | 0 | .5 | 1 | 1.5 | 2 | |

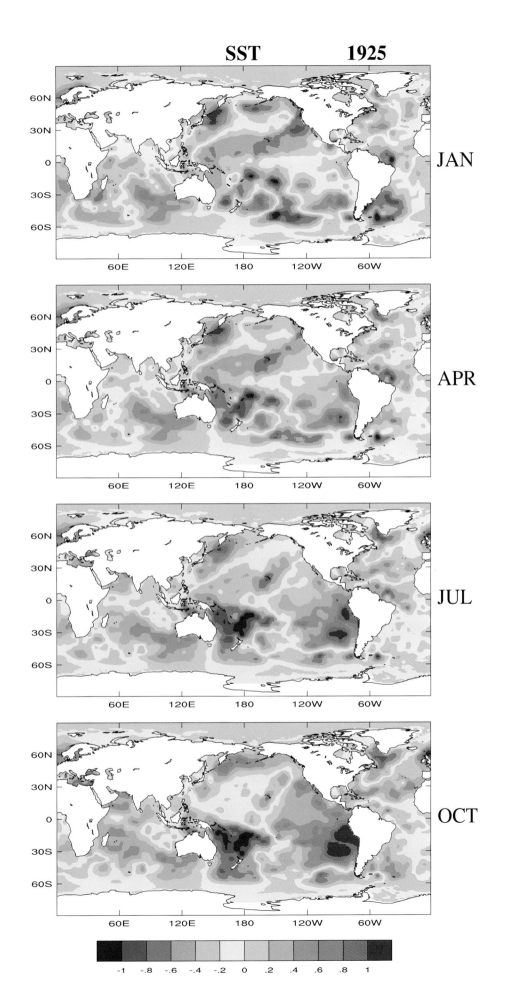

SST   1925

JAN

APR

JUL

OCT

-1   -.8   -.6   -.4   -.2   0   .2   .4   .6   .8   1

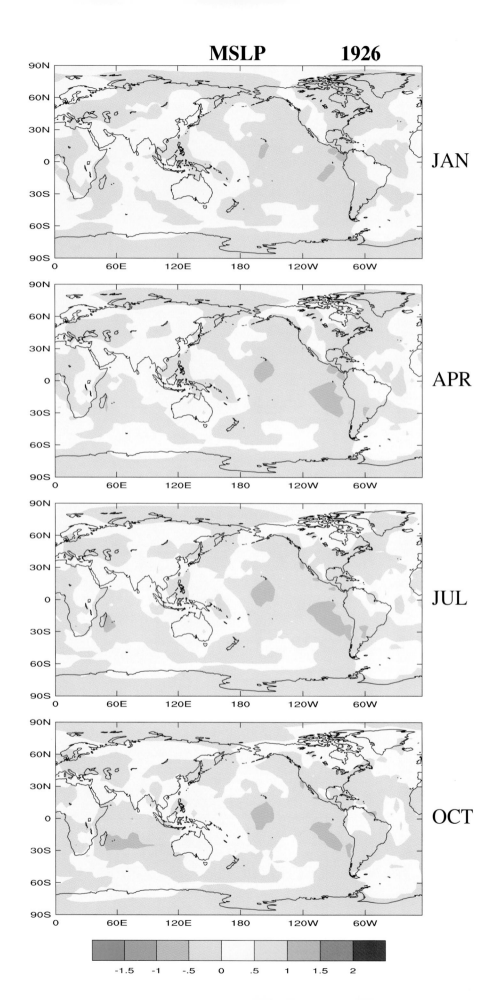

MSLP   **1926**

JAN

APR

JUL

OCT

-1.5   -1   -.5   0   .5   1   1.5   2

SST　　　**1926**

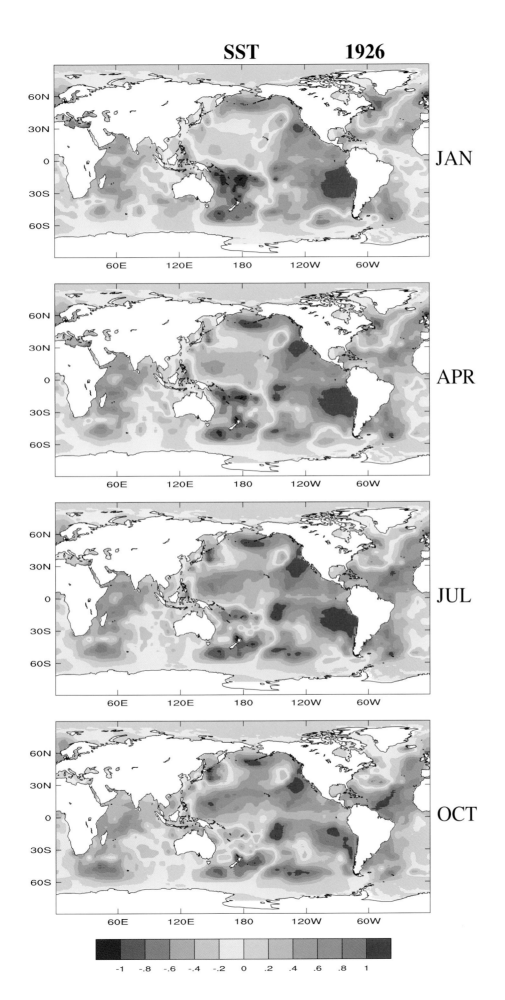

JAN

APR

JUL

OCT

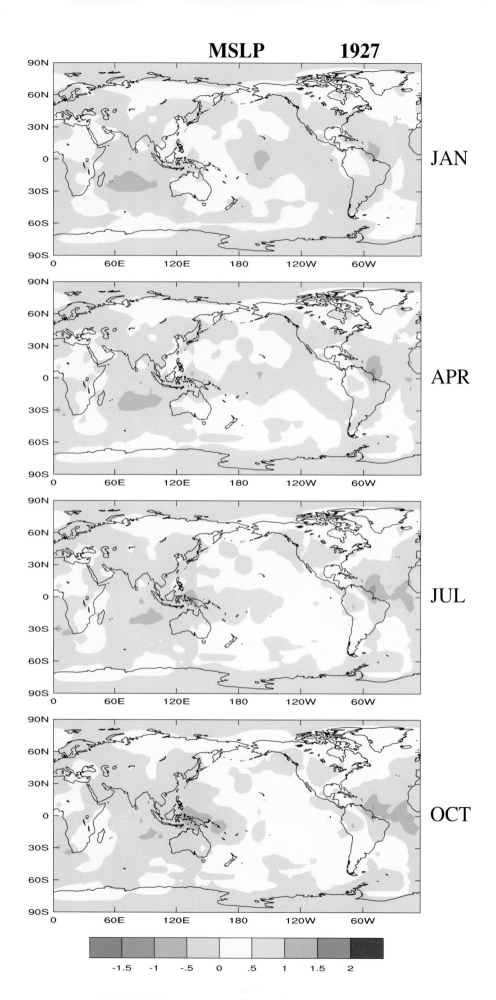

MSLP　　　　**1927**

JAN

APR

JUL

OCT

-1.5　-1　-.5　0　.5　1　1.5　2

SST **1927**

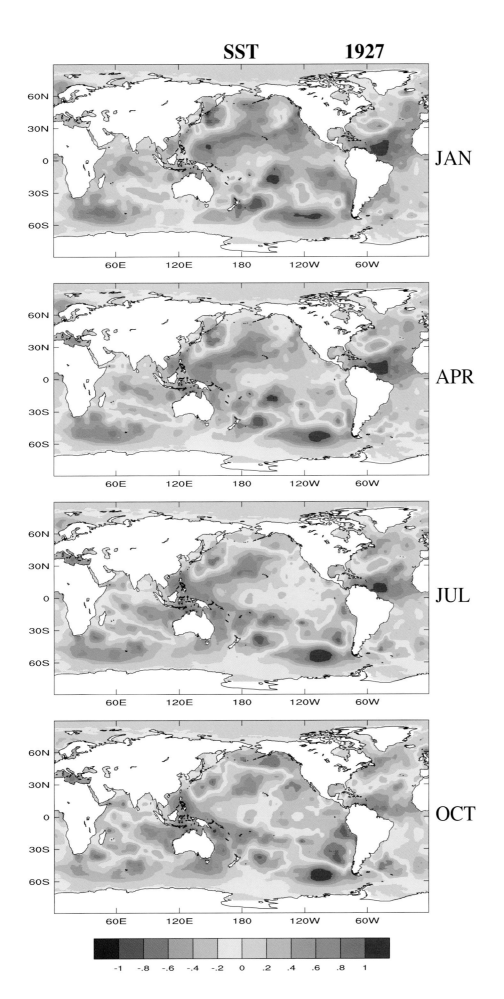

JAN

APR

JUL

OCT

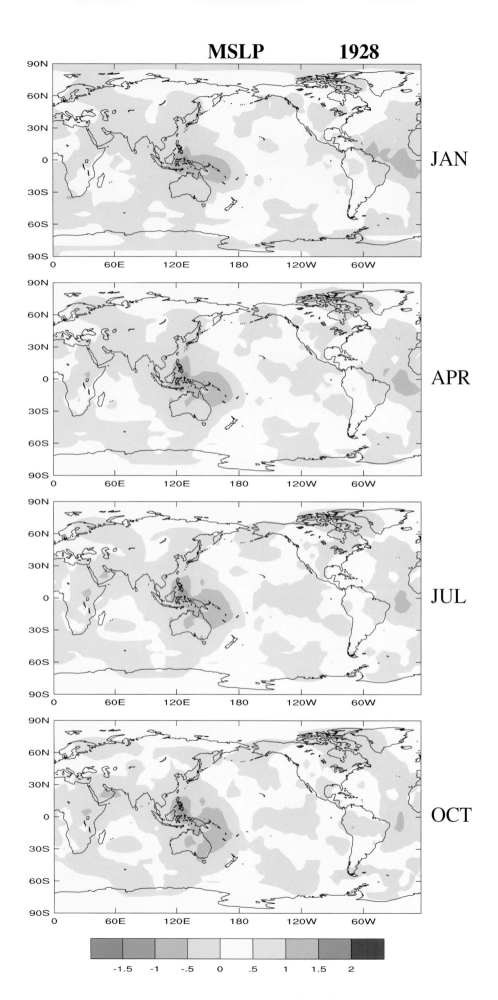

MSLP   1928

JAN

APR

JUL

OCT

-1.5   -1   -.5   0   .5   1   1.5   2

SST　　　**1928**

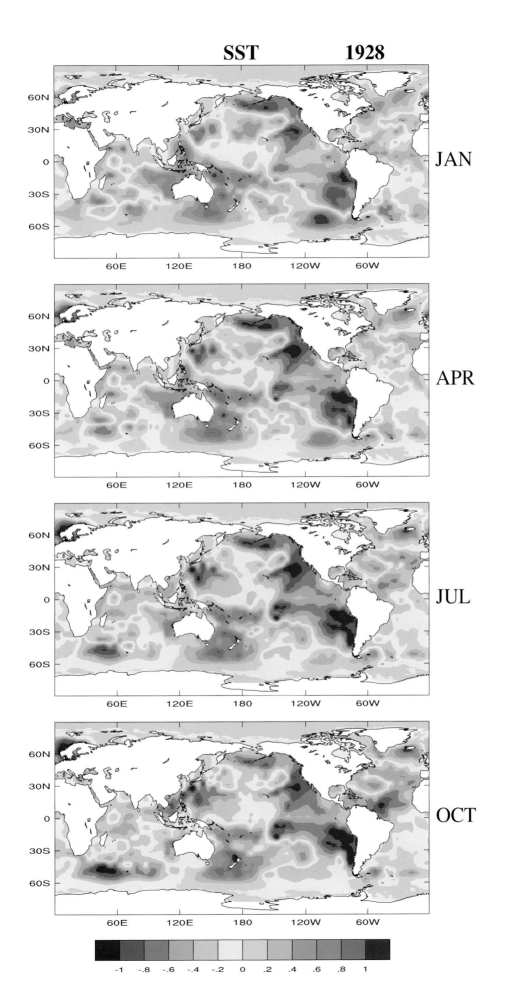

JAN

APR

JUL

OCT

-1 -.8 -.6 -.4 -.2 0 .2 .4 .6 .8 1

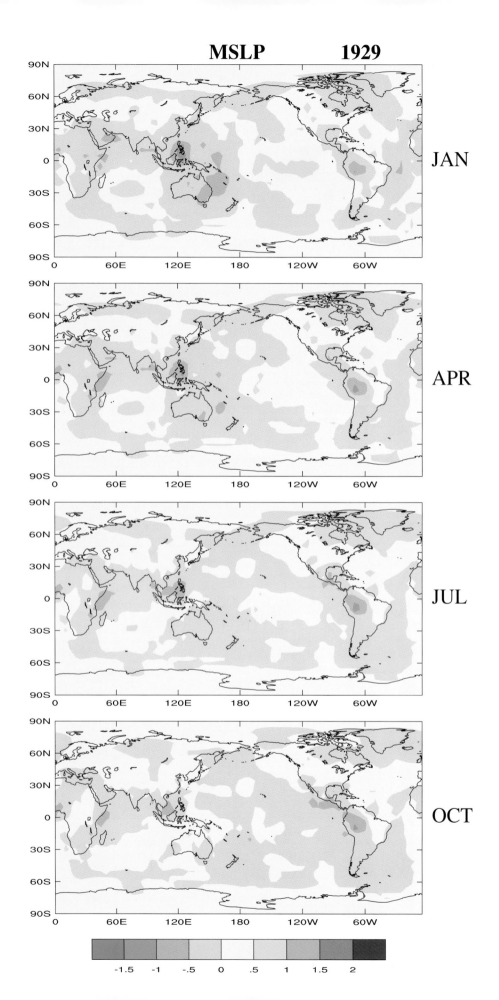

MSLP 1929

JAN

APR

JUL

OCT

-1.5 -1 -.5 0 .5 1 1.5 2

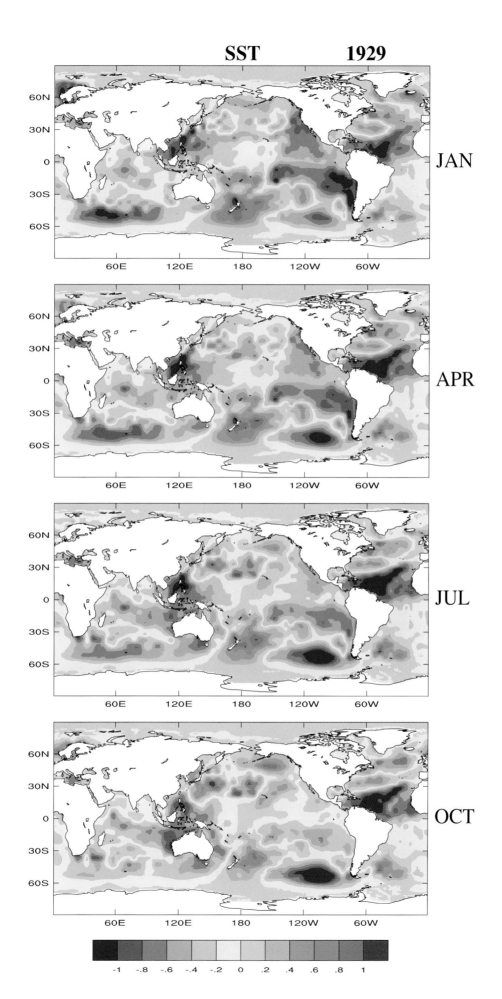

SST   **1929**

JAN

APR

JUL

OCT

-1   -.8   -.6   -.4   -.2   0   .2   .4   .6   .8   1

275

MSLP   **1930**

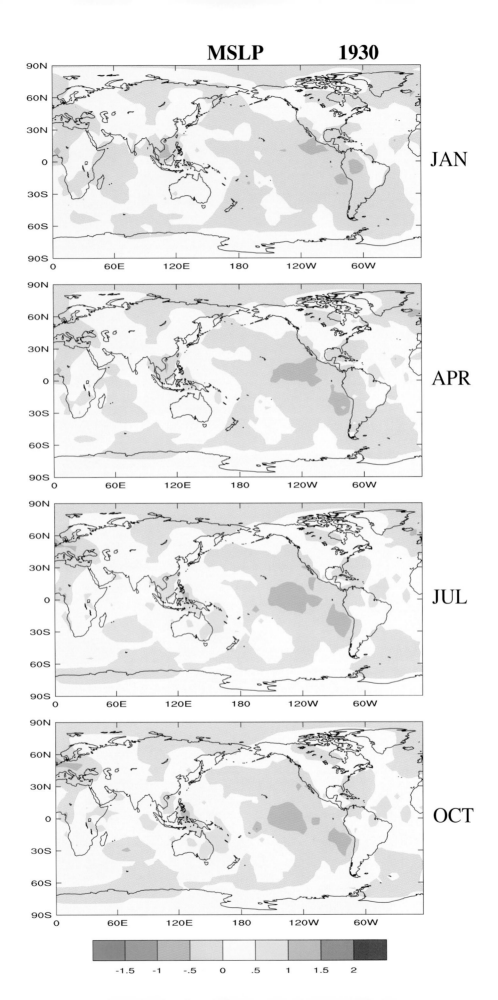

JAN

APR

JUL

OCT

SST 1930

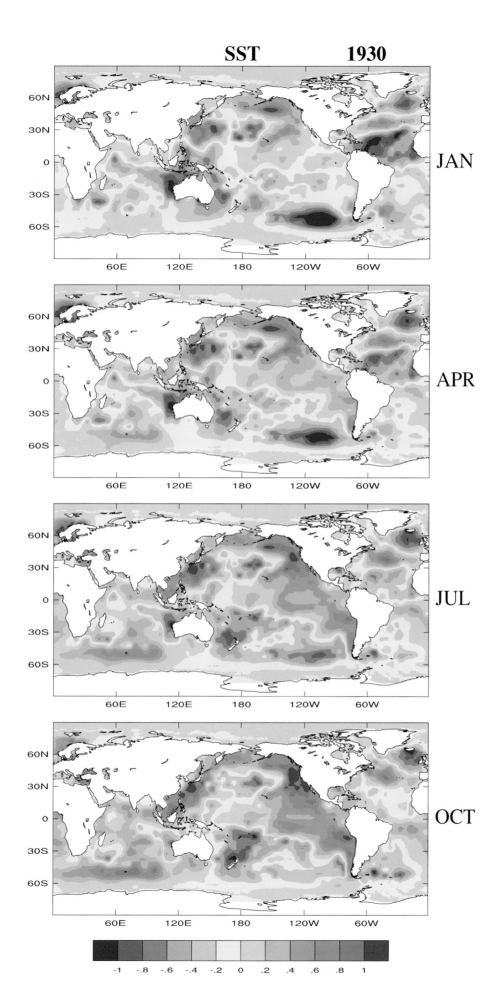

JAN

APR

JUL

OCT

277

MSLP   **1931**

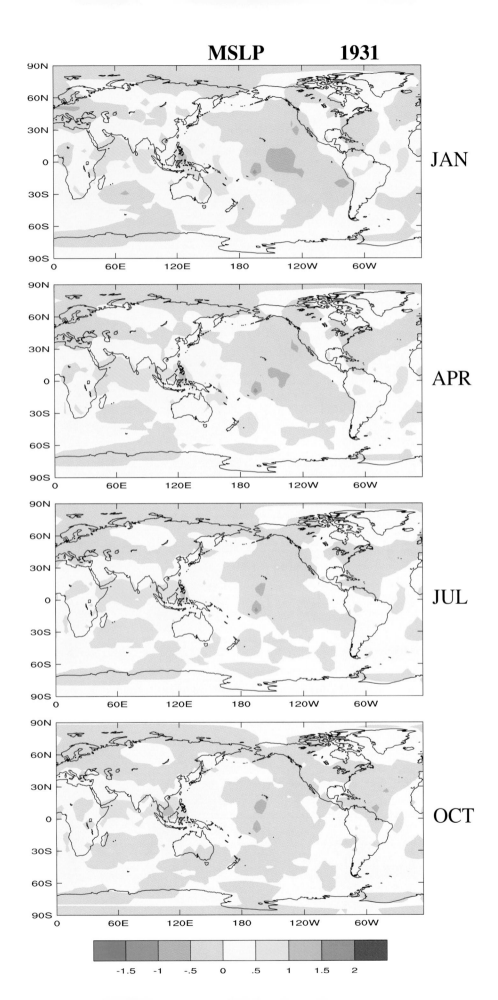

JAN

APR

JUL

OCT

SST 1931

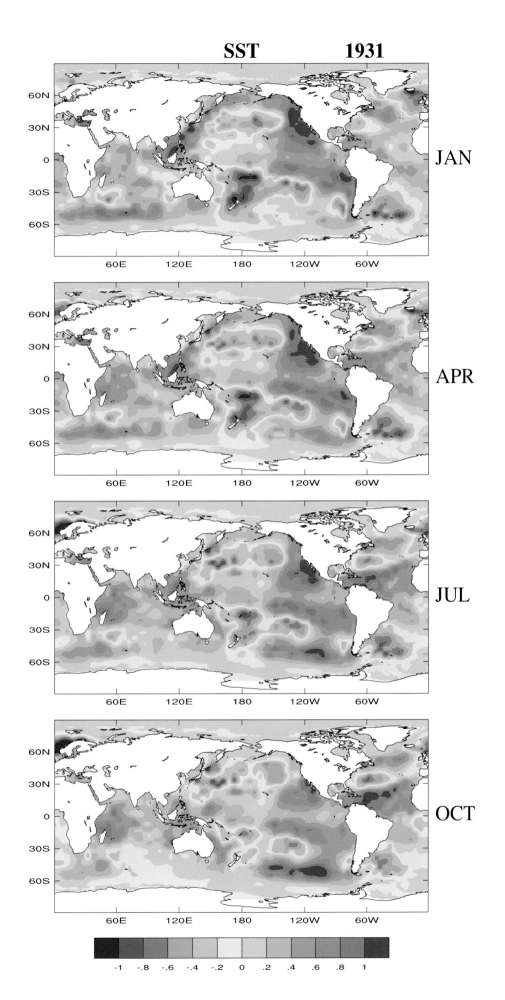

JAN

APR

JUL

OCT

-1 -.8 -.6 -.4 -.2 0 .2 .4 .6 .8 1

MSLP   1932

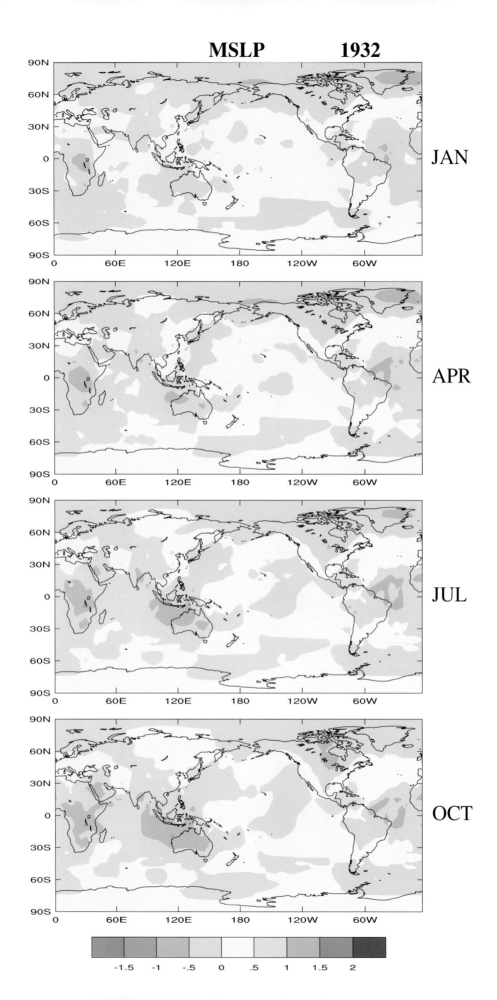

JAN

APR

JUL

OCT

-1.5   -1   -.5   0   .5   1   1.5   2

SST　　　**1932**

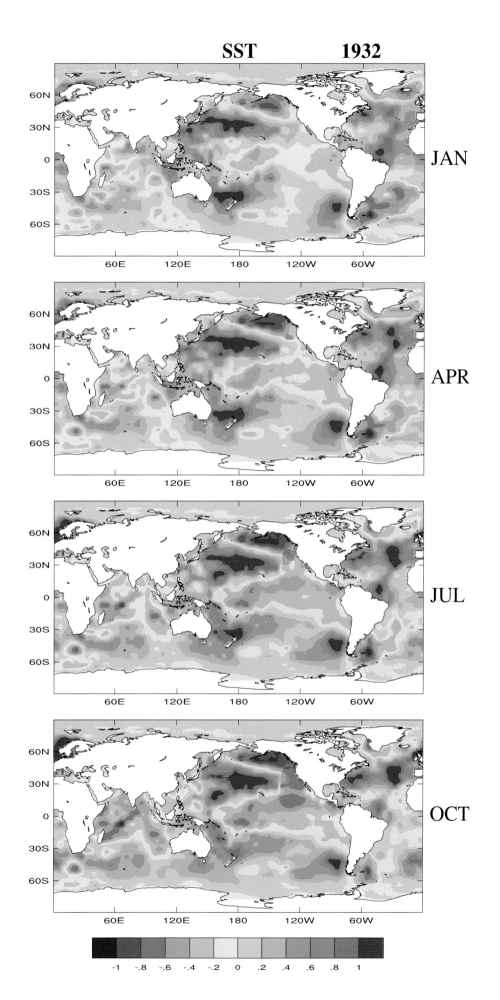

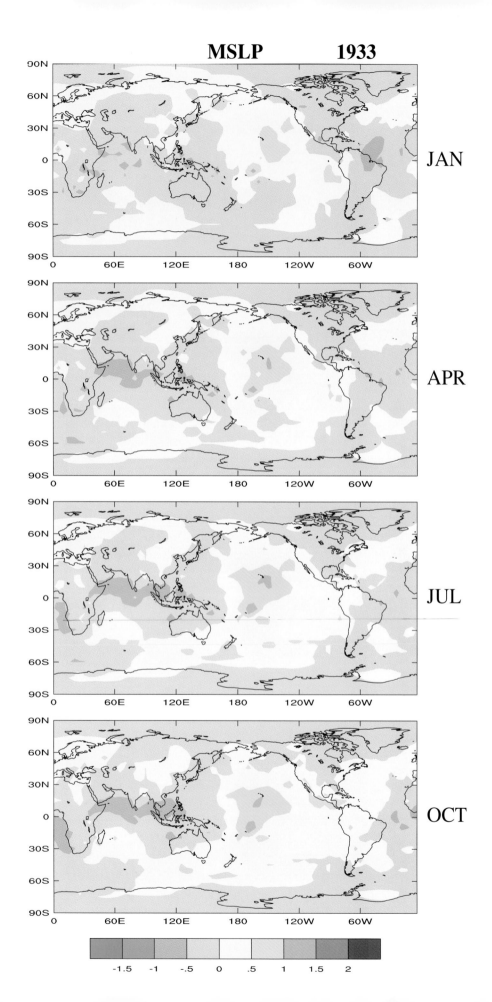

SST   **1933**

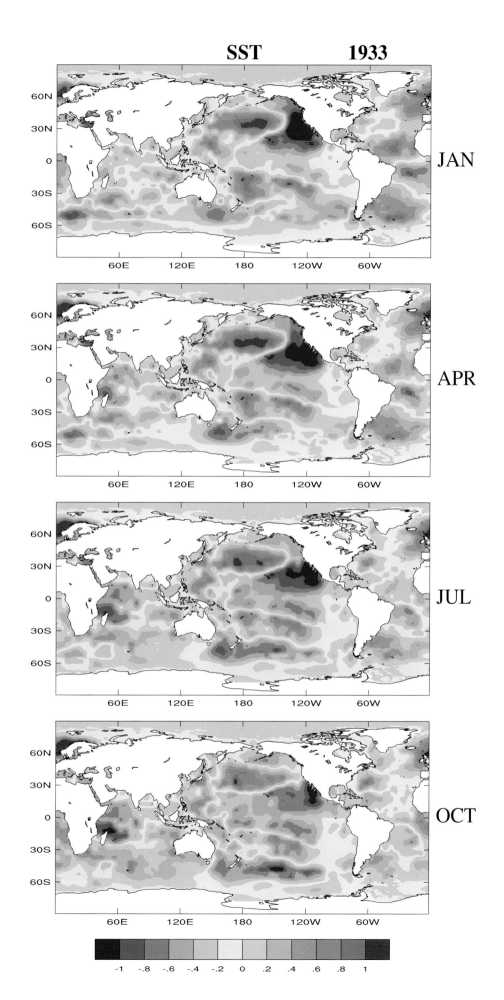

JAN

APR

JUL

OCT

-1   -.8   -.6   -.4   -.2   0   .2   .4   .6   .8   1

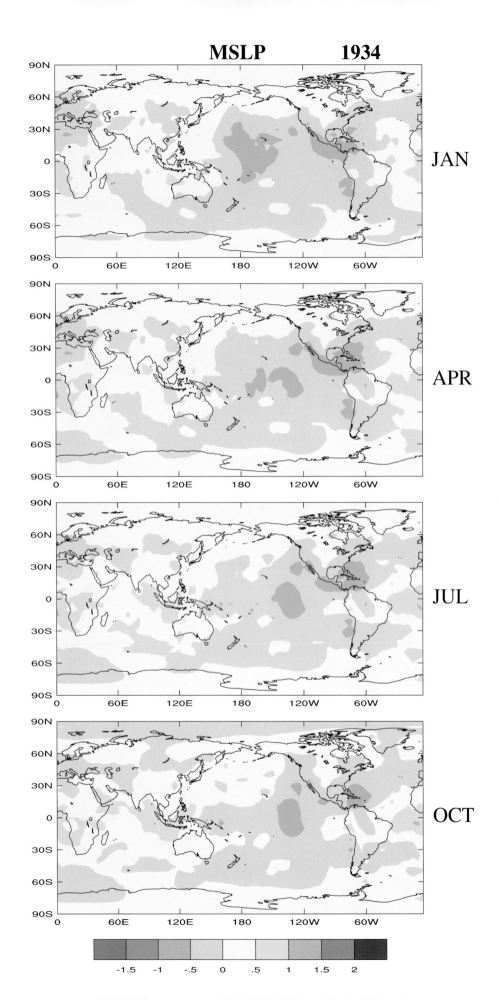

MSLP   **1934**

JAN

APR

JUL

OCT

-1.5   -1   -.5   0   .5   1   1.5   2

SST   **1934**

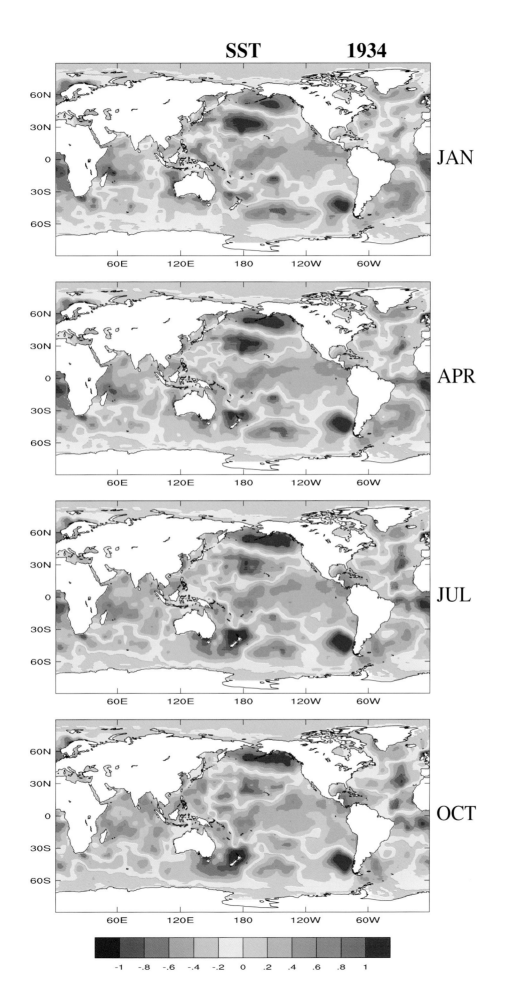

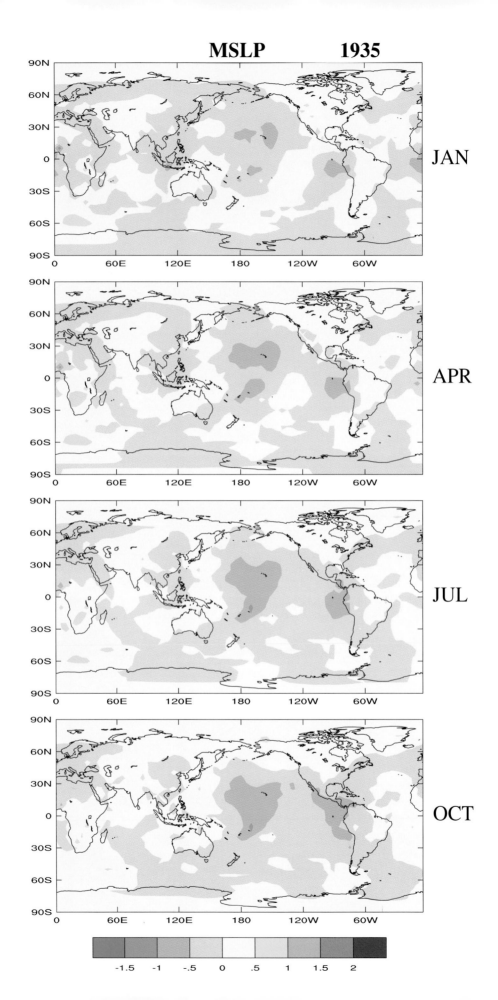

MSLP   1935

JAN

APR

JUL

OCT

-1.5   -1   -.5   0   .5   1   1.5   2

SST   1935

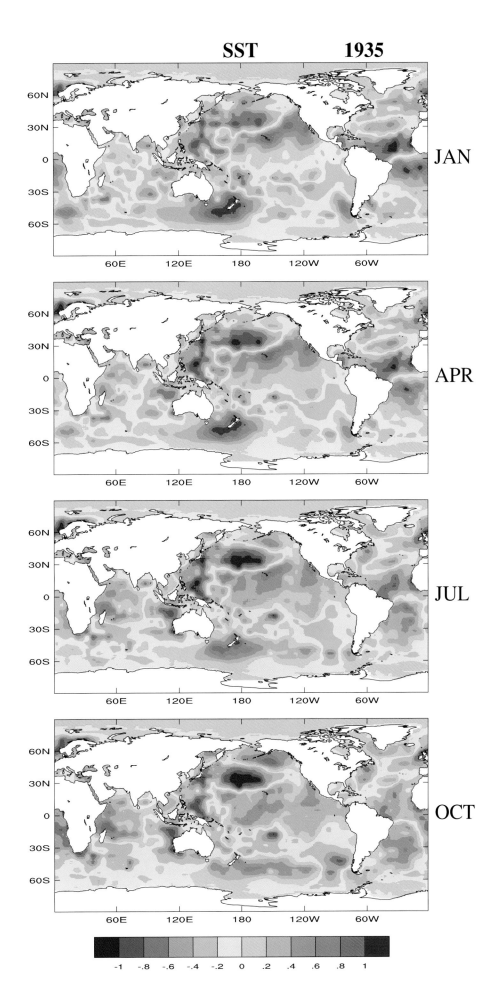

JAN

APR

JUL

OCT

-1   -.8   -.6   -.4   -.2   0   .2   .4   .6   .8   1

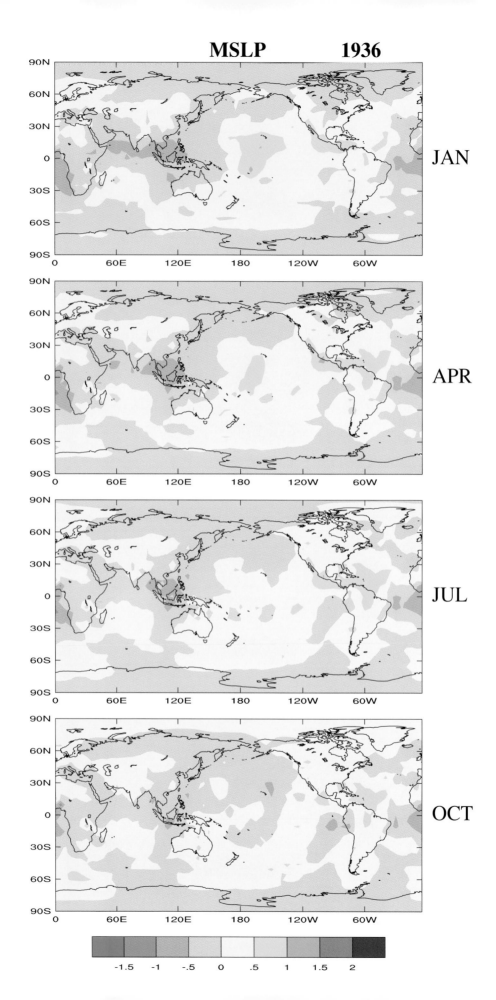

MSLP   **1936**

JAN

APR

JUL

OCT

-1.5   -1   -.5   0   .5   1   1.5   2

288

SST   **1936**

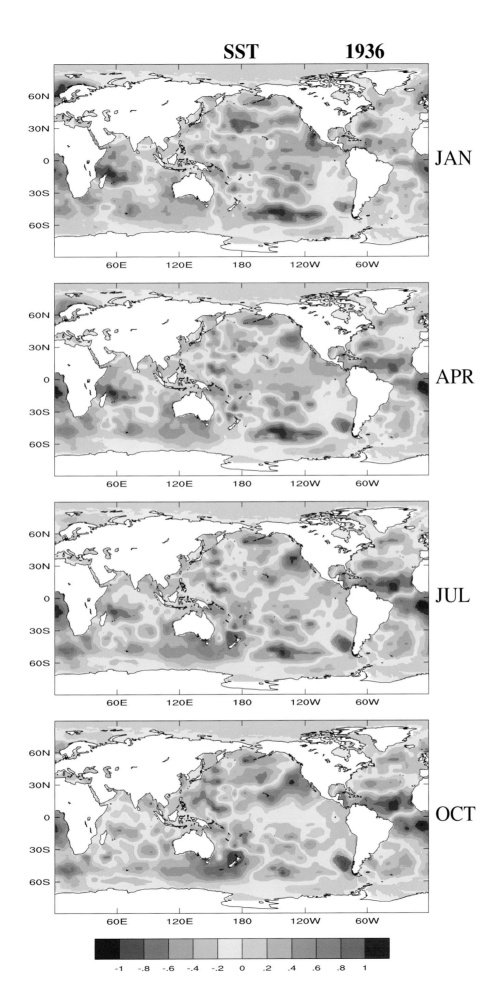

JAN

APR

JUL

OCT

-1   -.8   -.6   -.4   -.2   0   .2   .4   .6   .8   1

MSLP   **1937**

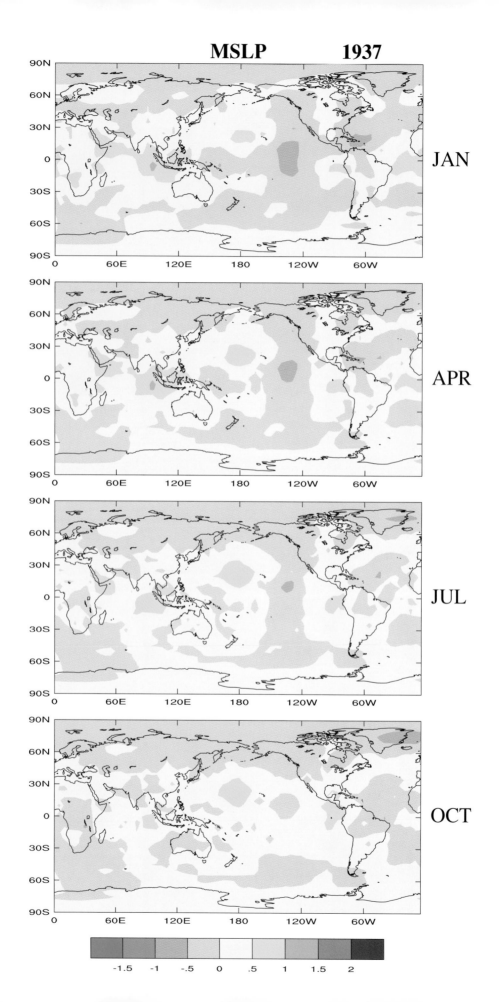

SST   **1937**

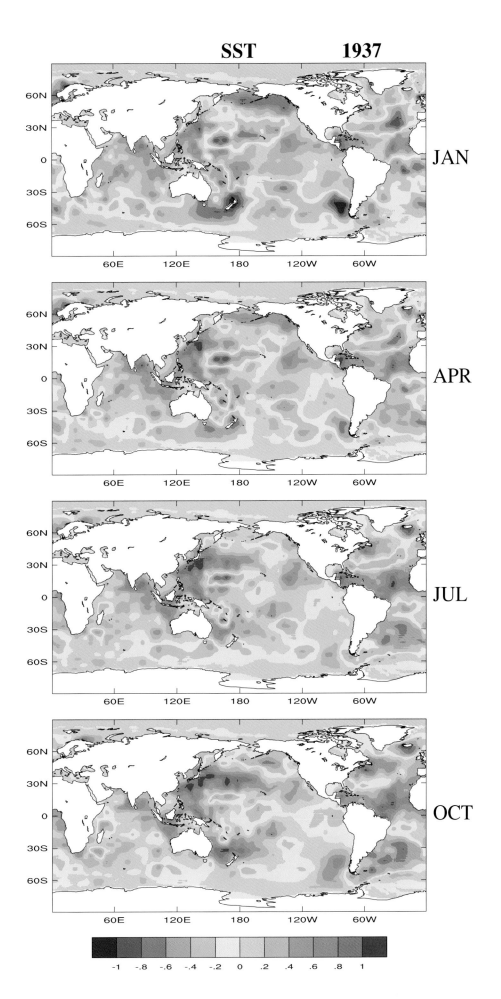

JAN

APR

JUL

OCT

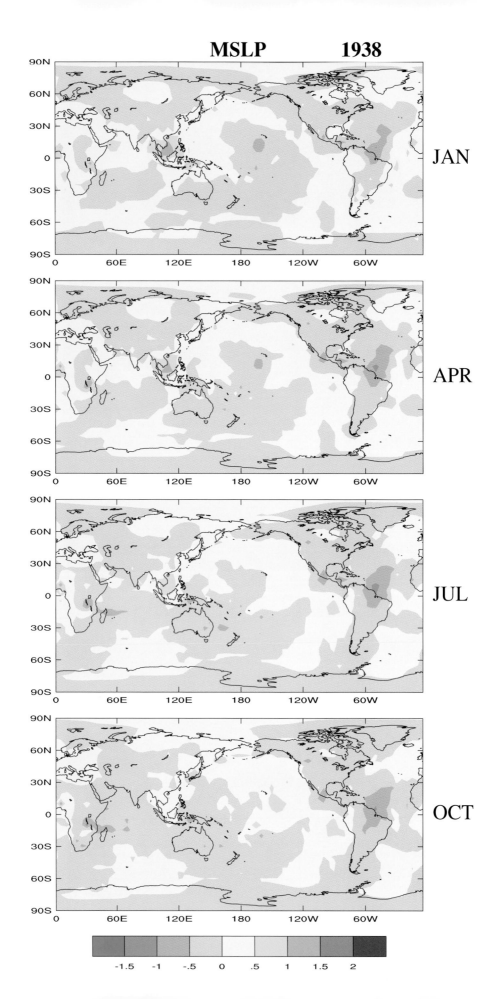

MSLP   **1938**

JAN

APR

JUL

OCT

-1.5   -1   -.5   0   .5   1   1.5   2

SST   1938

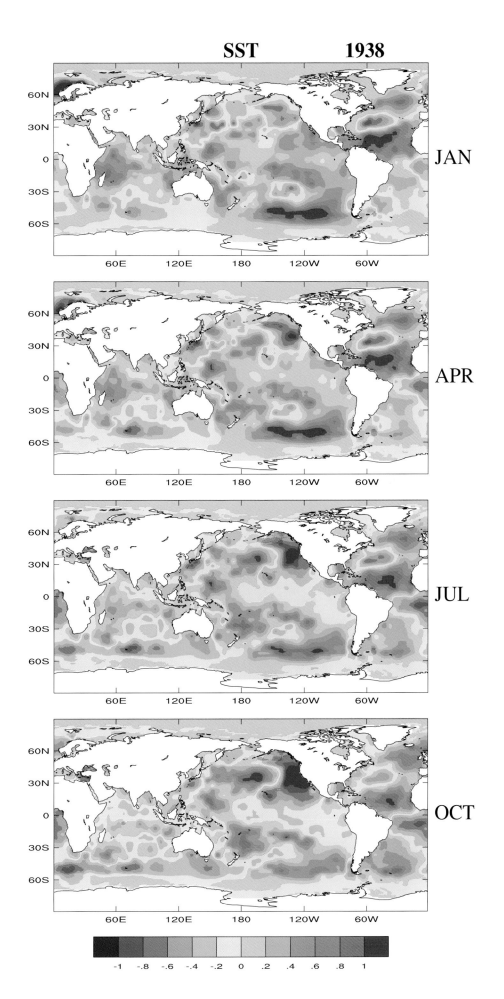

JAN

APR

JUL

OCT

-1 -.8 -.6 -.4 -.2 0 .2 .4 .6 .8 1

293

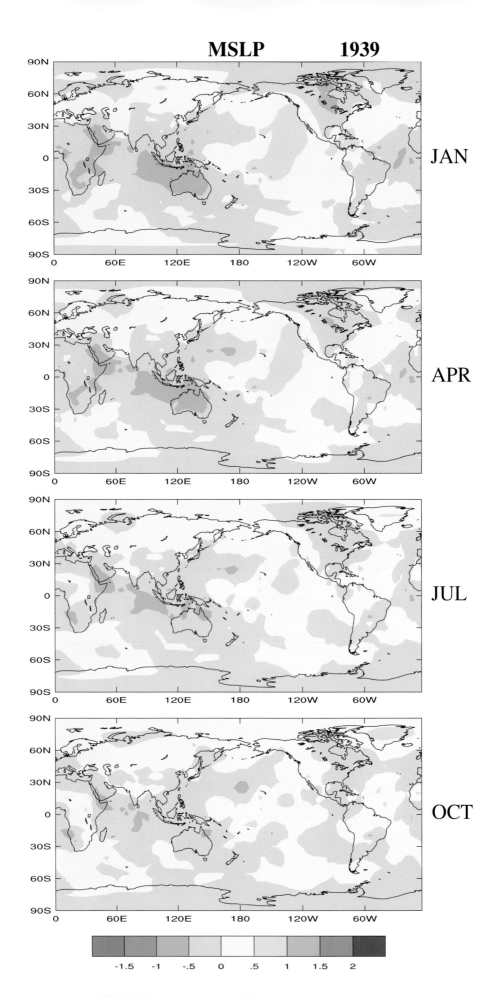

MSLP **1939**

JAN

APR

JUL

OCT

SST   1939

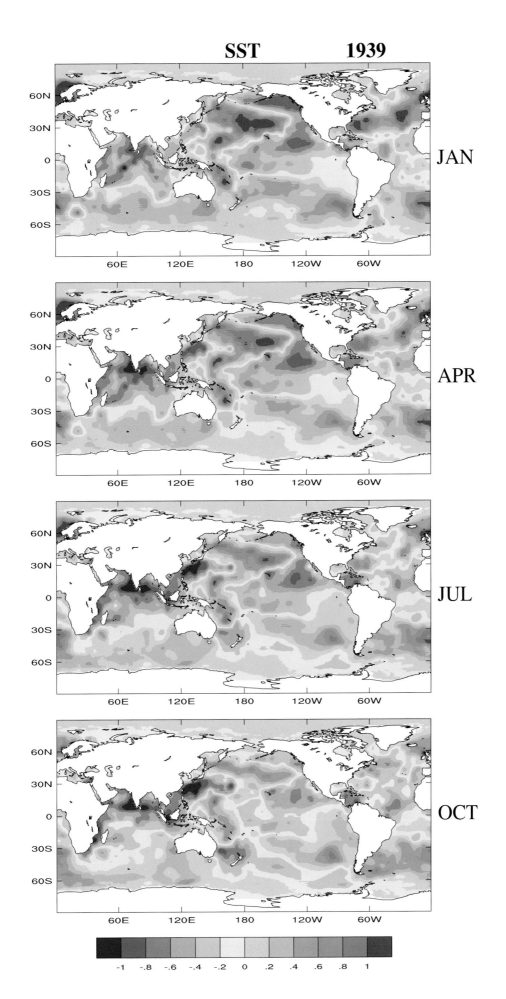

JAN

APR

JUL

OCT

MSLP   **1940**

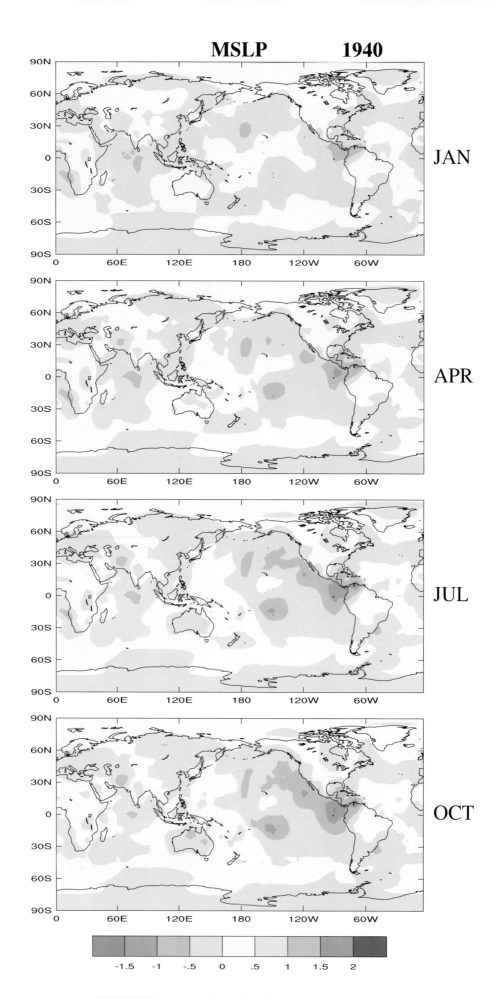

JAN

APR

JUL

OCT

-1.5   -1   -.5   0   .5   1   1.5   2

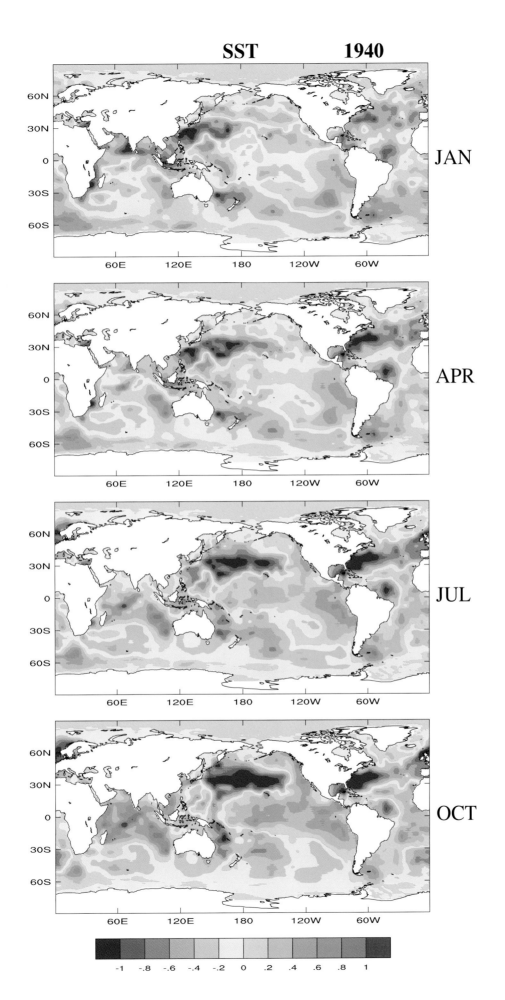

SST   1940

JAN

APR

JUL

OCT

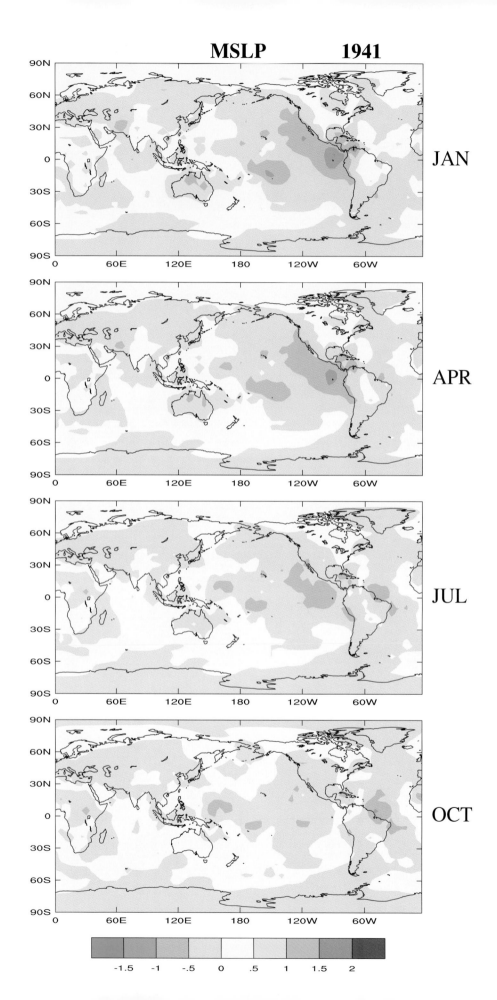

MSLP 1941

JAN

APR

JUL

OCT

-1.5   -1   -.5   0   .5   1   1.5   2

SST   1941

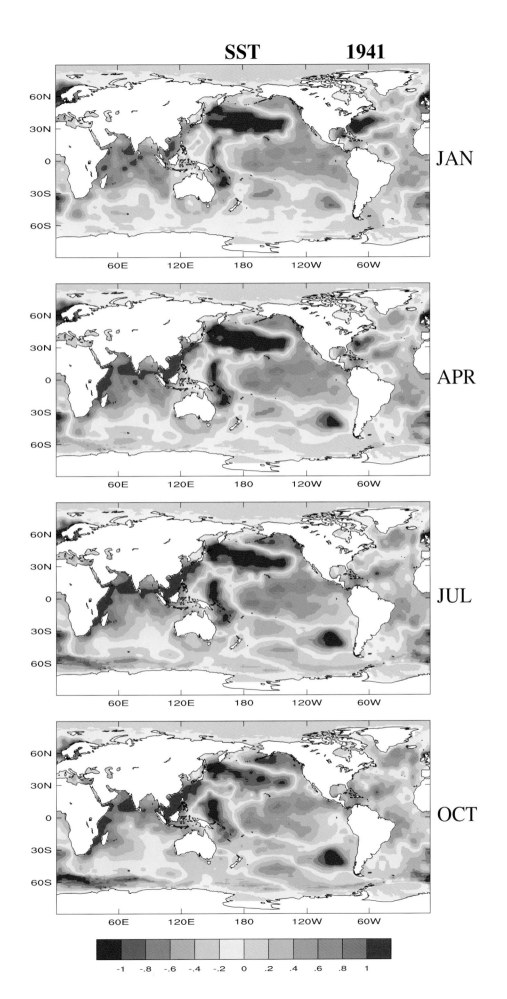

MSLP   **1942**

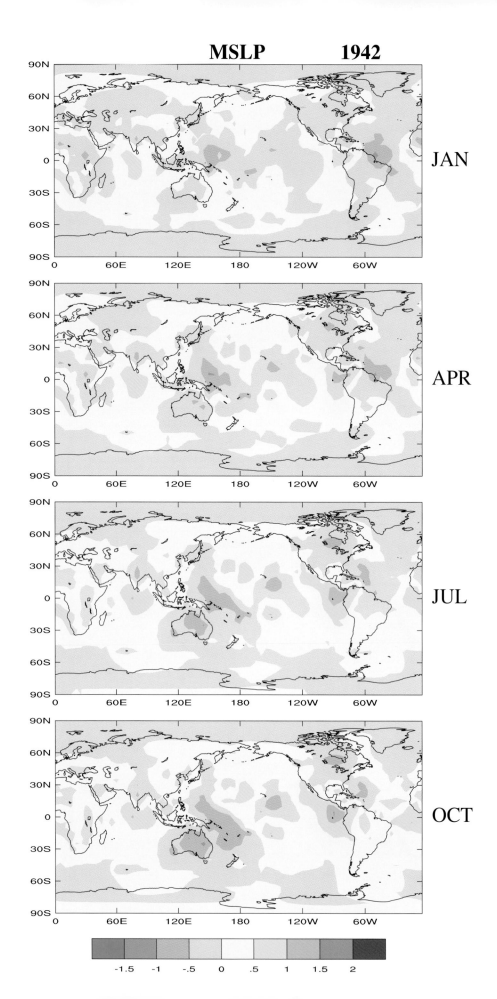

SST   **1942**

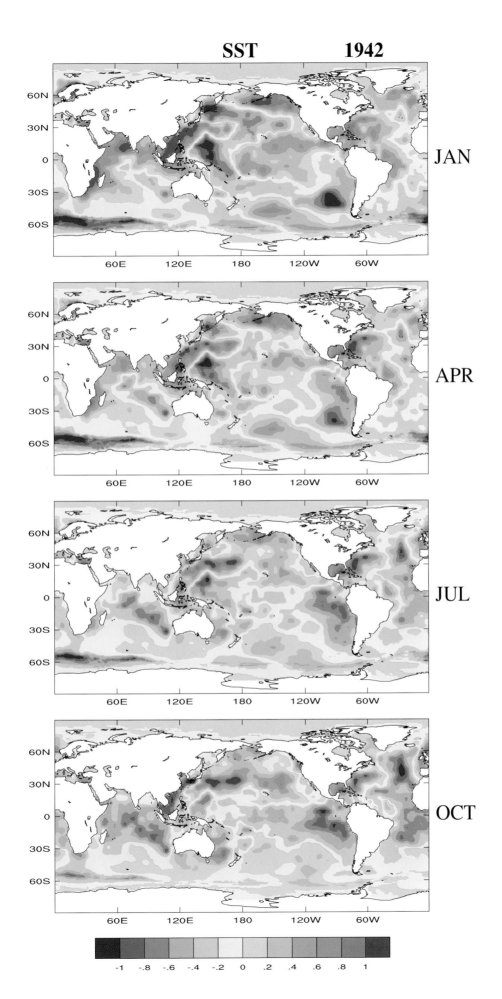

JAN

APR

JUL

OCT

MSLP   **1943**

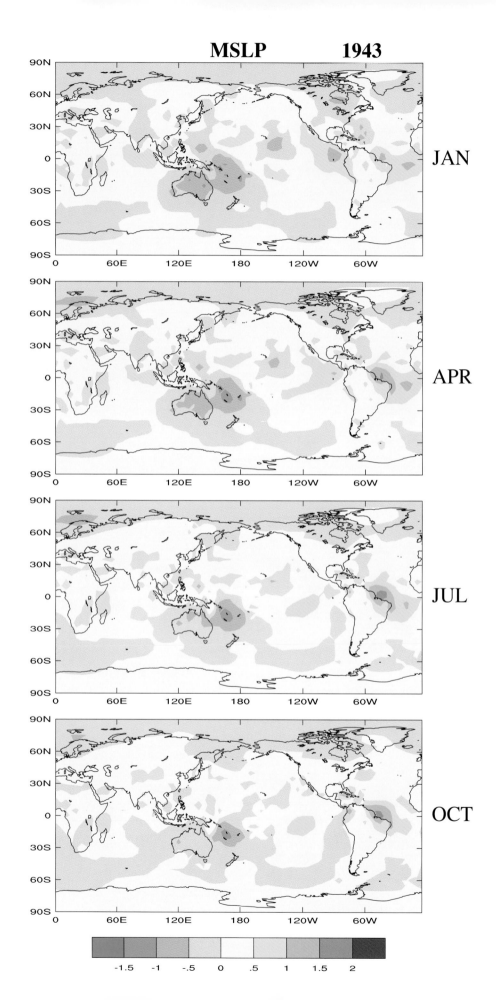

JAN

APR

JUL

OCT

-1.5   -1   -.5   0   .5   1   1.5   2

SST   1943

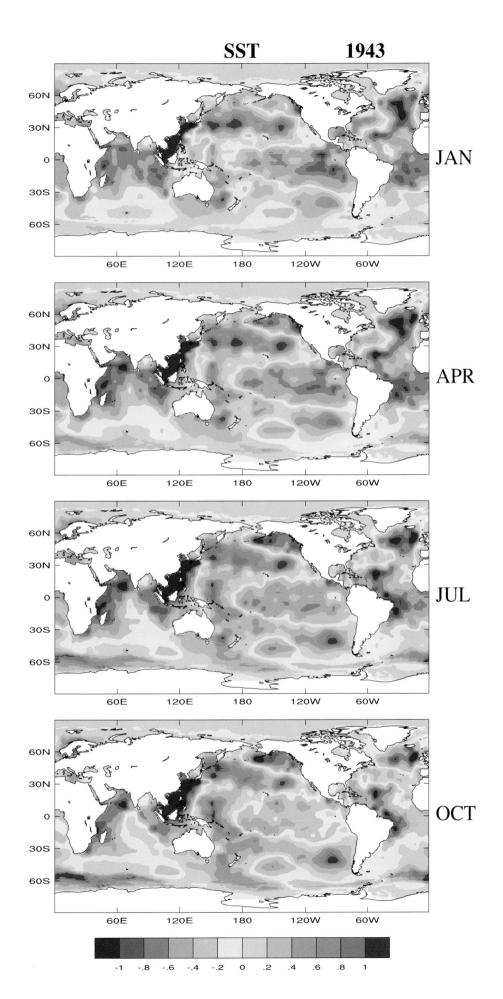

JAN

APR

JUL

OCT

MSLP   **1944**

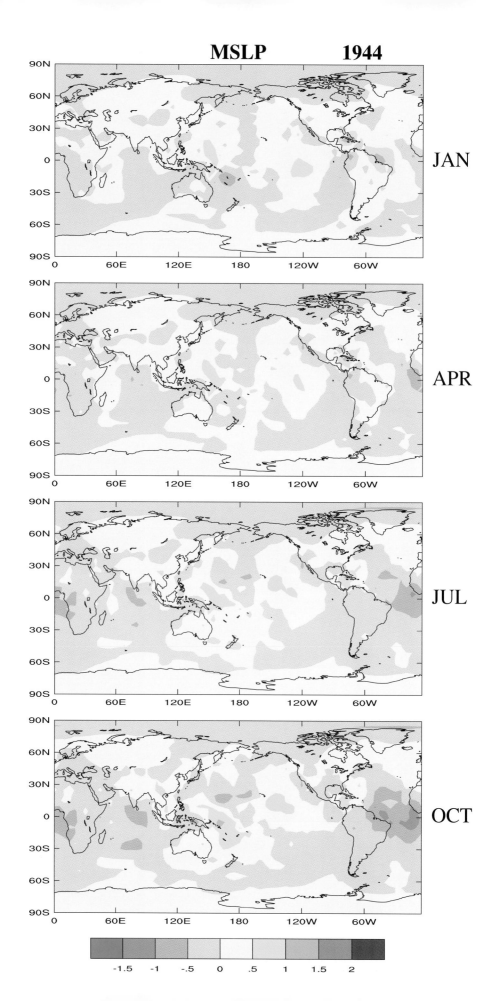

JAN

APR

JUL

OCT

-1.5   -1   -.5   0   .5   1   1.5   2

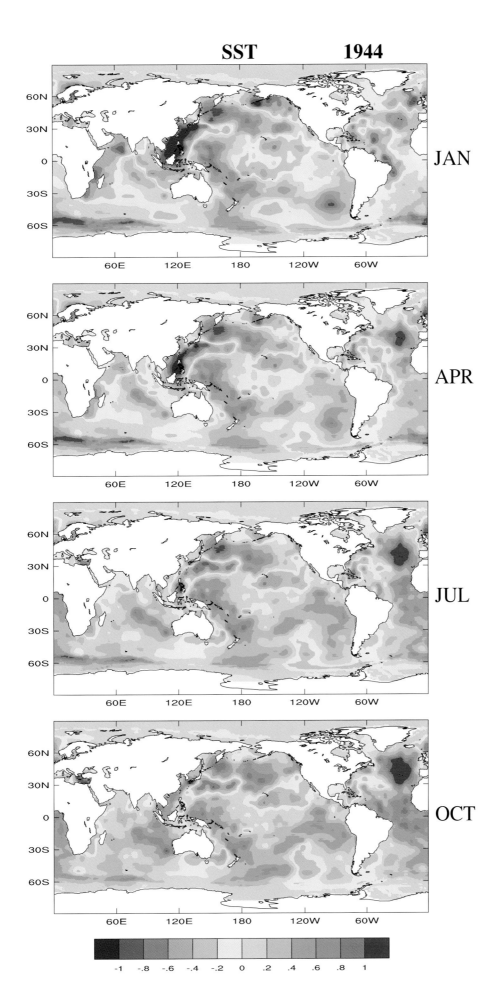

SST   1944

JAN

APR

JUL

OCT

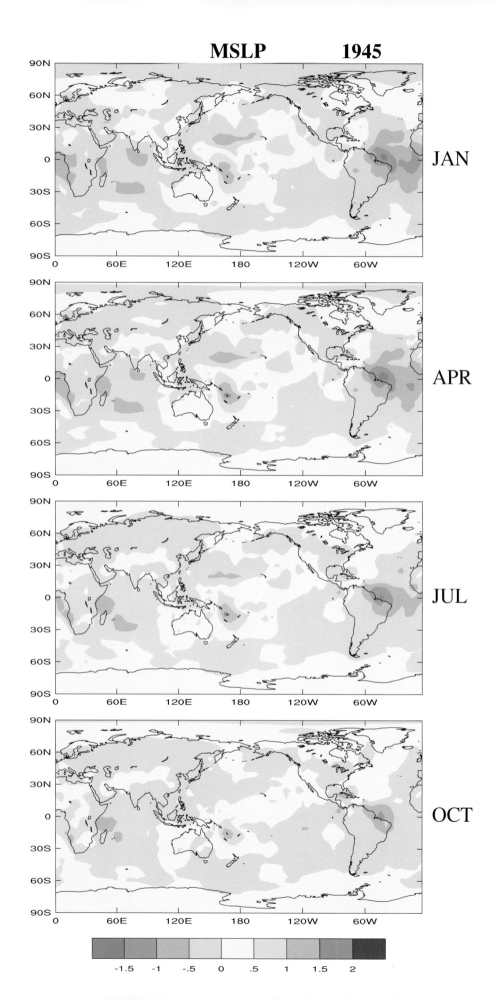

SST   **1945**

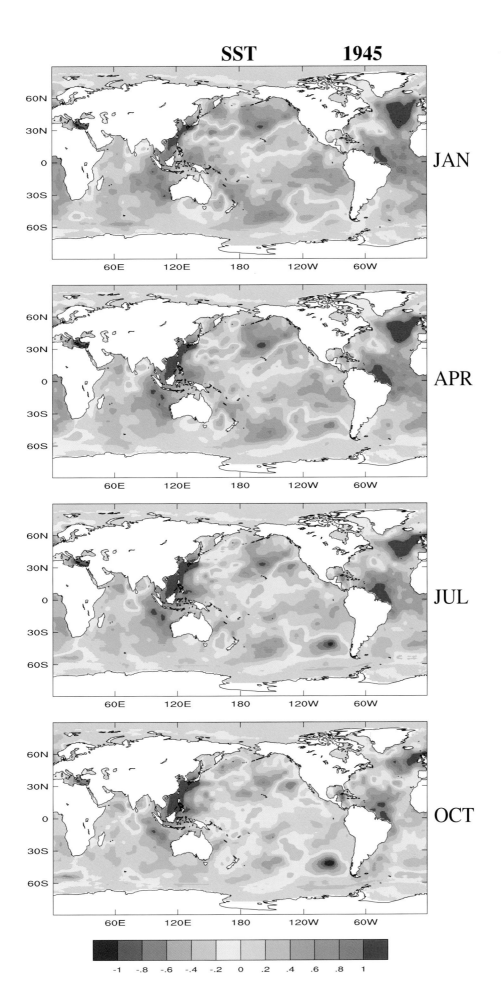

MSLP **1946**

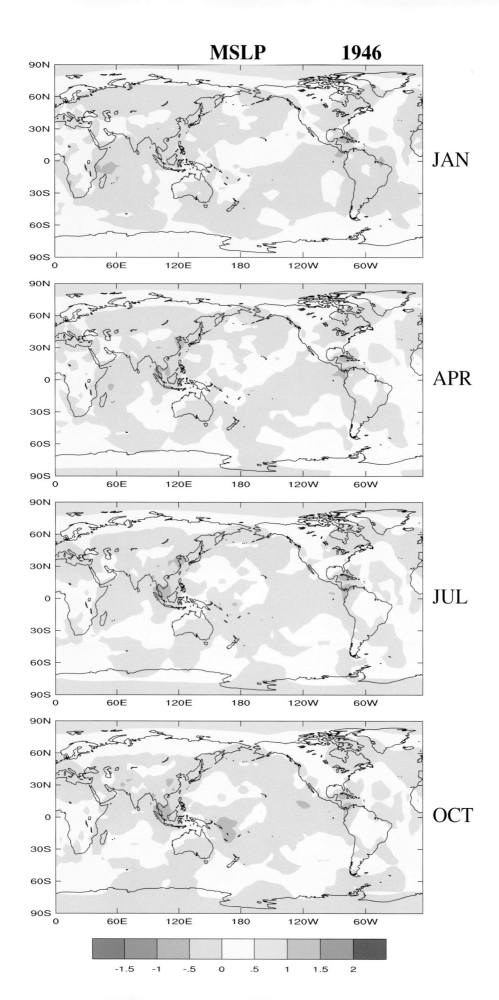

JAN

APR

JUL

OCT

-1.5 -1 -.5 0 .5 1 1.5 2

SST   **1946**

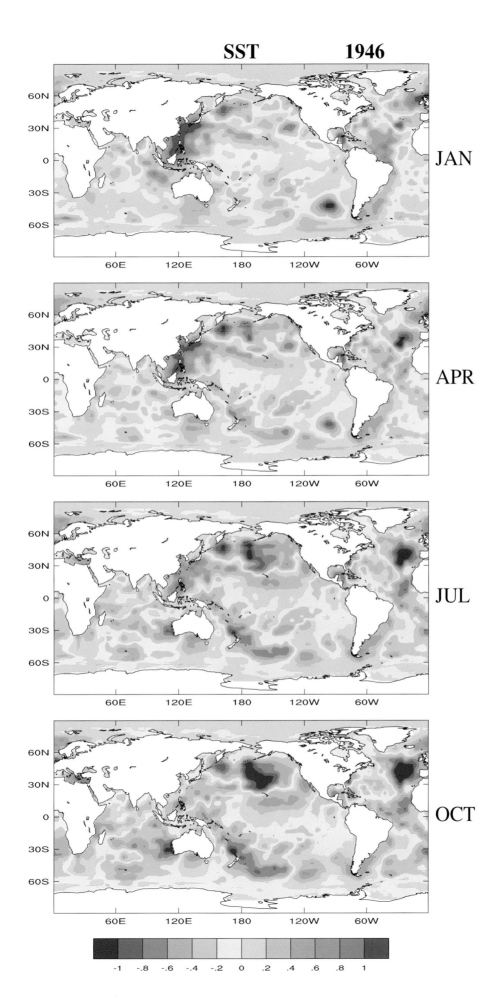

JAN

APR

JUL

OCT

-1   -.8   -.6   -.4   -.2   0   .2   .4   .6   .8   1

309

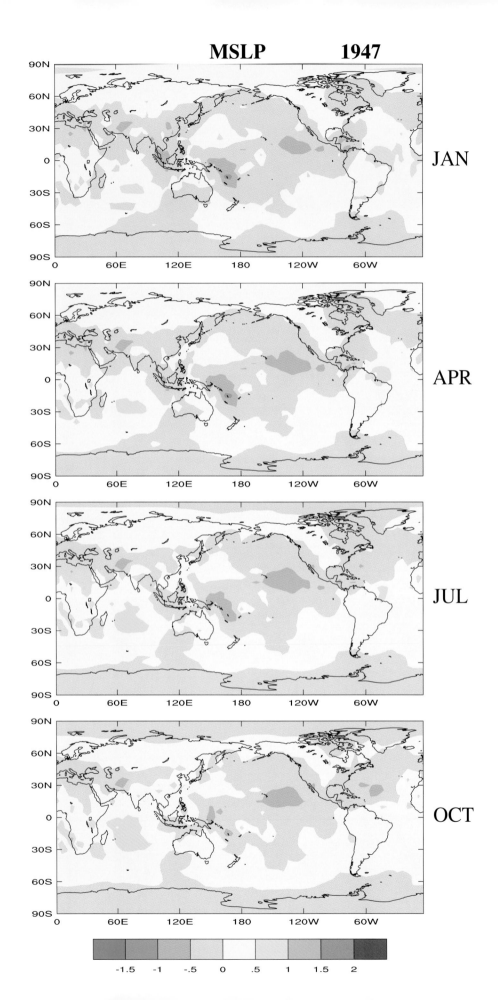

MSLP   1947

JAN

APR

JUL

OCT

-1.5   -1   -.5   0   .5   1   1.5   2

SST   1947

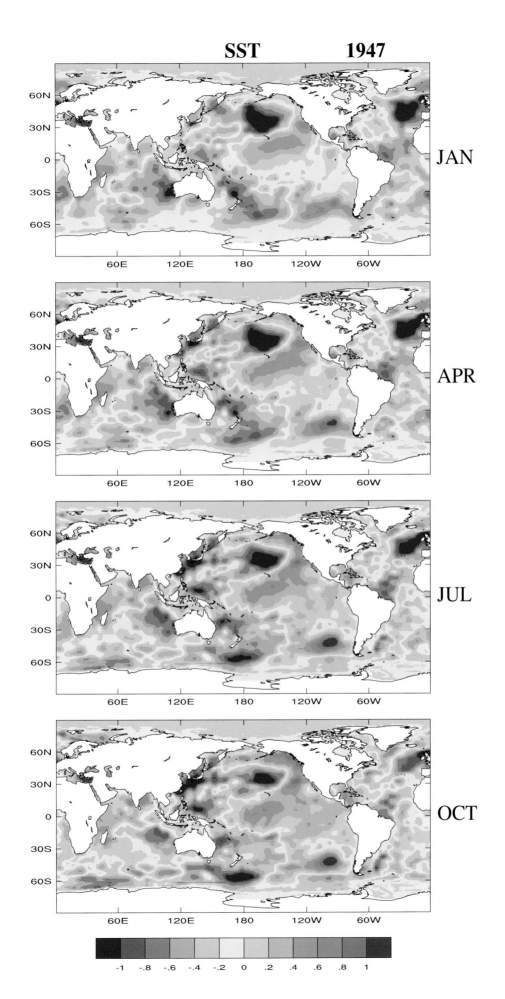

JAN

APR

JUL

OCT

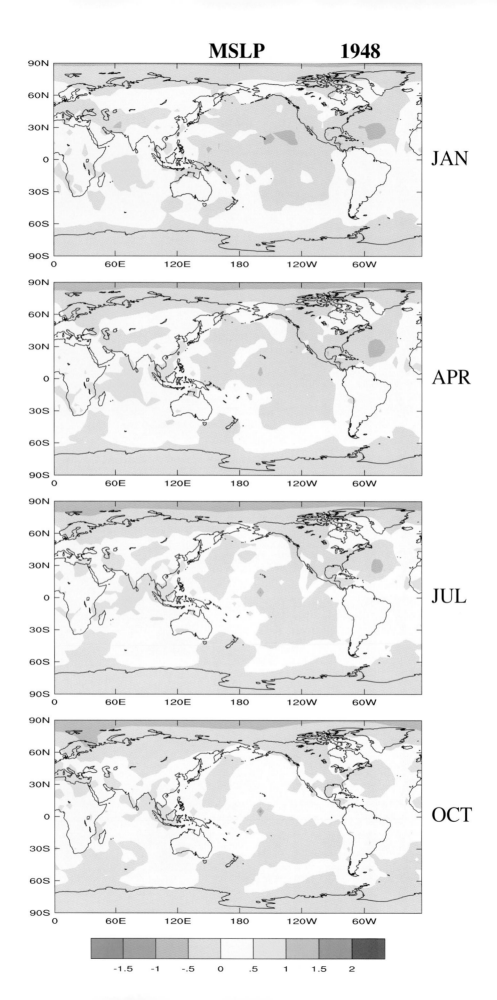

MSLP   1948

JAN

APR

JUL

OCT

-1.5   -1   -.5   0   .5   1   1.5   2

SST   1948

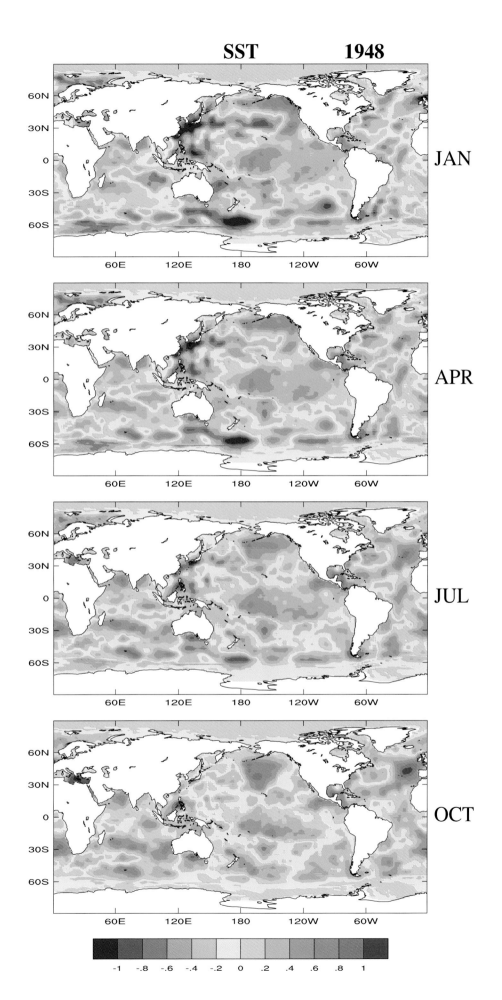

MSLP   **1949**

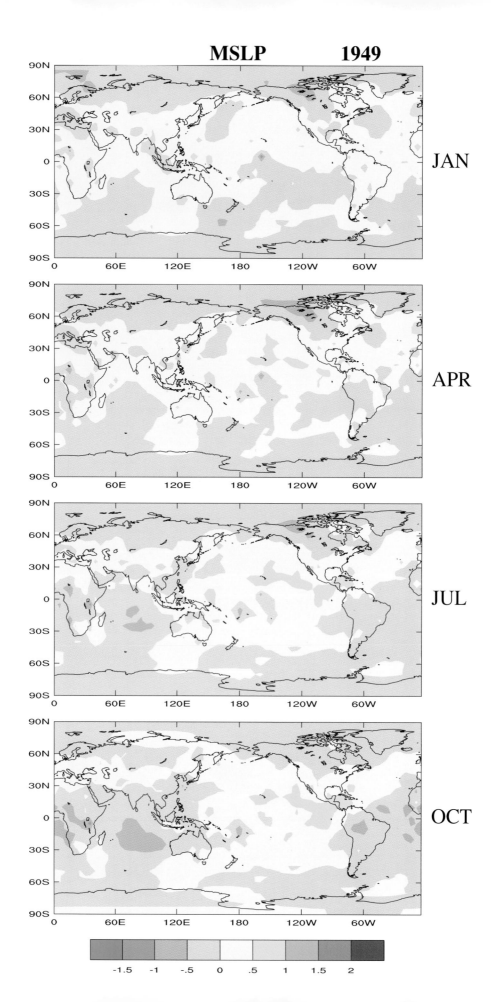

JAN

APR

JUL

OCT

-1.5   -1   -.5   0   .5   1   1.5   2

SST   **1949**

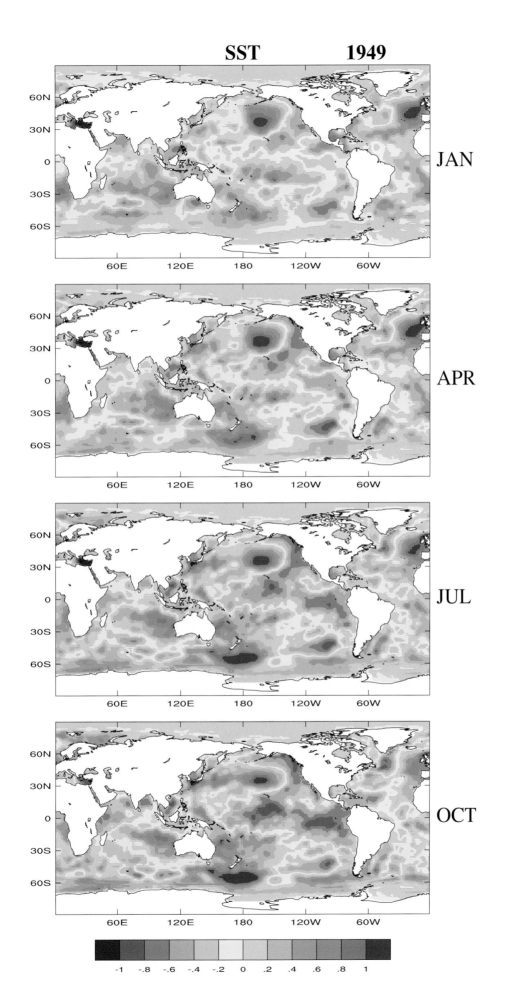

JAN

APR

JUL

OCT

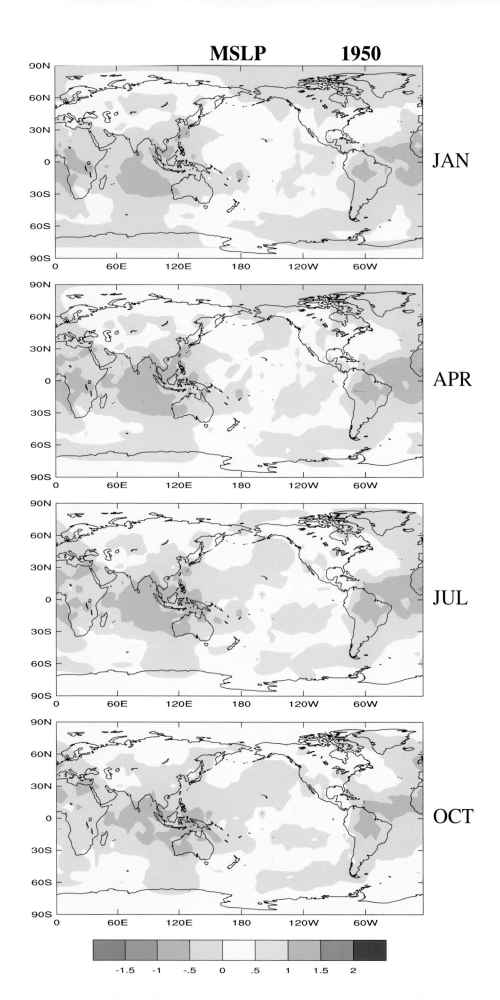

MSLP   1950

JAN

APR

JUL

OCT

-1.5   -1   -.5   0   .5   1   1.5   2

SST   **1950**

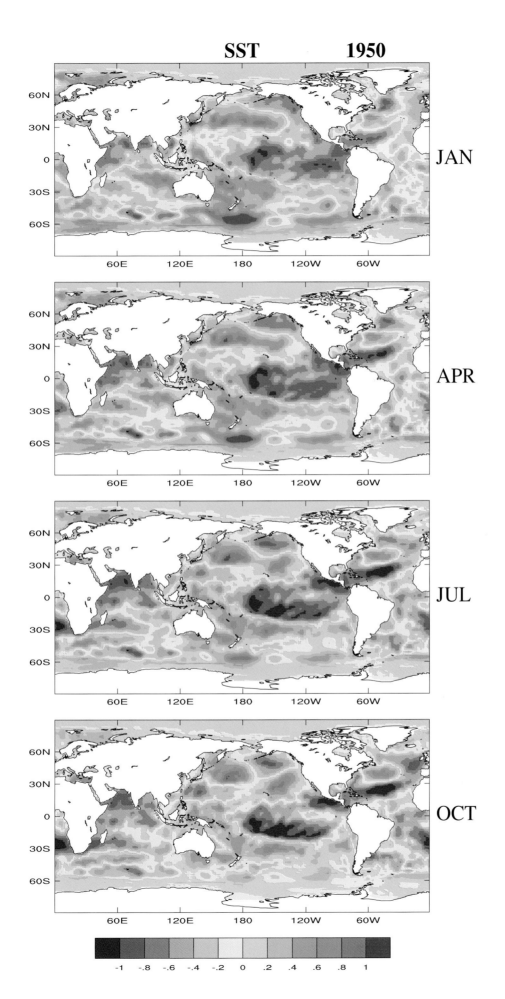

JAN

APR

JUL

OCT

-1   -.8   -.6   -.4   -.2   0   .2   .4   .6   .8   1

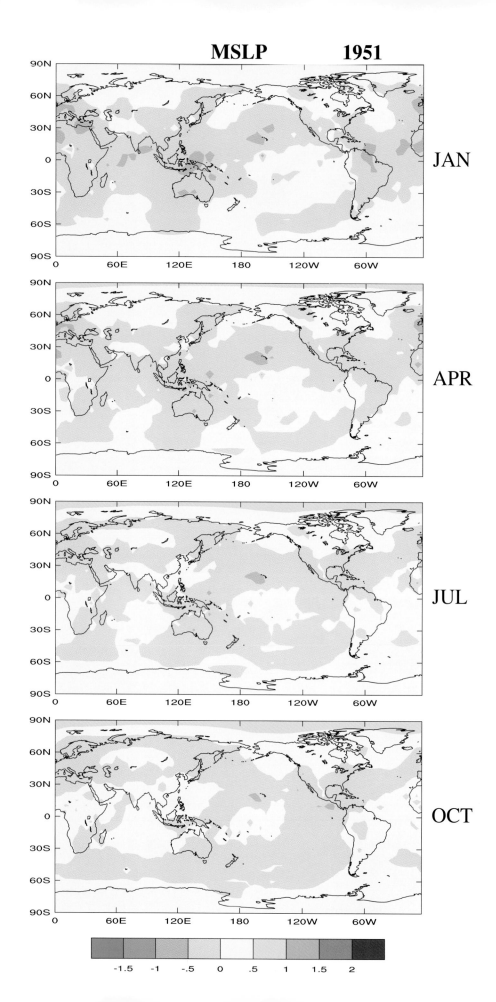

MSLP   1951

JAN

APR

JUL

OCT

-1.5 -1 -.5 0 .5 1 1.5 2

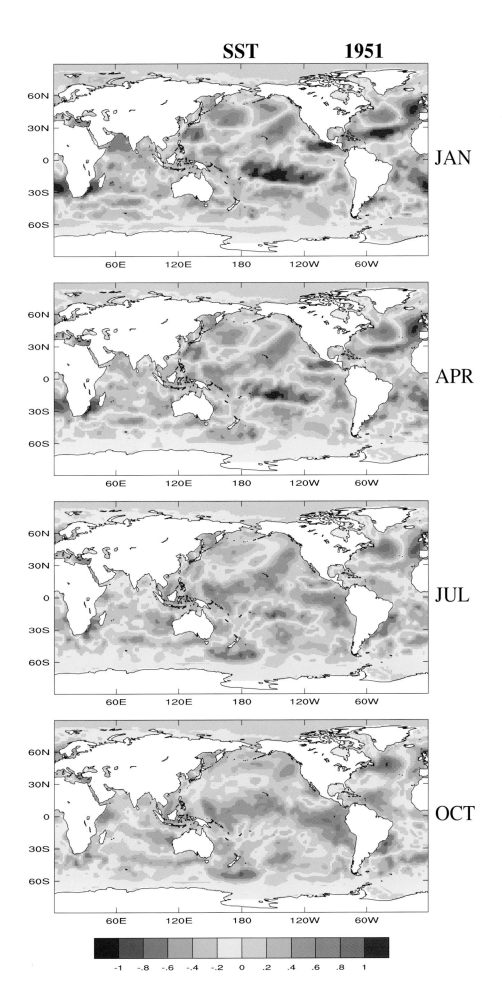

SST   **1951**

JAN

APR

JUL

OCT

-1   -.8   -.6   -.4   -.2   0   .2   .4   .6   .8   1

MSLP   **1952**

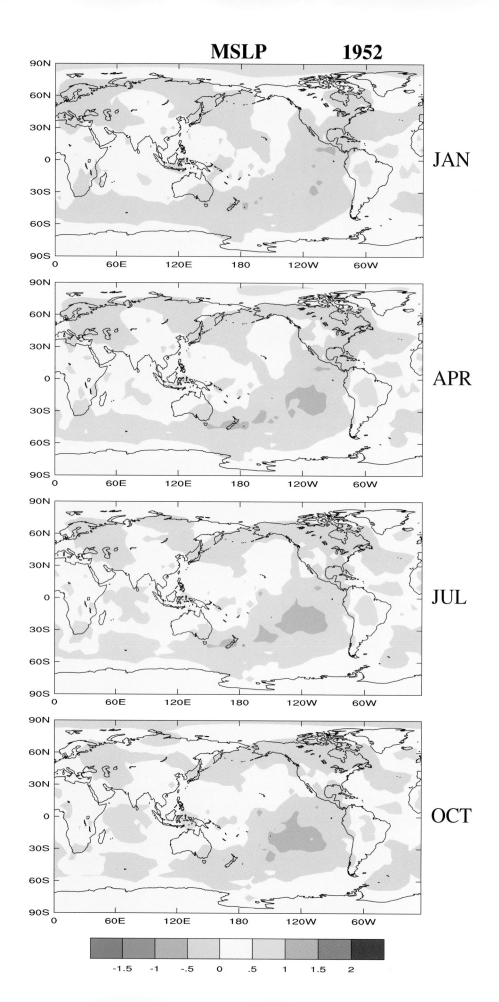

JAN

APR

JUL

OCT

-1.5   -1   -.5   0   .5   1   1.5   2

SST　　**1952**

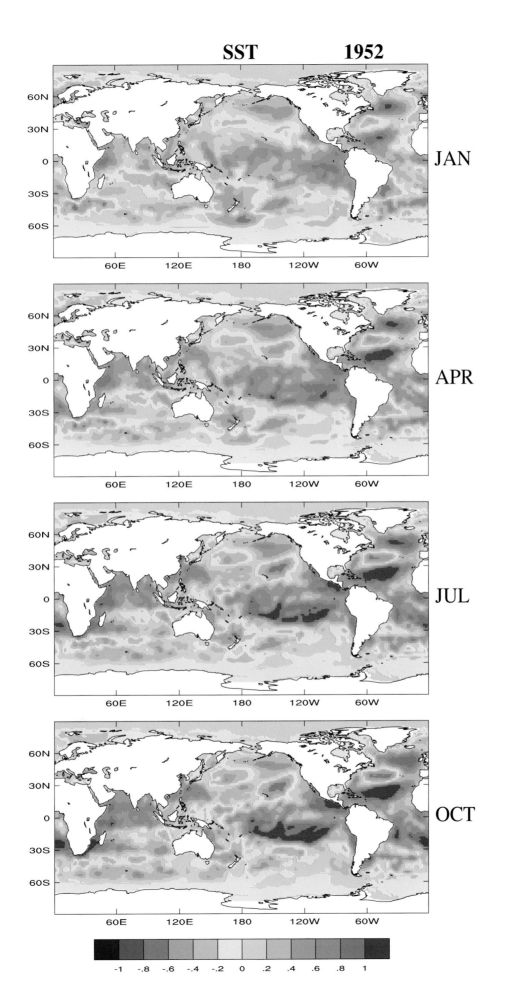

JAN

APR

JUL

OCT

-1　-.8　-.6　-.4　-.2　0　.2　.4　.6　.8　1

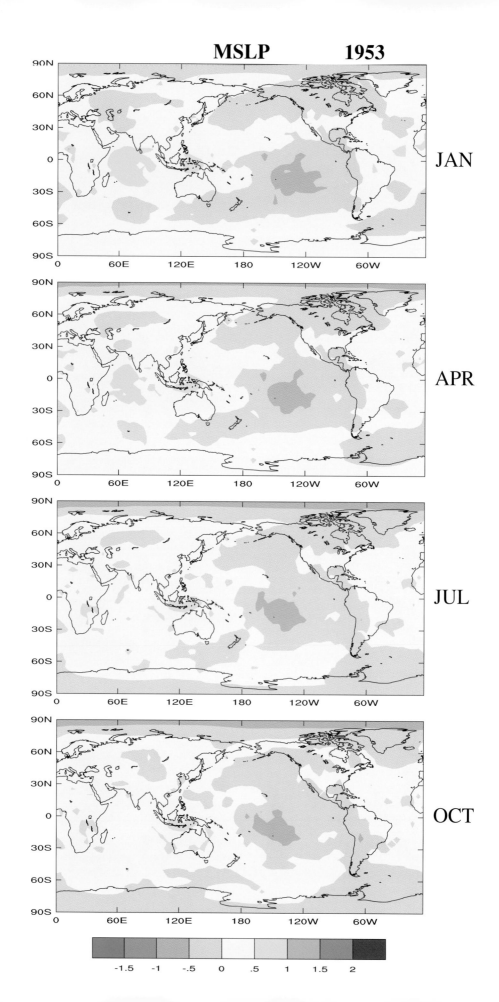

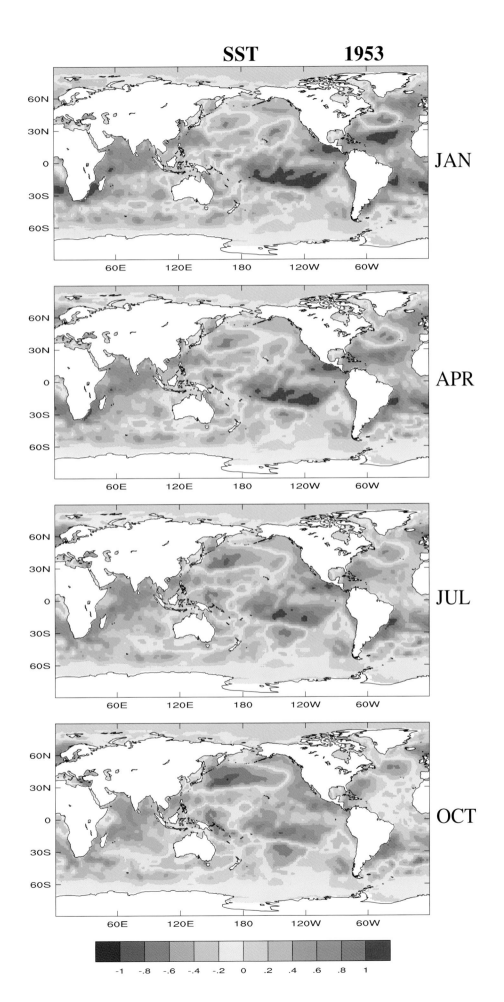

SST   1953

JAN

APR

JUL

OCT

-1 -.8 -.6 -.4 -.2 0 .2 .4 .6 .8 1

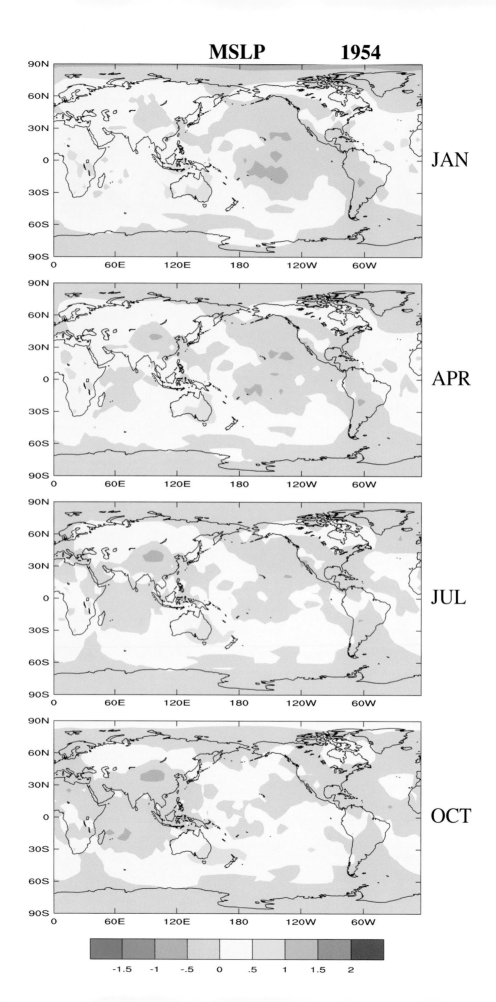

MSLP   **1954**

JAN

APR

JUL

OCT

-1.5   -1   -.5   0   .5   1   1.5   2

SST   1954

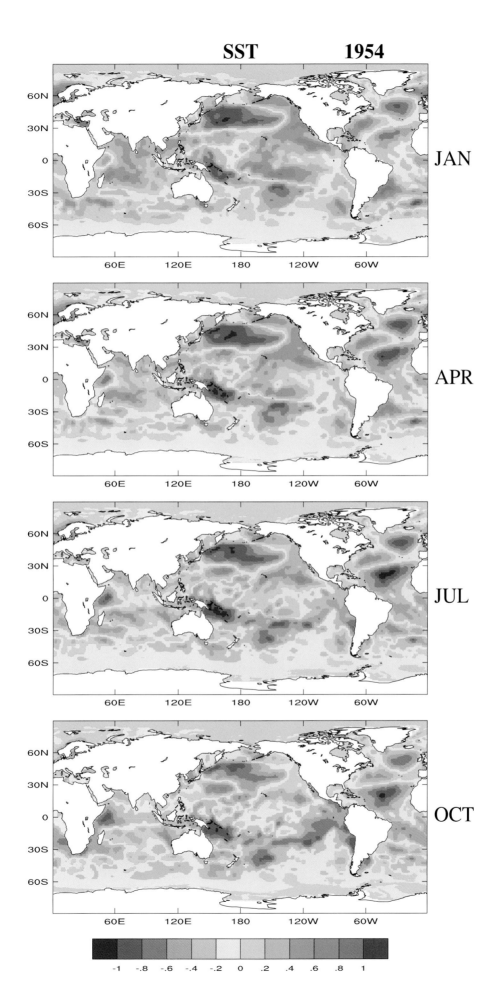

JAN

APR

JUL

OCT

-1 -.8 -.6 -.4 -.2 0 .2 .4 .6 .8 1

MSLP   **1955**

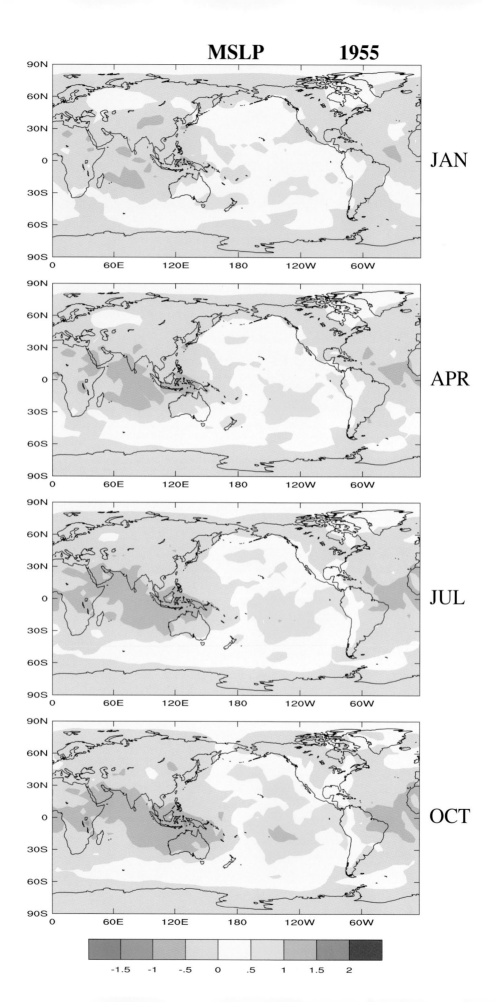

JAN

APR

JUL

OCT

-1.5   -1   -.5   0   .5   1   1.5   2

SST   1955

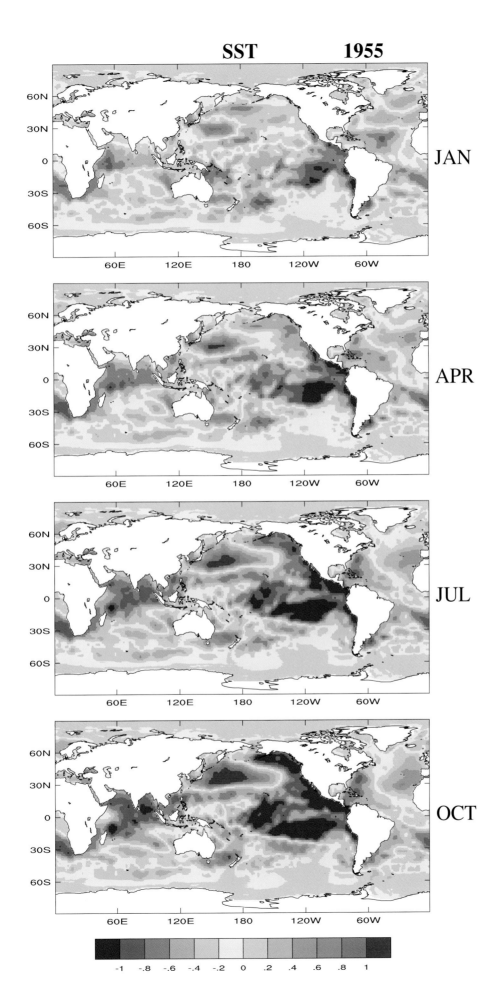

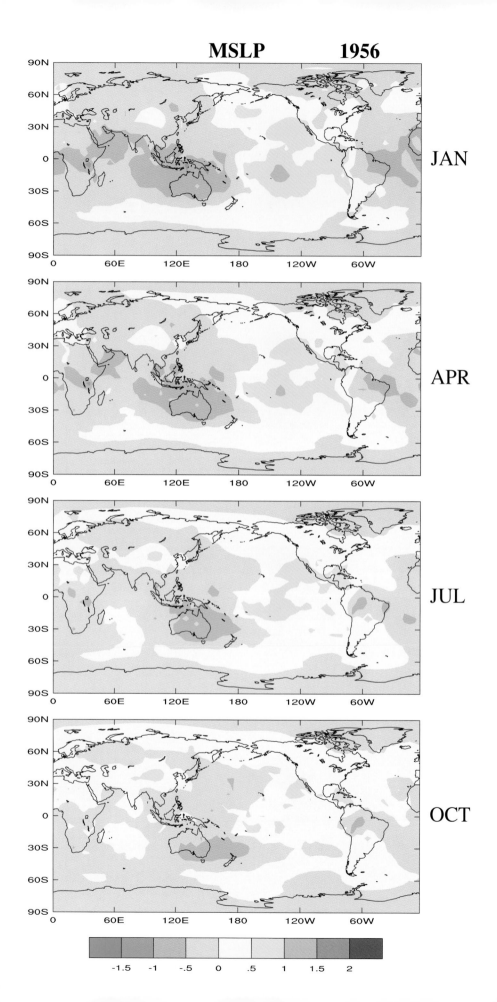

MSLP 1956

JAN

APR

JUL

OCT

-1.5 -1 -.5 0 .5 1 1.5 2

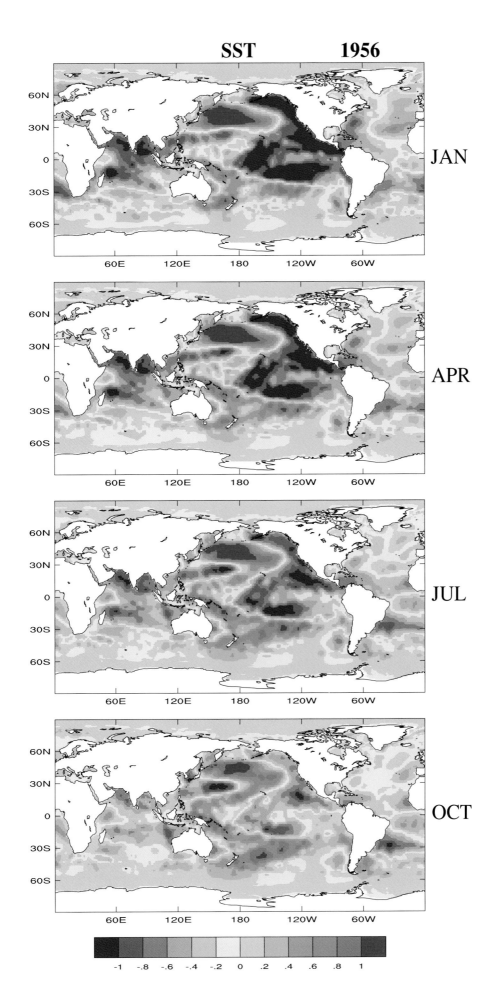

SST   1956

JAN

APR

JUL

OCT

-1 -.8 -.6 -.4 -.2 0 .2 .4 .6 .8 1

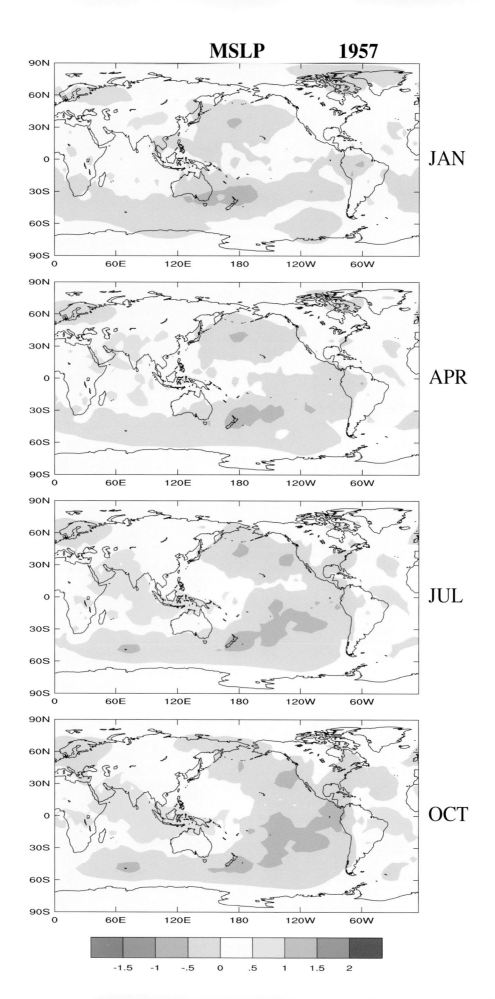

SST   1957

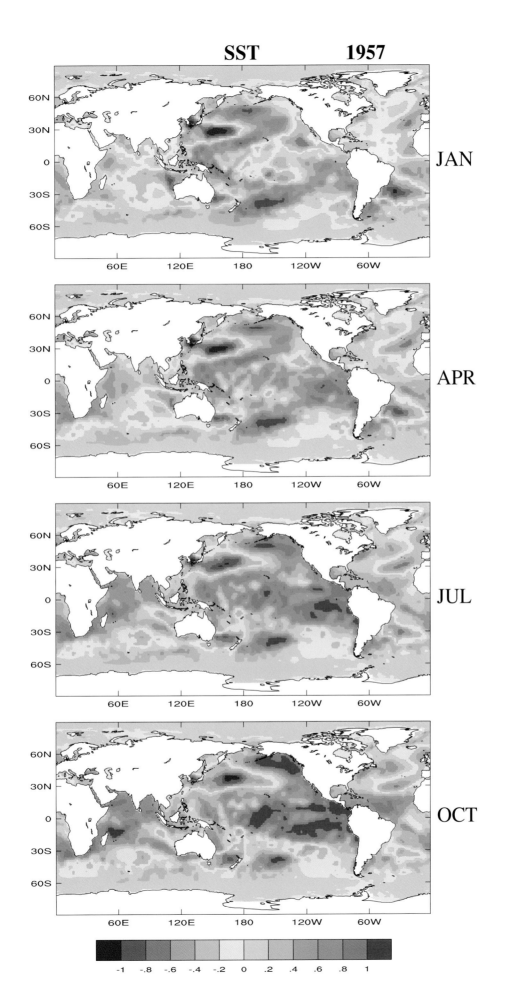

JAN

APR

JUL

OCT

MSLP   1958

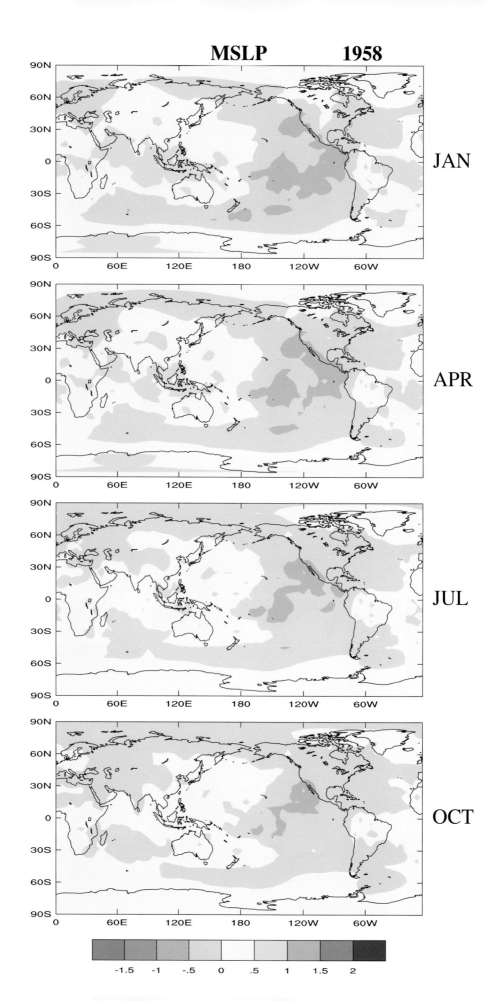

JAN

APR

JUL

OCT

-1.5   -1   -.5   0   .5   1   1.5   2

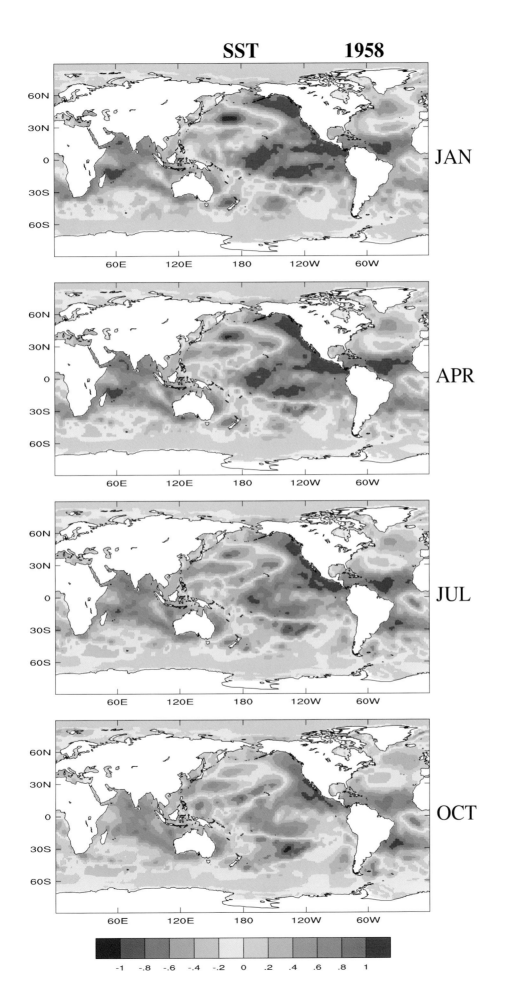

SST   **1958**

JAN

APR

JUL

OCT

-1   -.8   -.6   -.4   -.2   0   .2   .4   .6   .8   1

MSLP   **1959**

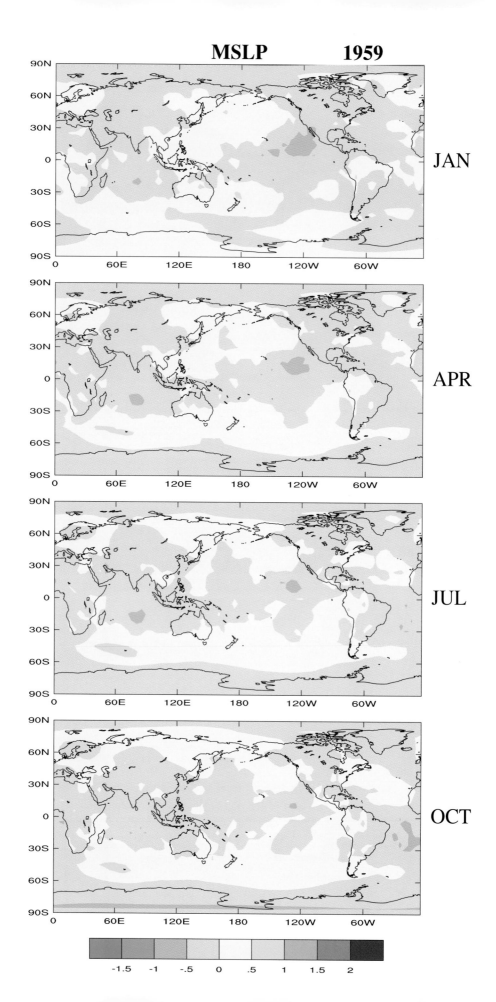

JAN

APR

JUL

OCT

-1.5   -1   -.5   0   .5   1   1.5   2

SST   1959

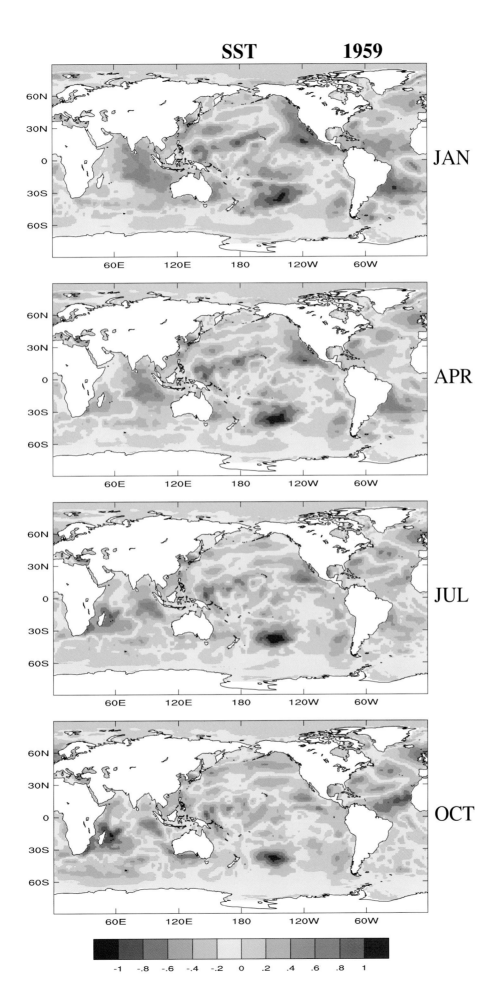

JAN

APR

JUL

OCT

-1 -.8 -.6 -.4 -.2 0 .2 .4 .6 .8 1

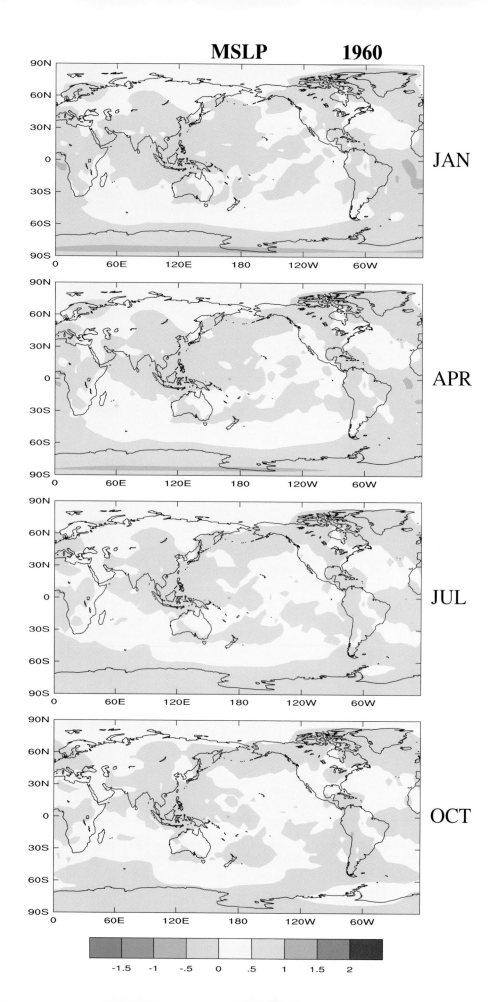

SST   **1960**

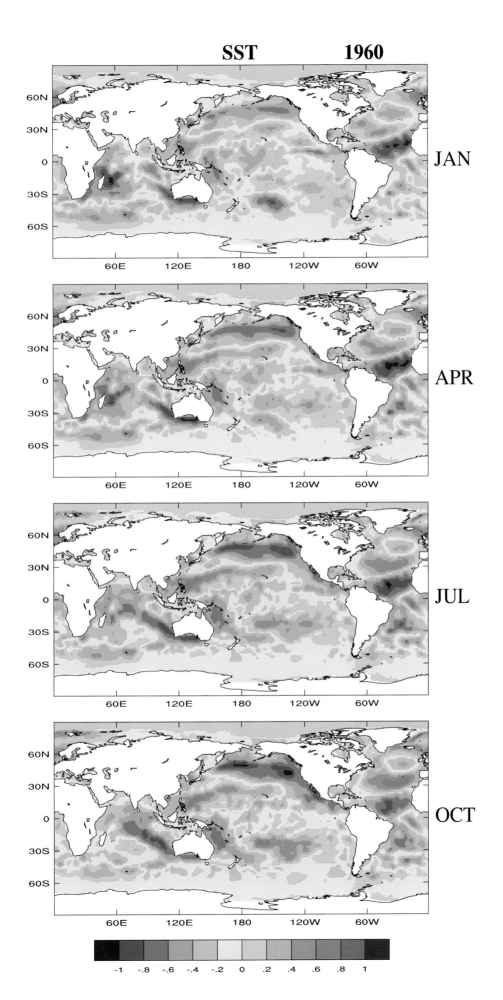

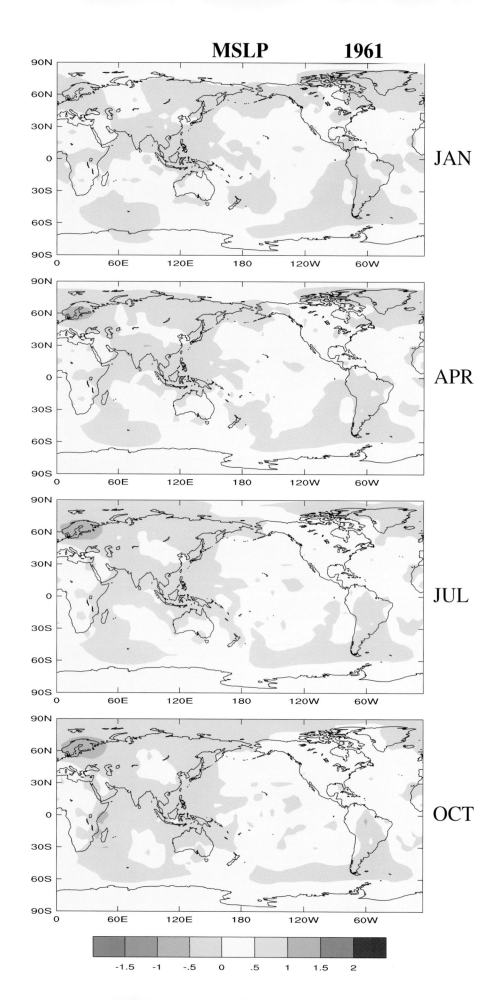

SST   **1961**

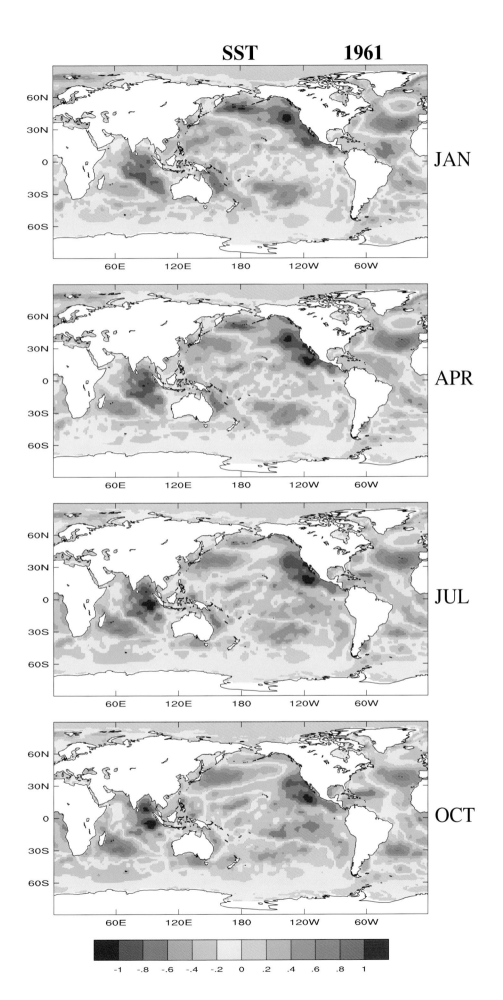

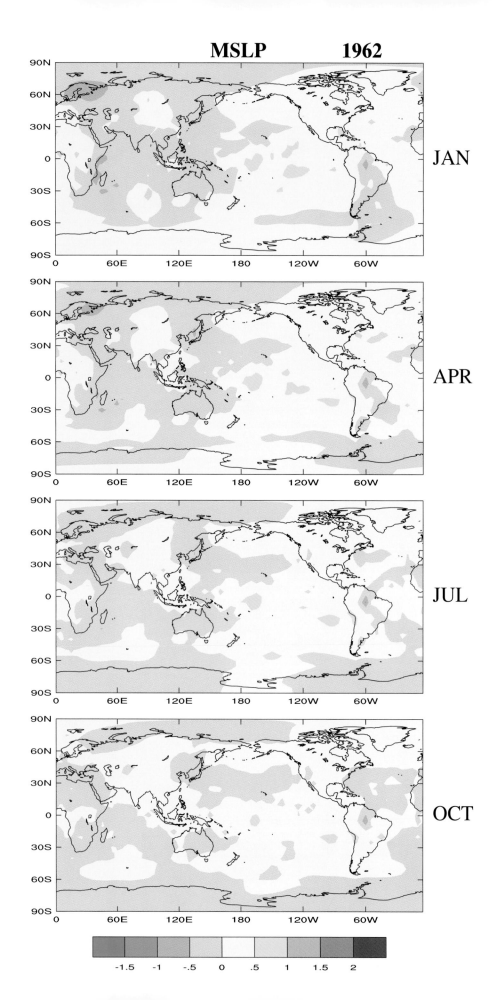

MSLP   1962

JAN

APR

JUL

OCT

-1.5   -1   -.5   0   .5   1   1.5   2

340

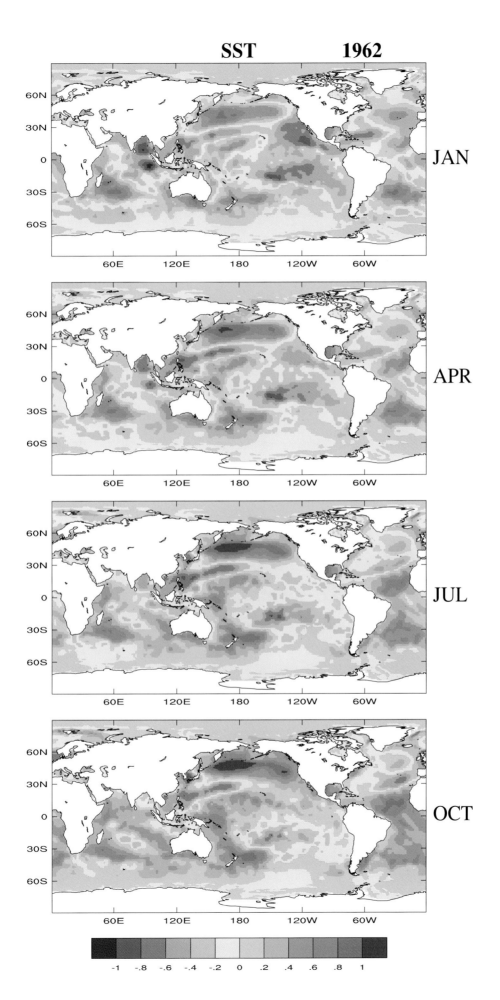

SST   1962

JAN

APR

JUL

OCT

-1 -.8 -.6 -.4 -.2 0 .2 .4 .6 .8 1

MSLP 1963

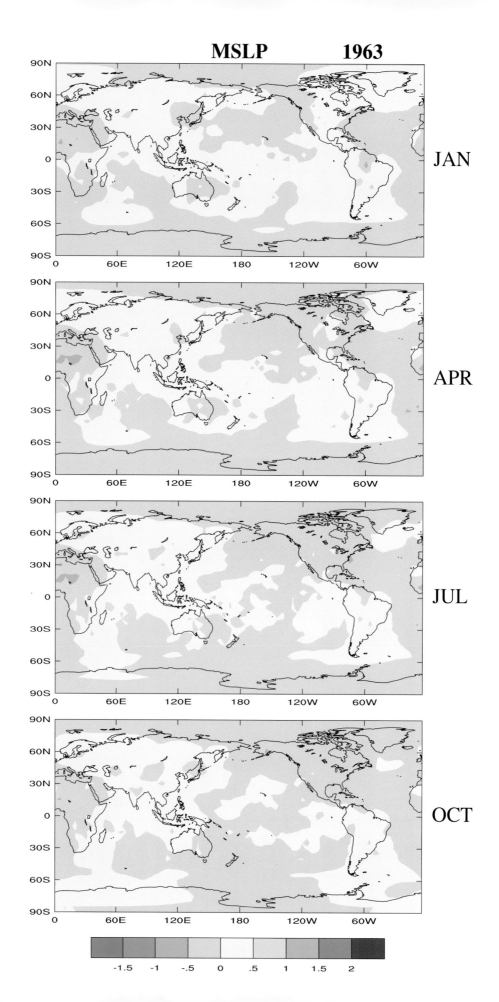

JAN

APR

JUL

OCT

-1.5 -1 -.5 0 .5 1 1.5 2

342

SST   1963

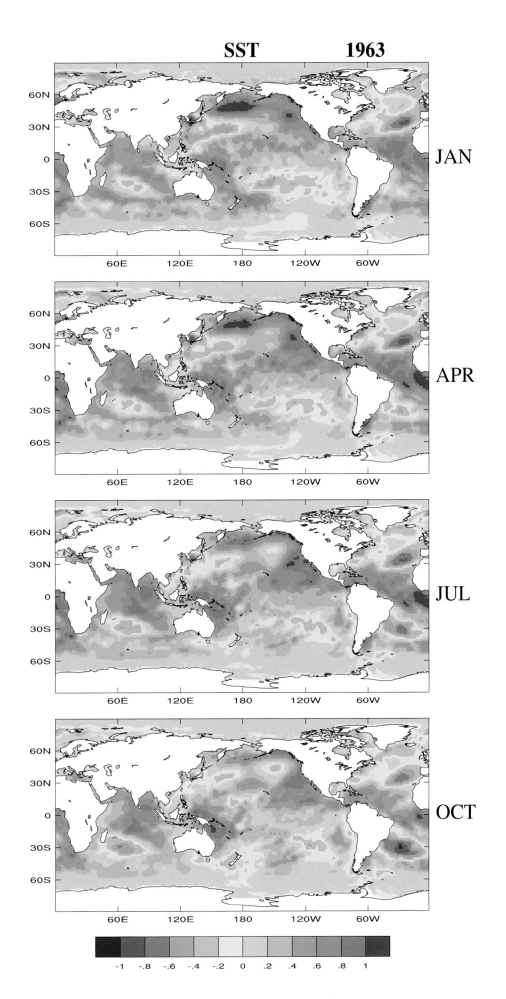

JAN

APR

JUL

OCT

-1 -.8 -.6 -.4 -.2 0 .2 .4 .6 .8 1

343

MSLP   1964

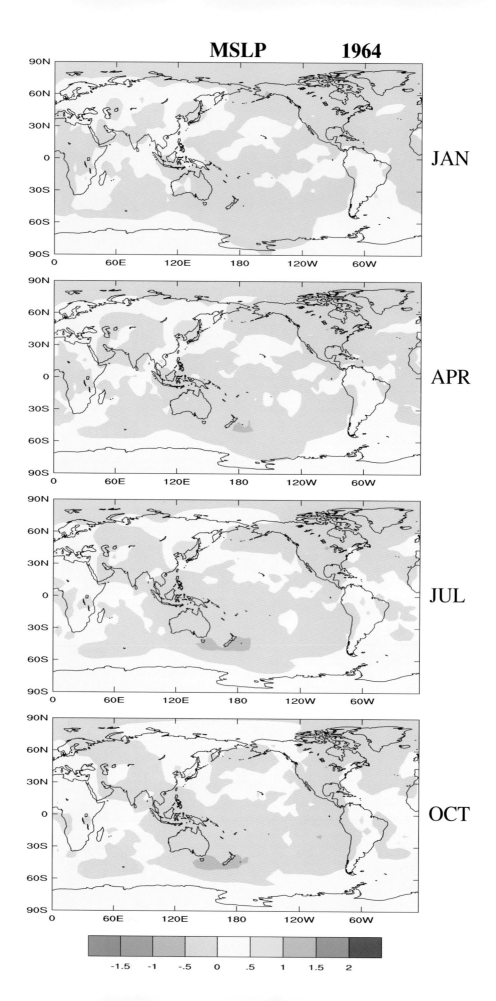

JAN

APR

JUL

OCT

-1.5   -1   -.5   0   .5   1   1.5   2

SST 1964

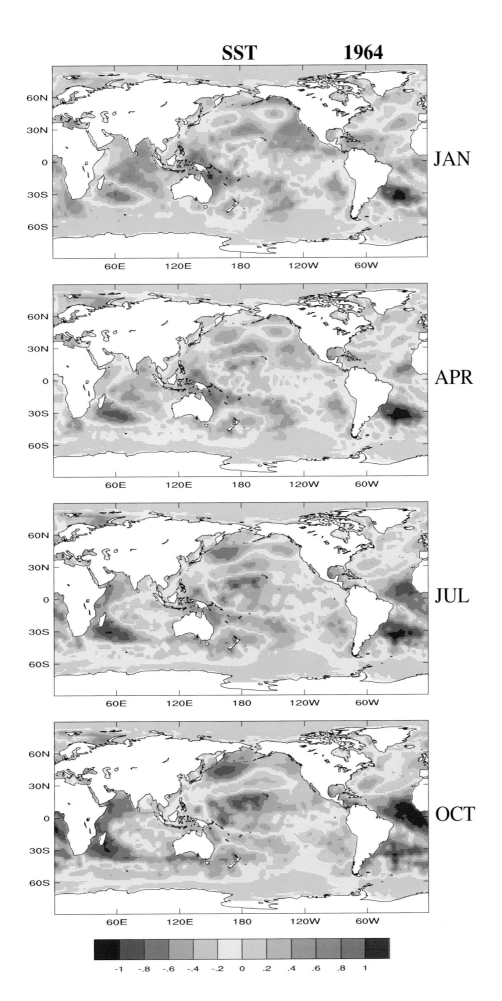

JAN

APR

JUL

OCT

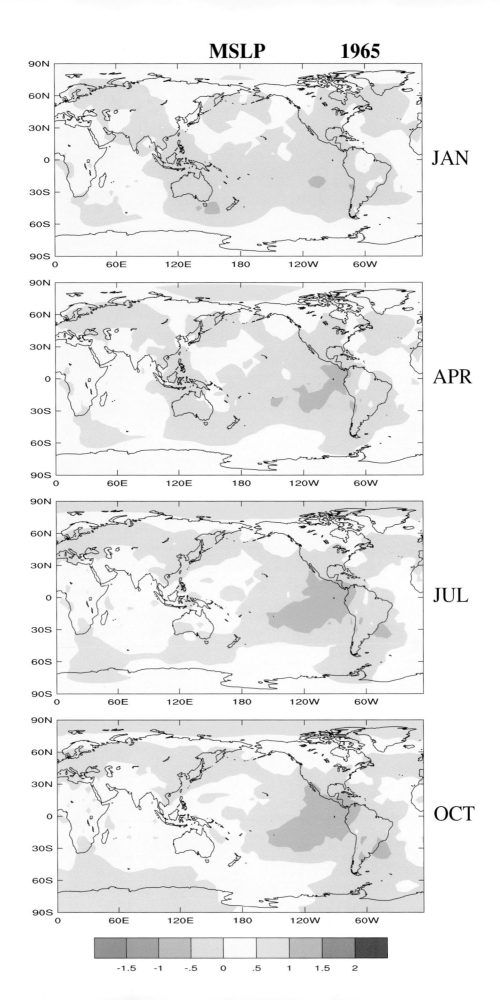

SST　　　**1965**

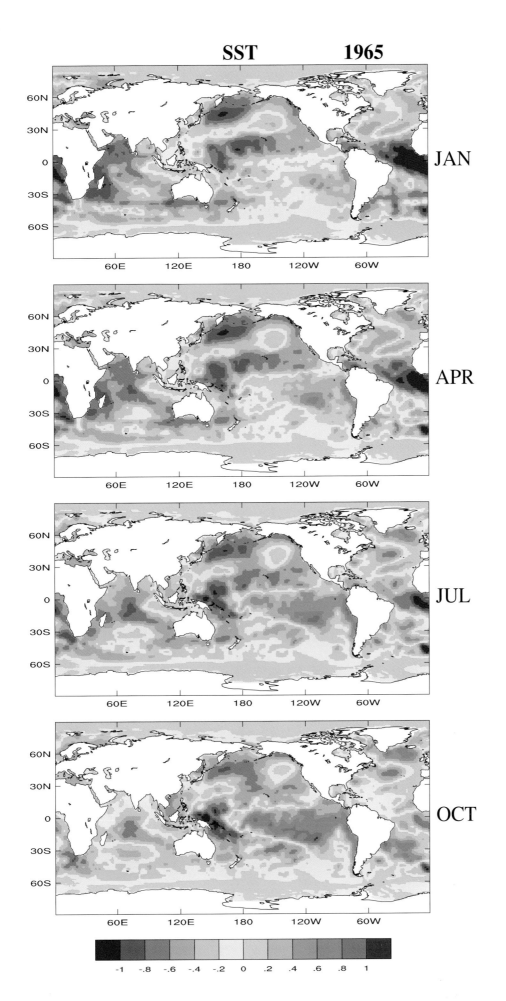

JAN

APR

JUL

OCT

$$-1 \quad -.8 \quad -.6 \quad -.4 \quad -.2 \quad 0 \quad .2 \quad .4 \quad .6 \quad .8 \quad 1$$

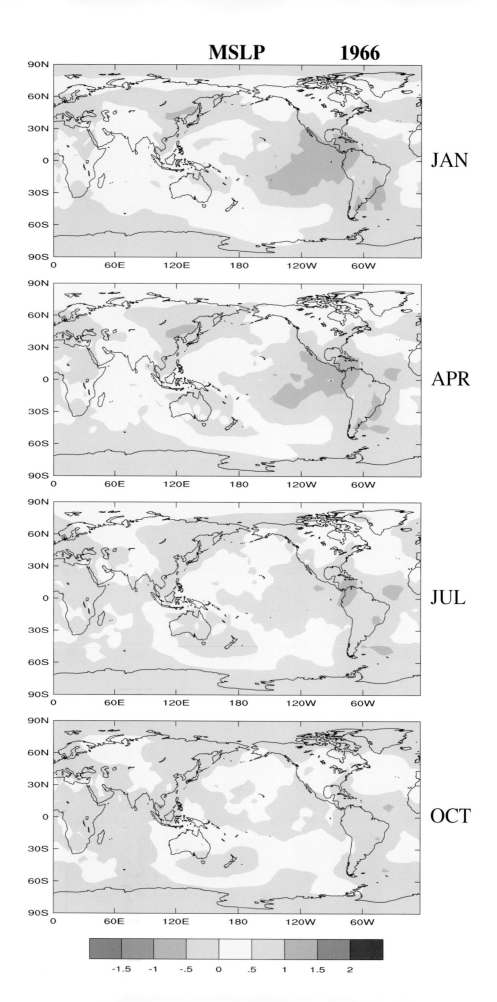

MSLP   1966

JAN

APR

JUL

OCT

-1.5   -1   -.5   0   .5   1   1.5   2

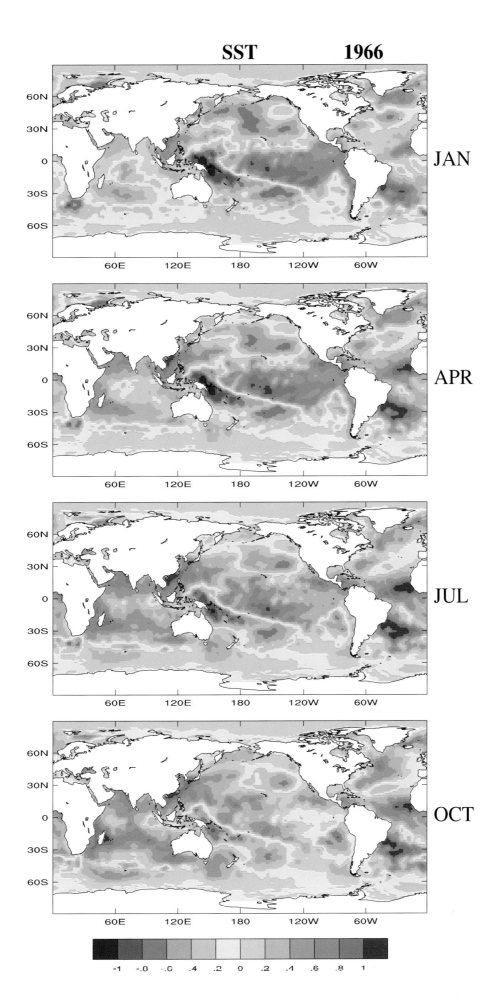

SST   **1966**

JAN

APR

JUL

OCT

-1   -.0   -.6   .4   .2   0   .2   .4   .6   .8   1

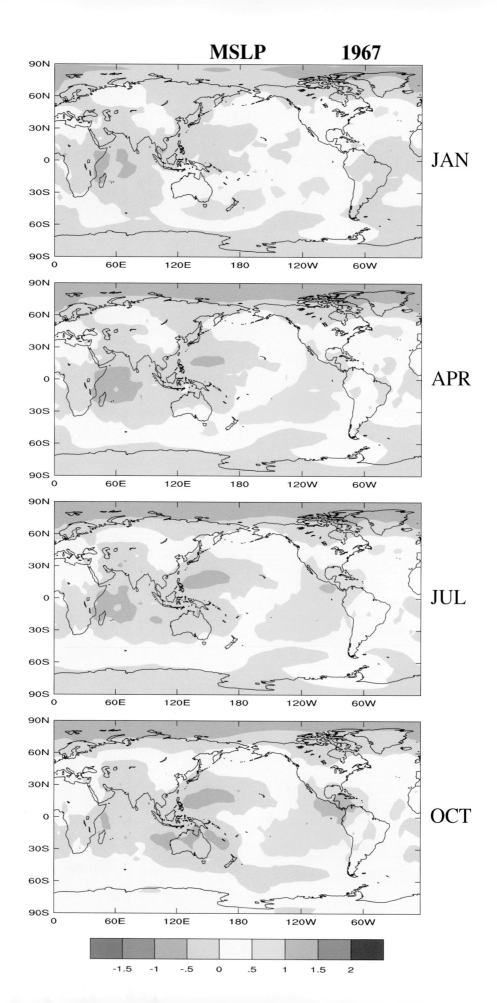

MSLP　　　　1967

JAN

APR

JUL

OCT

-1.5　-1　-.5　0　.5　1　1.5　2

350

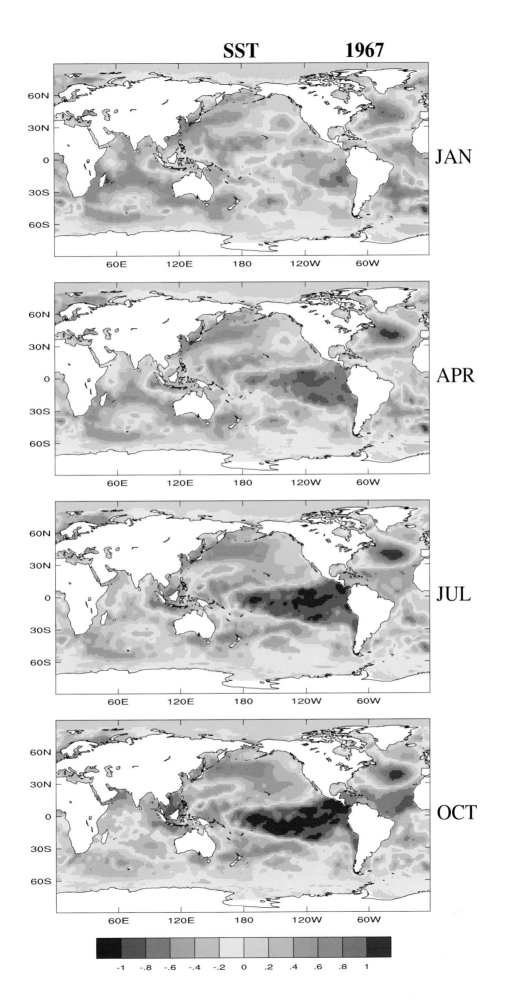

SST   1967

JAN

APR

JUL

OCT

MSLP   **1968**

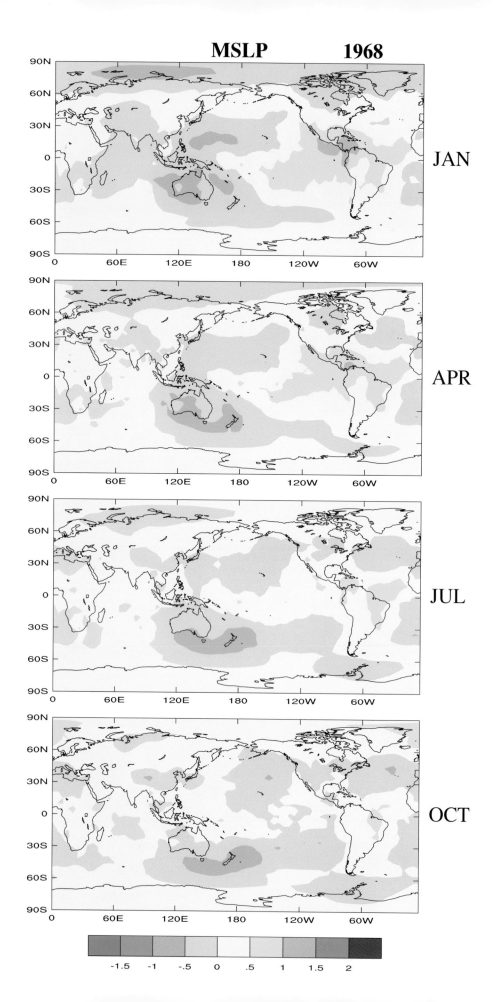

JAN

APR

JUL

OCT

-1.5   -1   -.5   0   .5   1   1.5   2

SST   **1968**

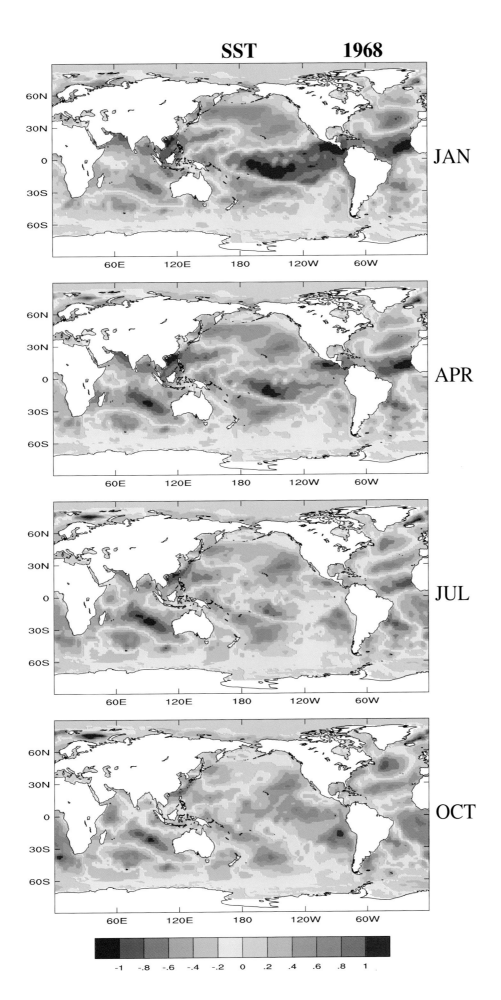

JAN

APR

JUL

OCT

-1 -.8 -.6 -.4 -.2 0 .2 .4 .6 .8 1

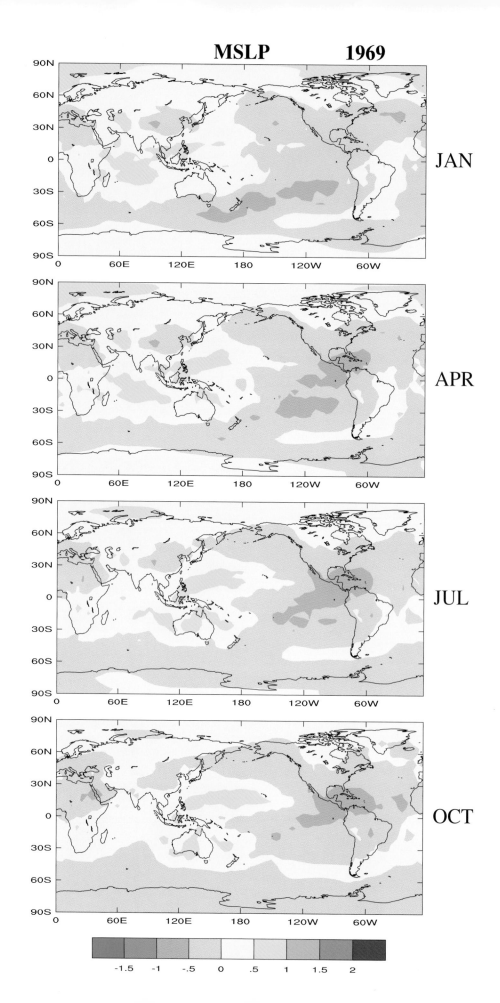

MSLP   **1969**

JAN

APR

JUL

OCT

-1.5 -1 -.5 0 .5 1 1.5 2

354

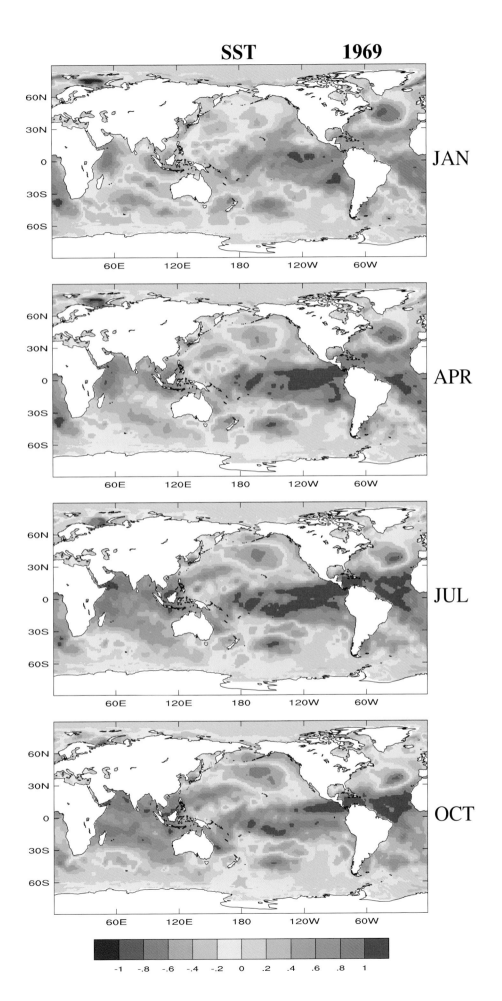

SST   1969

JAN

APR

JUL

OCT

-1 -.8 -.6 -.4 -.2   0   .2   .4   .6   .8   1

MSLP   1970

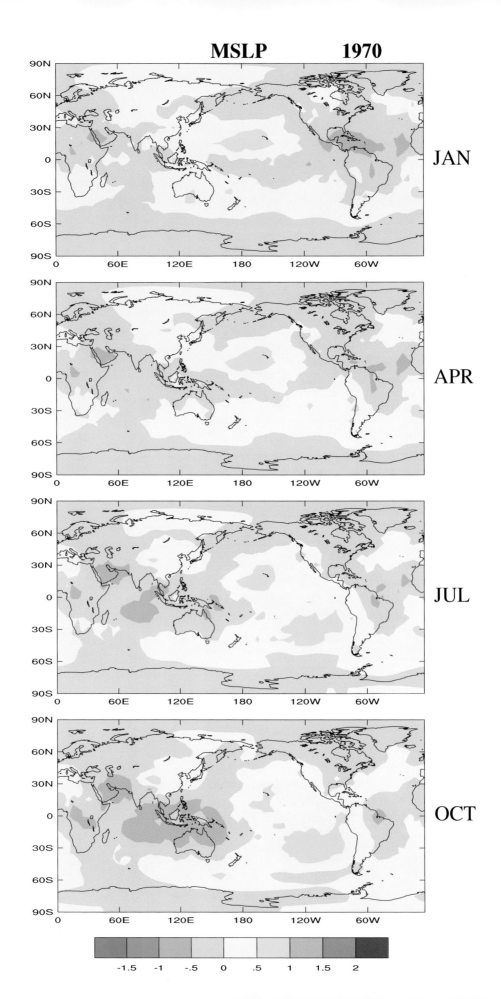

JAN

APR

JUL

OCT

SST   1970

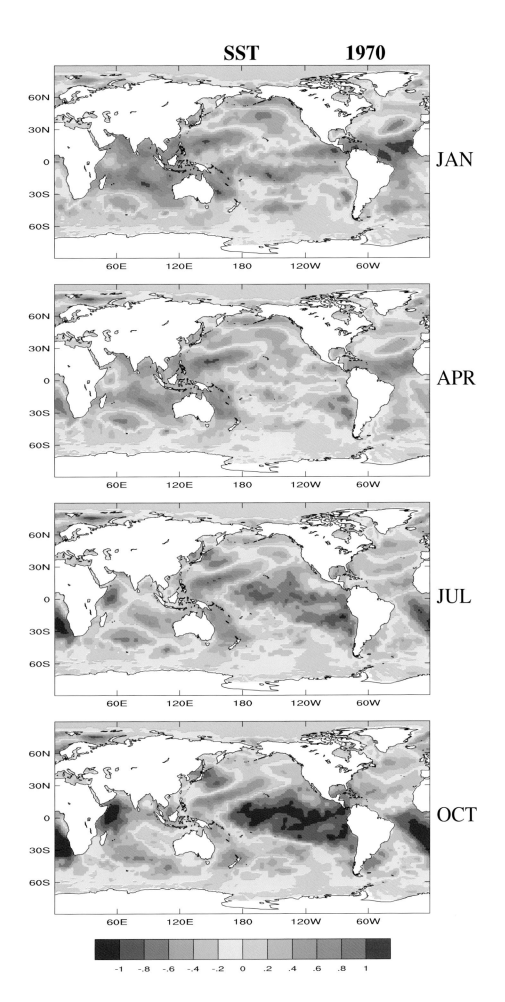

JAN

APR

JUL

OCT

MSLP   **1971**

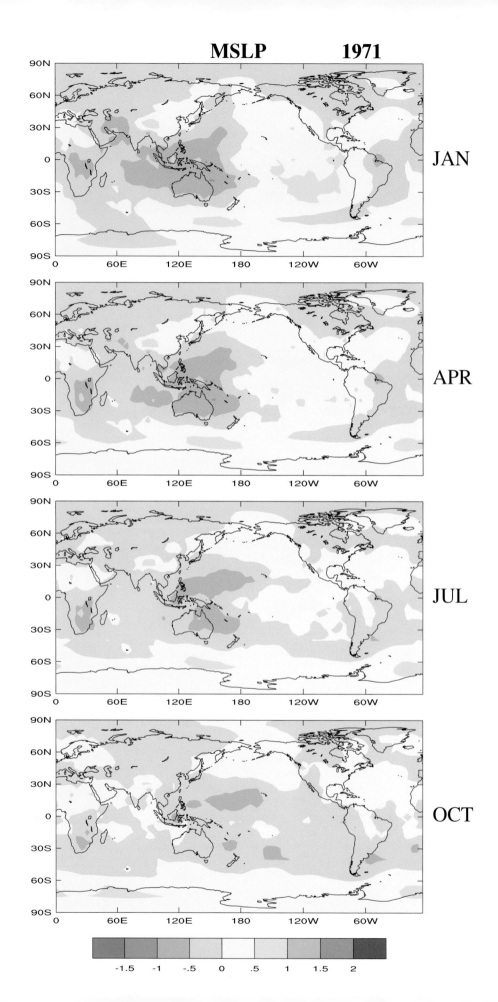

JAN

APR

JUL

OCT

-1.5   -1   -.5   0   .5   1   1.5   2

SST   1971

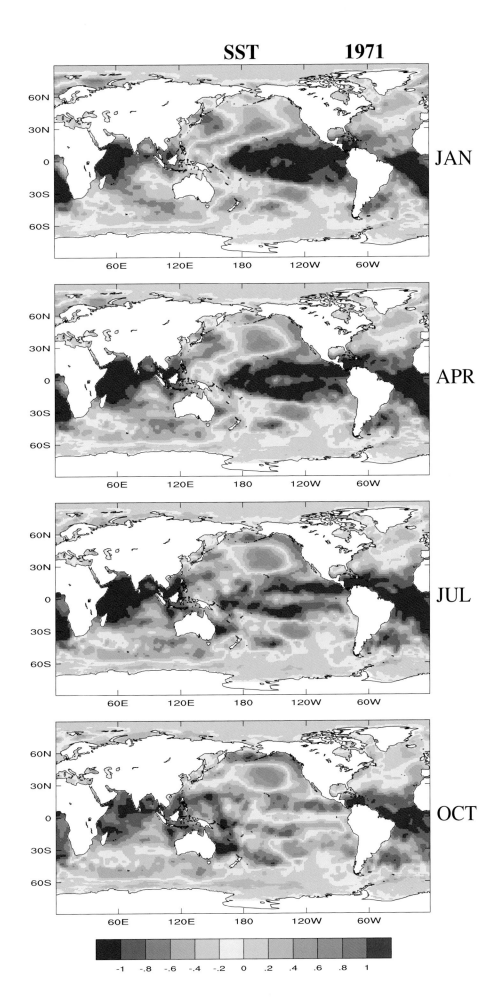

JAN

APR

JUL

OCT

MSLP   1972

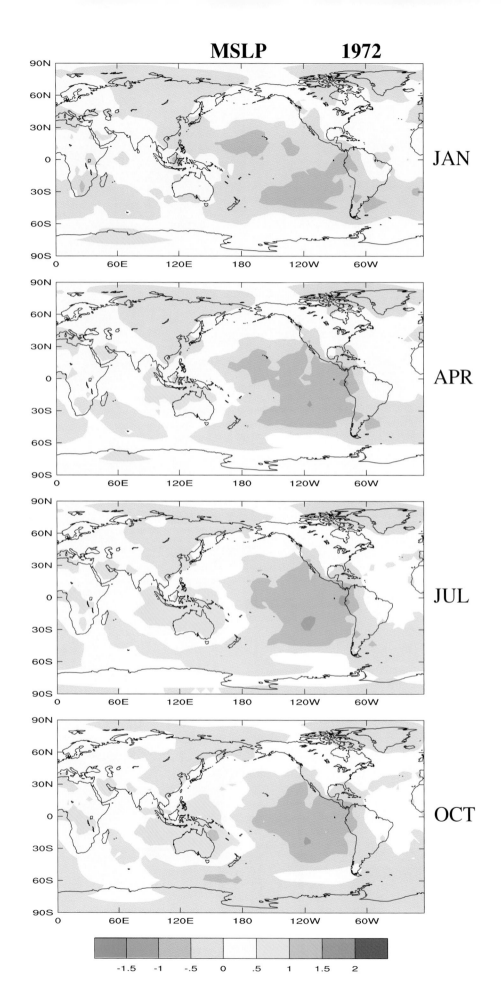

JAN

APR

JUL

OCT

SST   1972

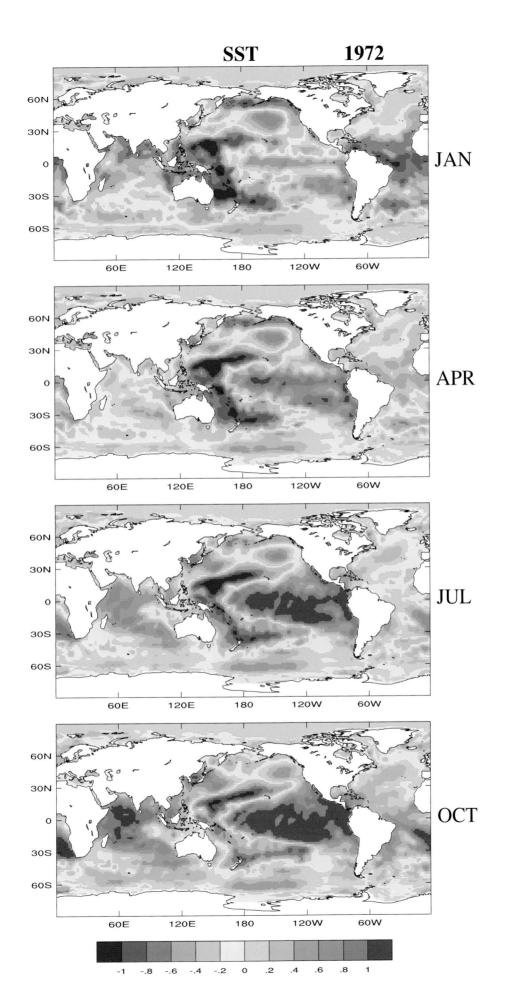

JAN

APR

JUL

OCT

-1   -.8   -.6   -.4   -.2   0   .2   .4   .6   .8   1

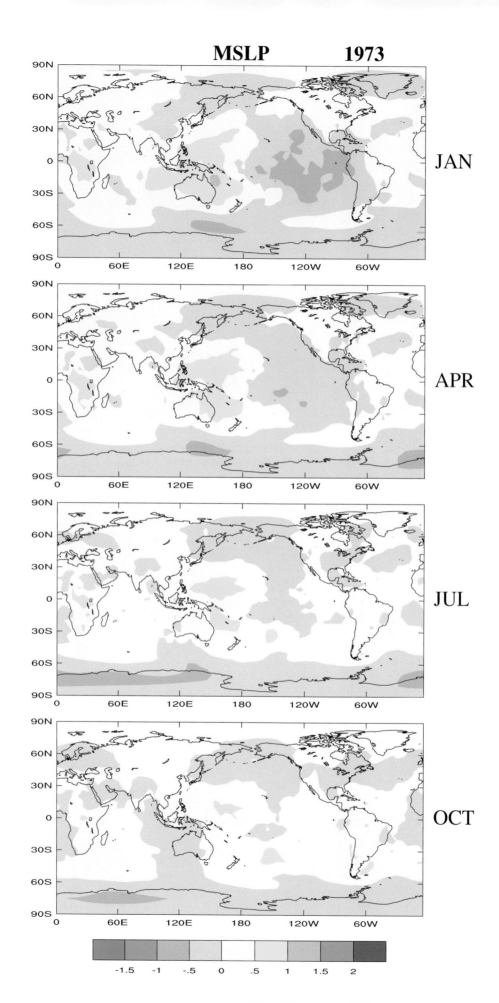

MSLP   1973

JAN

APR

JUL

OCT

SST   1973

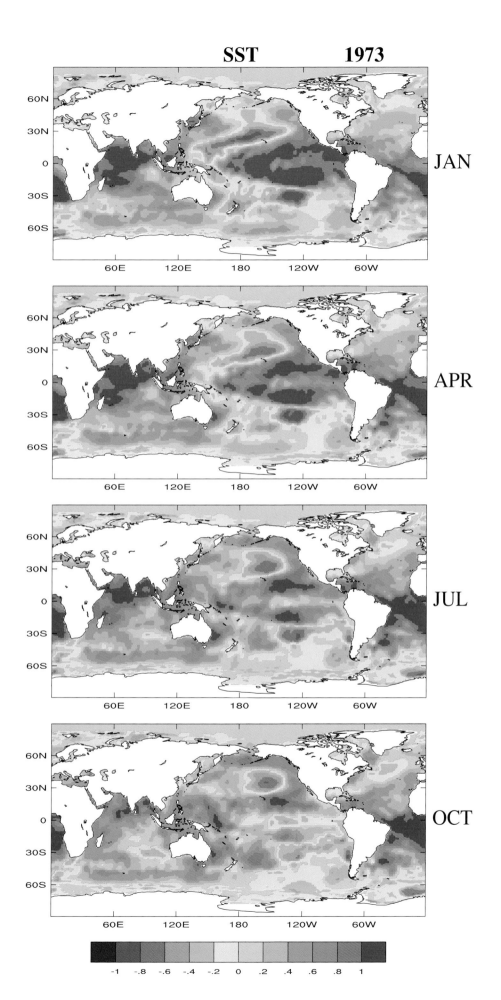

JAN

APR

JUL

OCT

-1 -.8 -.6 -.4 -.2 0 .2 .4 .6 .8 1

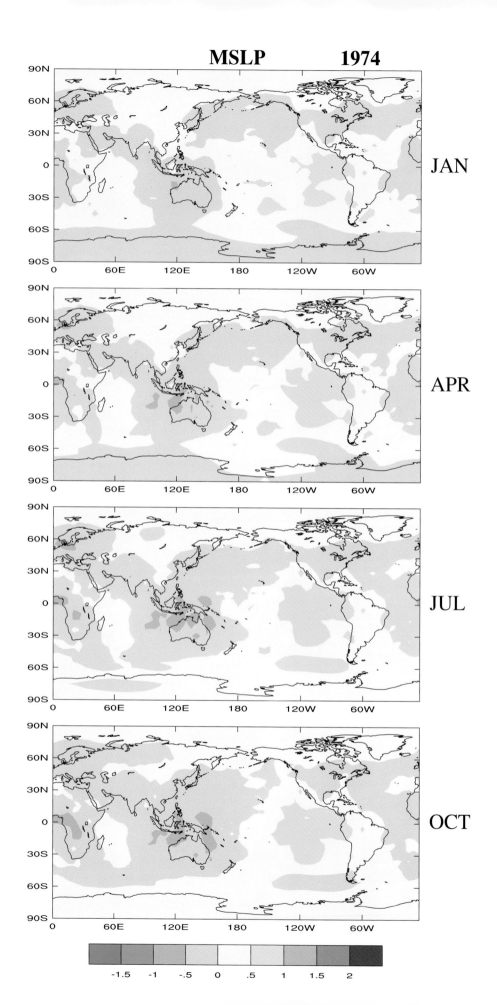

MSLP   **1974**

JAN

APR

JUL

OCT

SST   **1974**

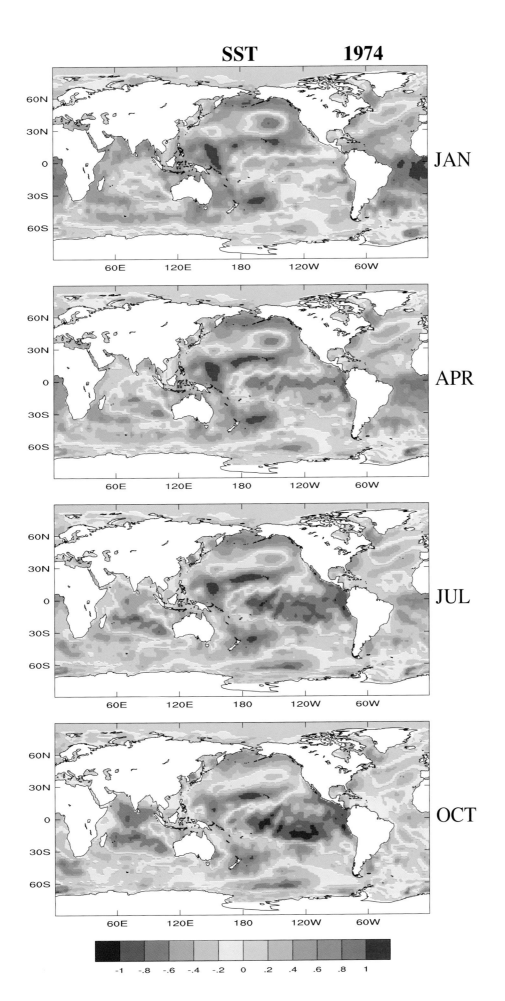

JAN

APR

JUL

OCT

-1 -.8 -.6 -.4 -.2 0 .2 .4 .6 .8 1

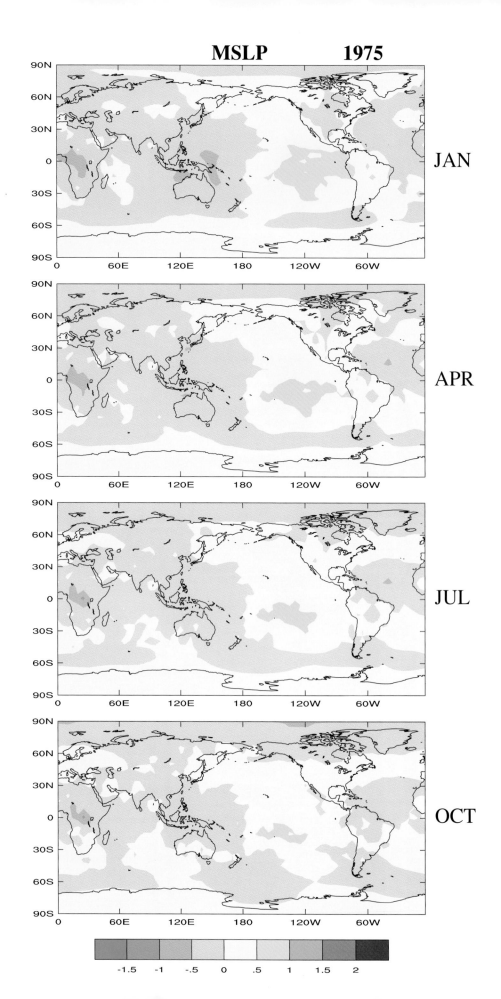

MSLP   **1975**

JAN

APR

JUL

OCT

-1.5   -1   -.5   0   .5   1   1.5   2

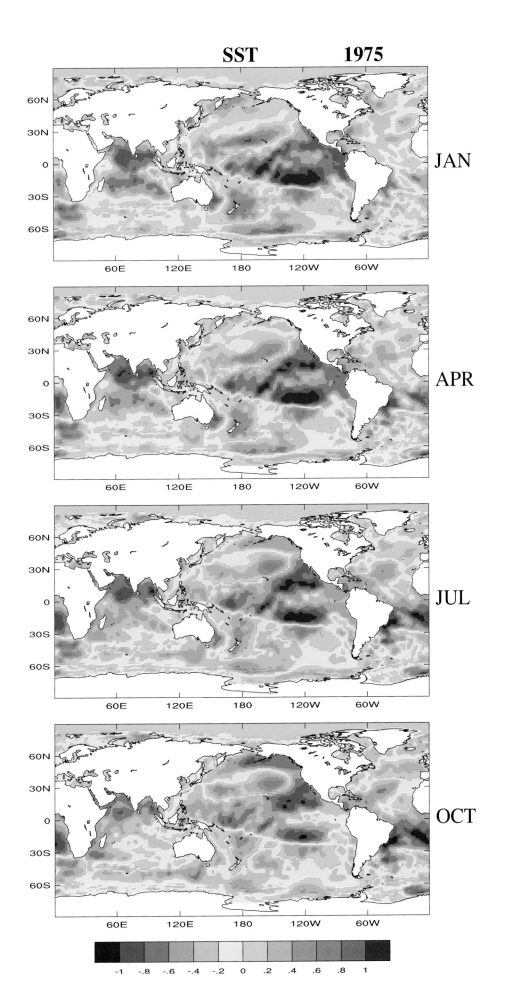

SST   1975

JAN

APR

JUL

OCT

-1 -.8 -.6 -.4 -.2 0 .2 .4 .6 .8 1

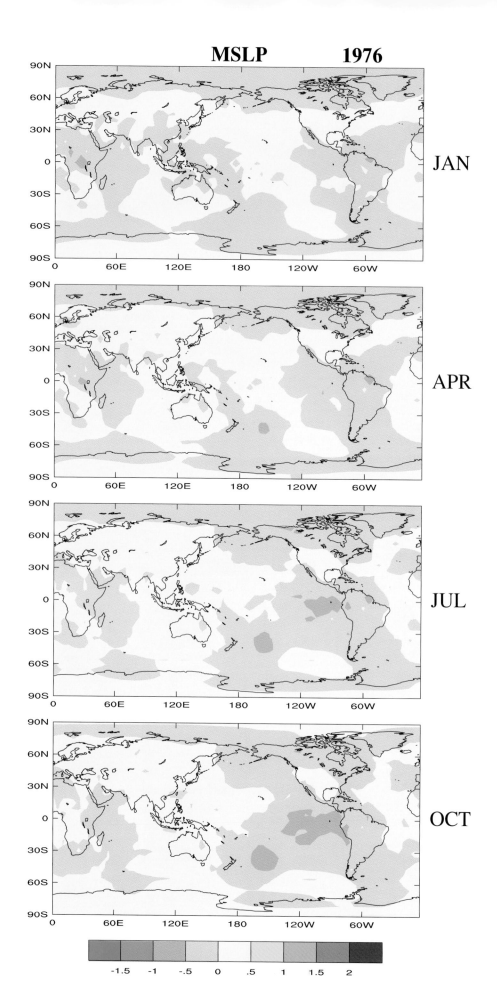

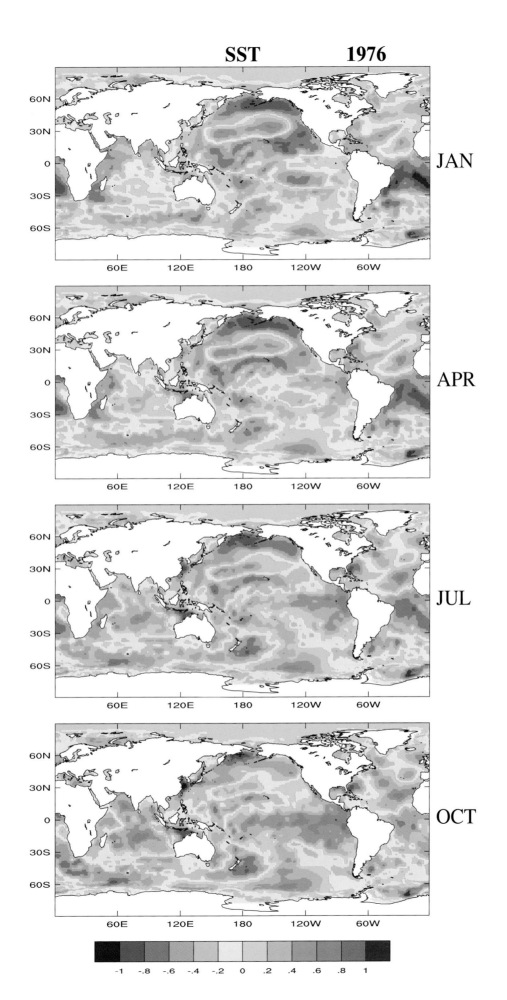

SST   1976

JAN

APR

JUL

OCT

-1 -.8 -.6 -.4 -.2 0 .2 .4 .6 .8 1

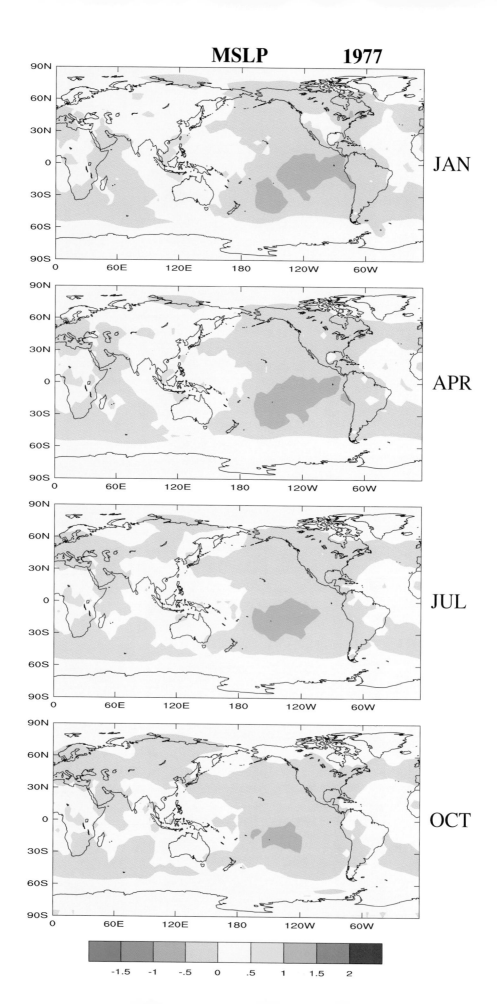

MSLP   **1977**

JAN

APR

JUL

OCT

-1.5   -1   -.5   0   .5   1   1.5   2

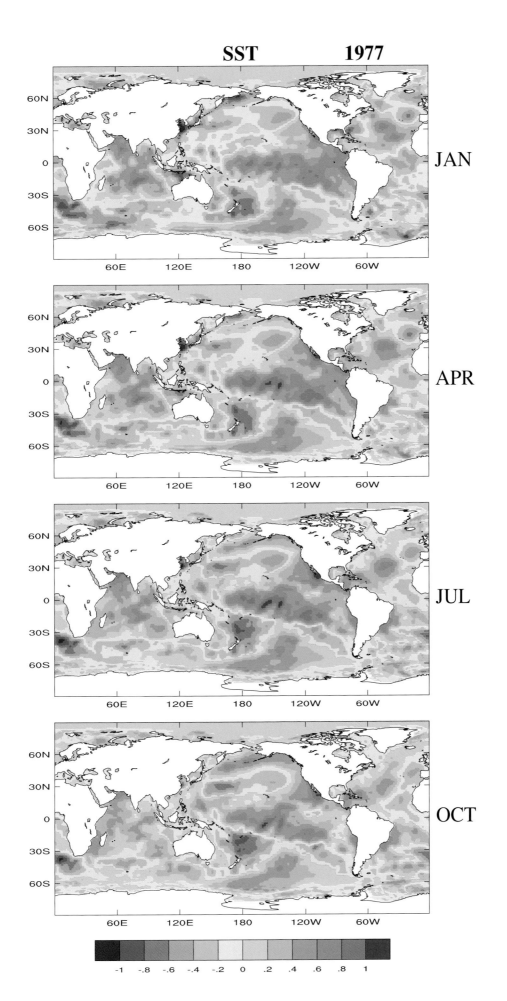

SST   **1977**

JAN

APR

JUL

OCT

-1   -.8   -.6   -.4   -.2   0   .2   .4   .6   .8   1

371

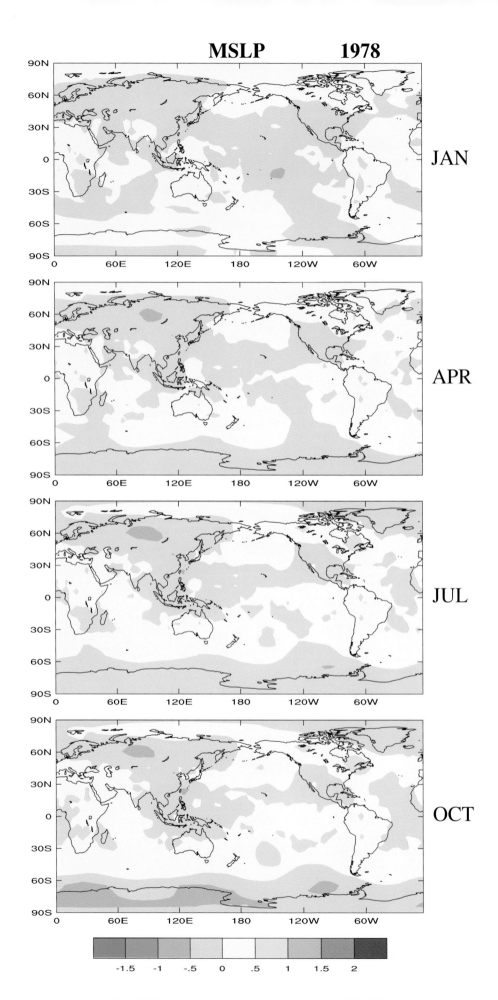

MSLP   1978

JAN

APR

JUL

OCT

-1.5   -1   -.5   0   .5   1   1.5   2

SST   **1978**

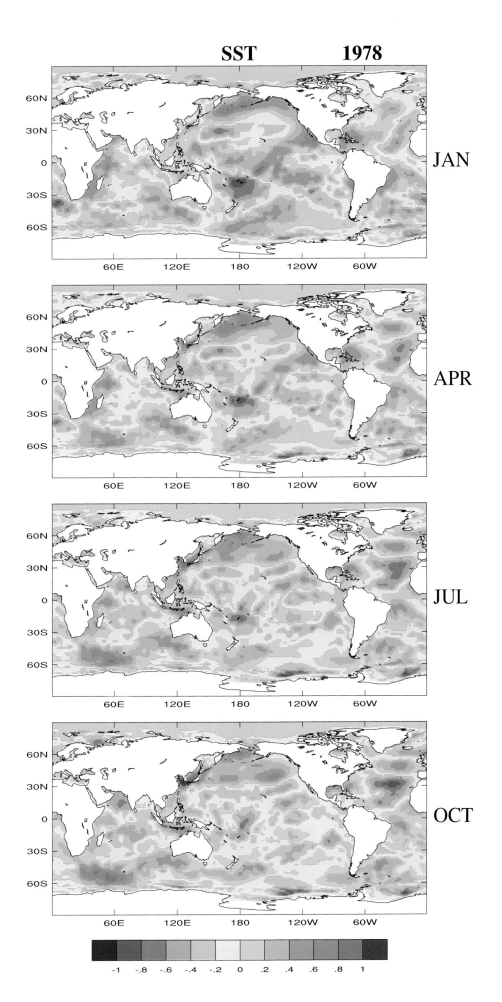

JAN

APR

JUL

OCT

-1   -.8   -.6   -.4   -.2   0   .2   .4   .6   .8   1

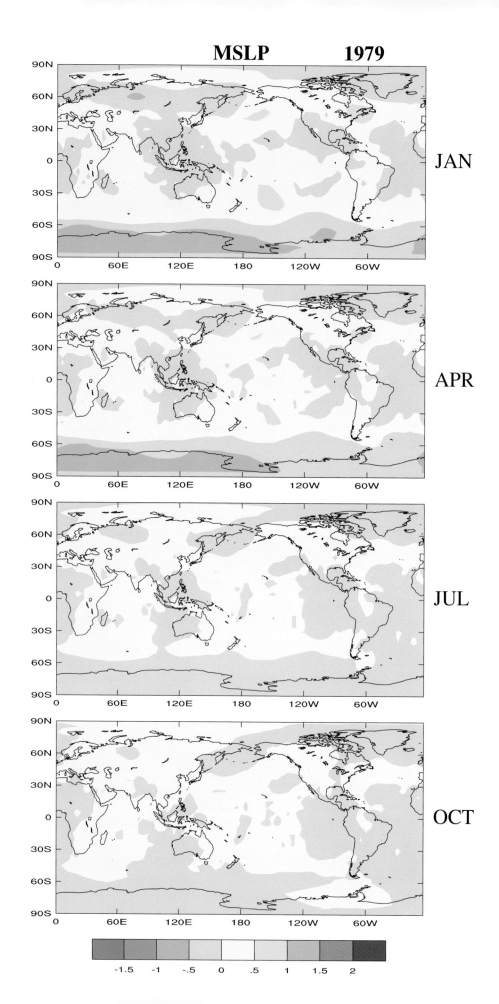

MSLP   **1979**

JAN

APR

JUL

OCT

SST   **1979**

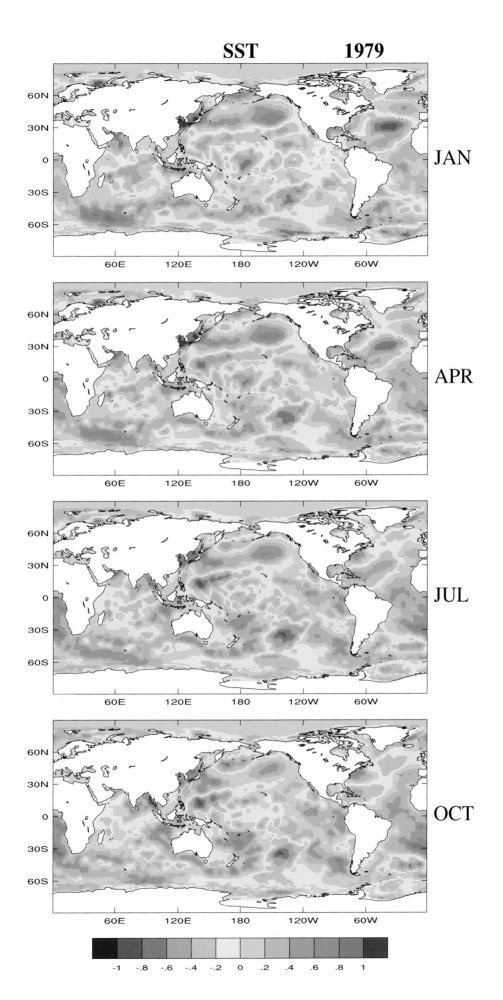

JAN

APR

JUL

OCT

-1   -.8   -.6   -.4   -.2   0   .2   .4   .6   .8   1

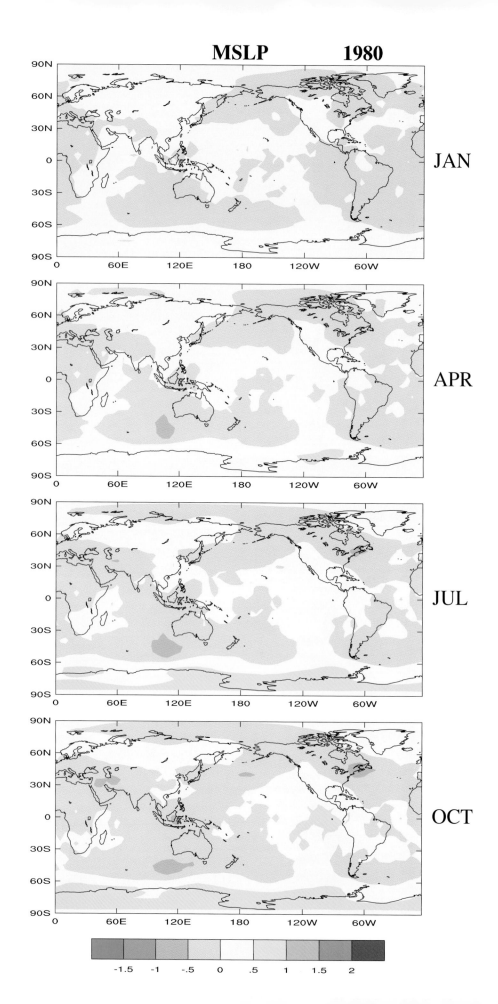

MSLP   1980

JAN

APR

JUL

OCT

-1.5   -1   -.5   0   .5   1   1.5   2

376

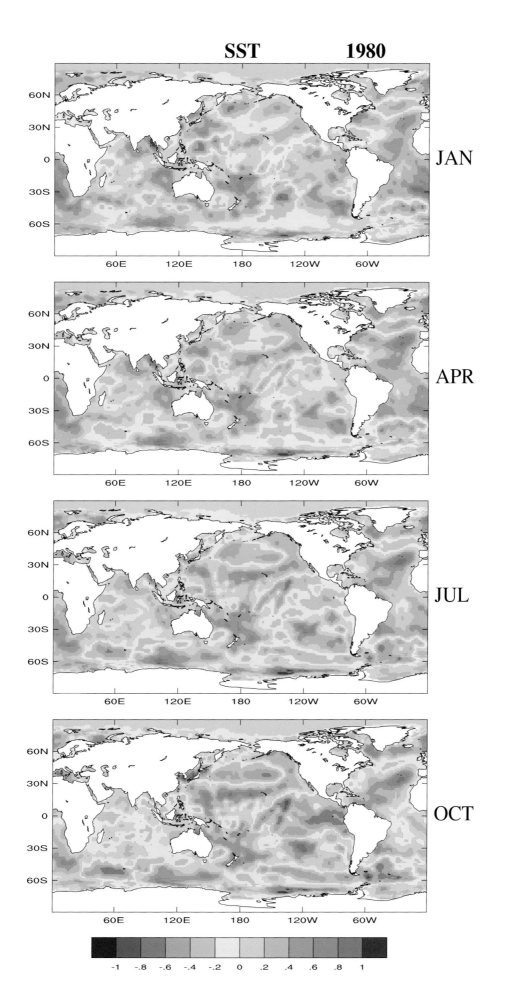

SST   1980

JAN

APR

JUL

OCT

-1 -.8 -.6 -.4 -.2 0 .2 .4 .6 .8 1

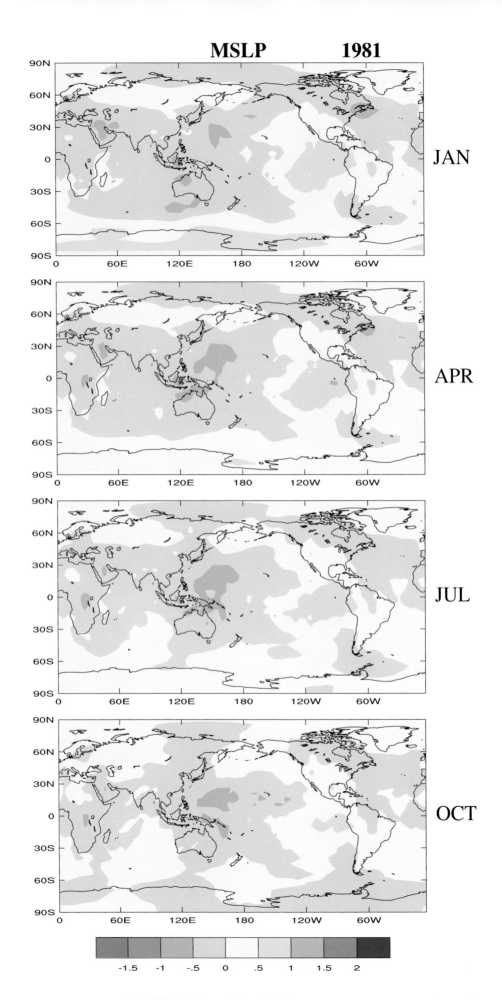

SST   **1981**

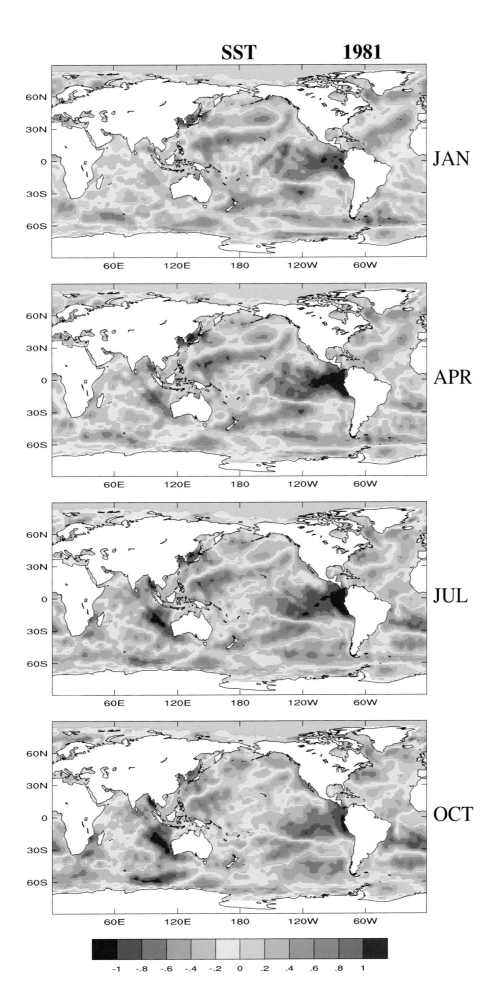

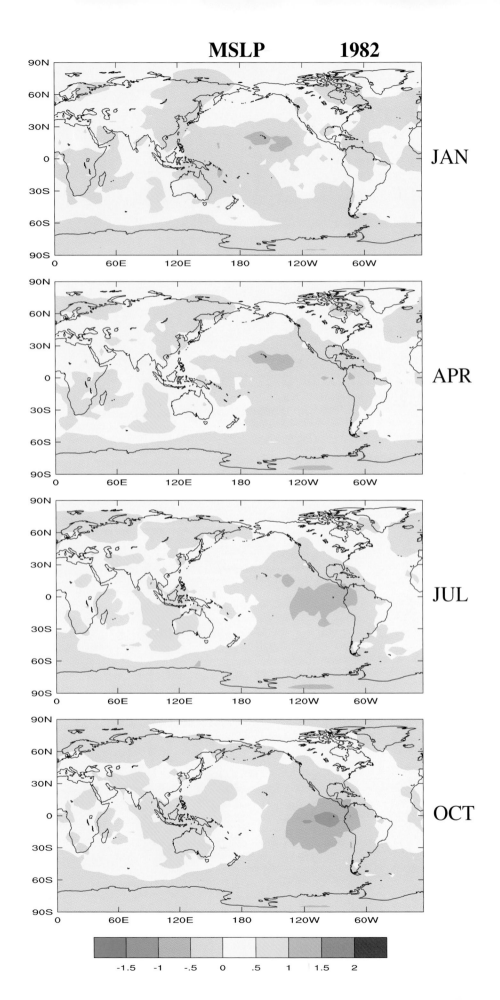

MSLP **1982**

JAN

APR

JUL

OCT

-1.5   -1   -.5   0   .5   1   1.5   2

SST   **1982**

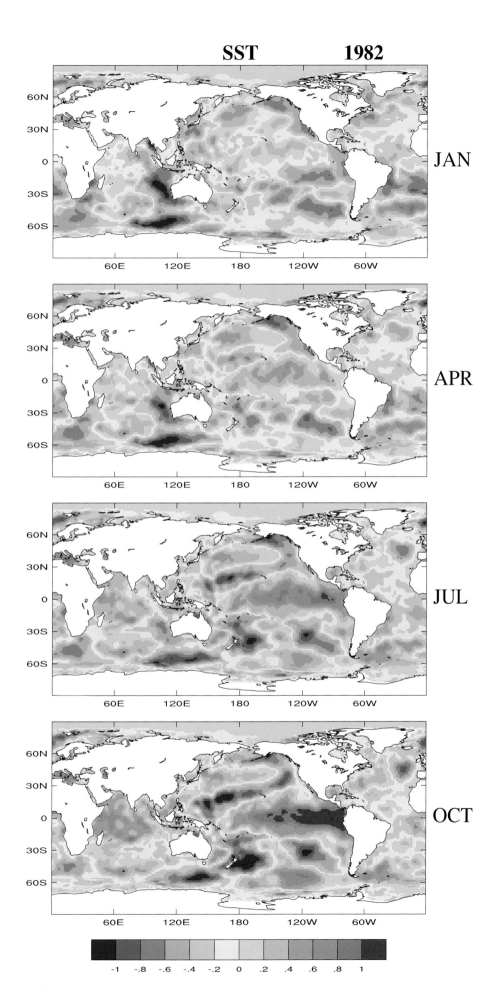

JAN

APR

JUL

OCT

-1   -.8   -.6   -.4   -.2   0   .2   .4   .6   .8   1

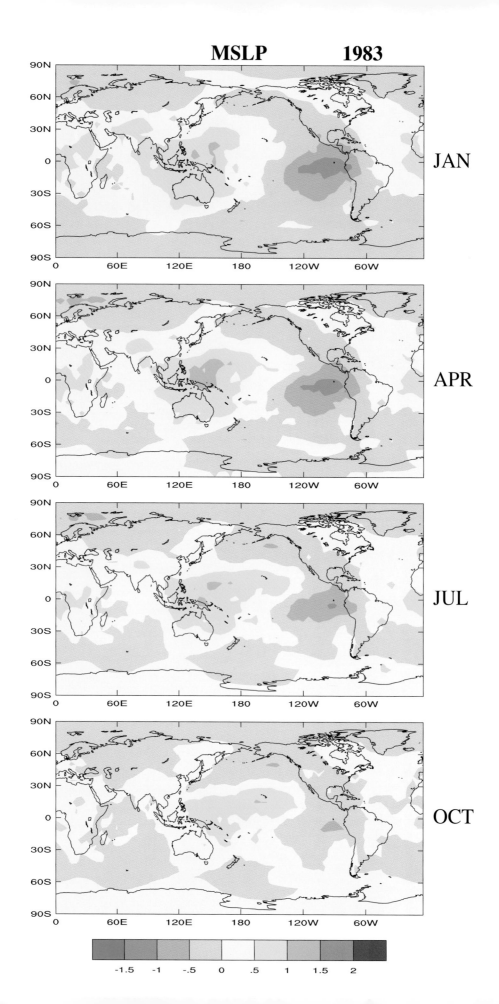

SST   **1983**

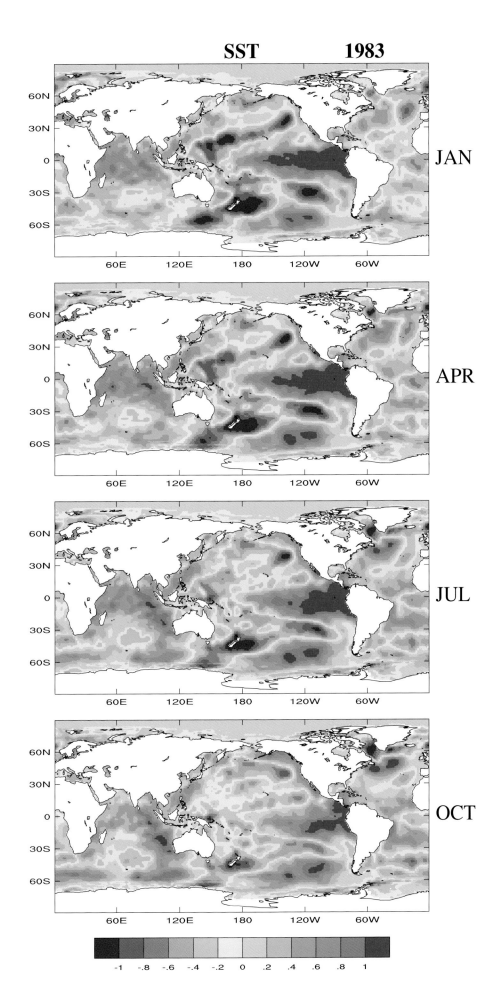

JAN

APR

JUL

OCT

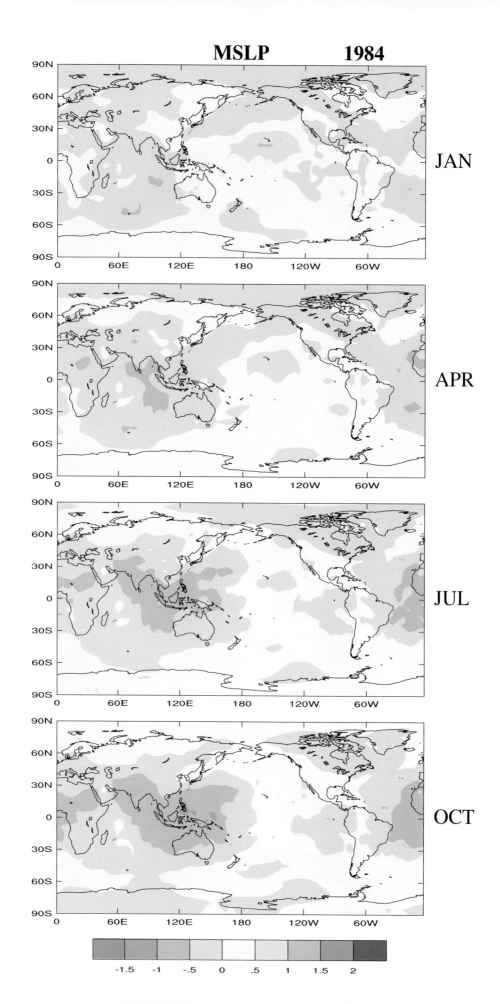

SST   **1984**

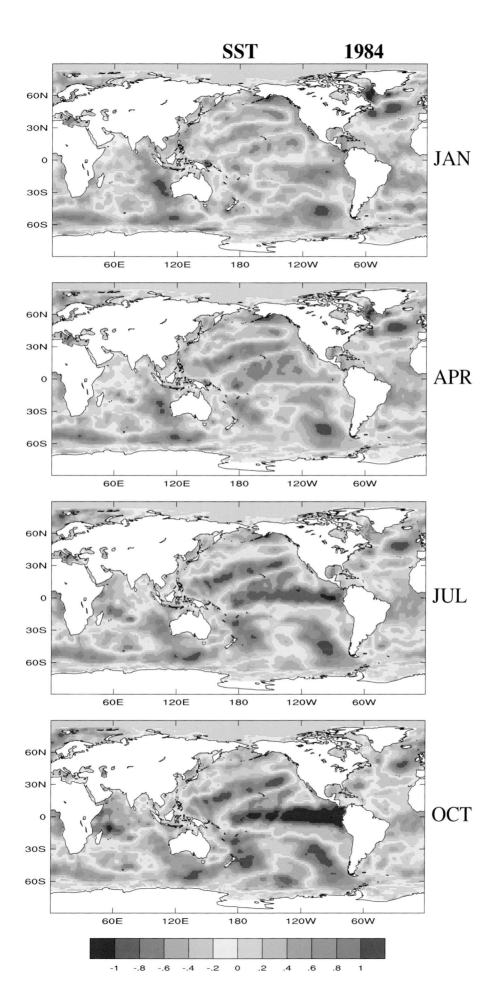

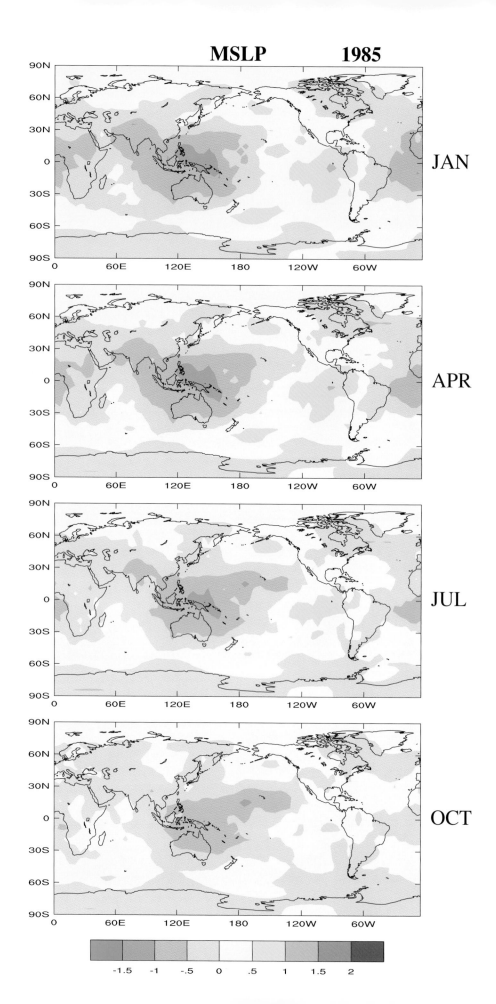

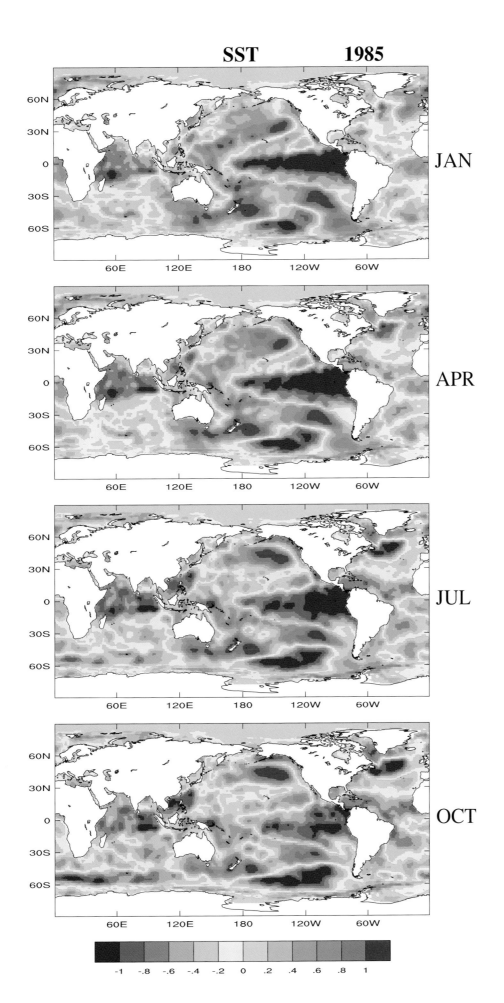

SST   **1985**

JAN

APR

JUL

OCT

-1   -.8   -.6   -.4   -.2   0   .2   .4   .6   .8   1

MSLP   **1986**

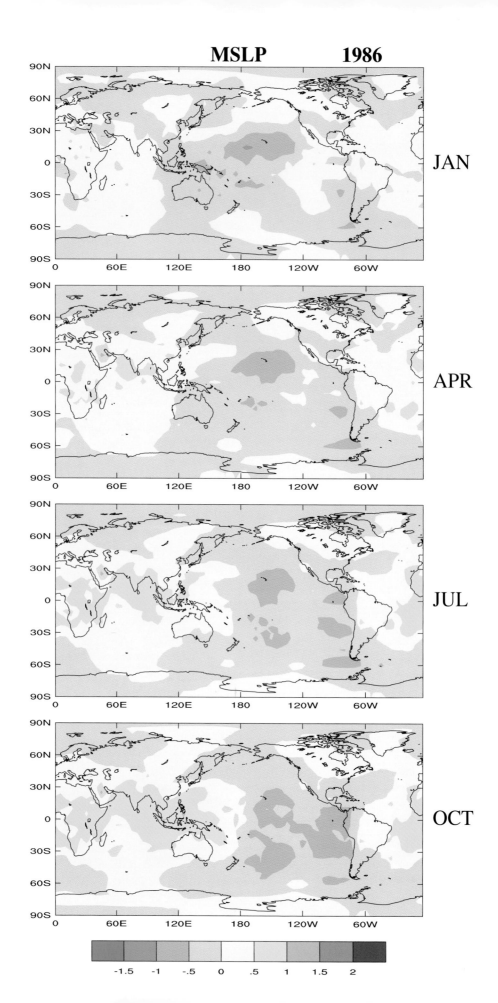

JAN

APR

JUL

OCT

-1.5   -1   -.5   0   .5   1   1.5   2

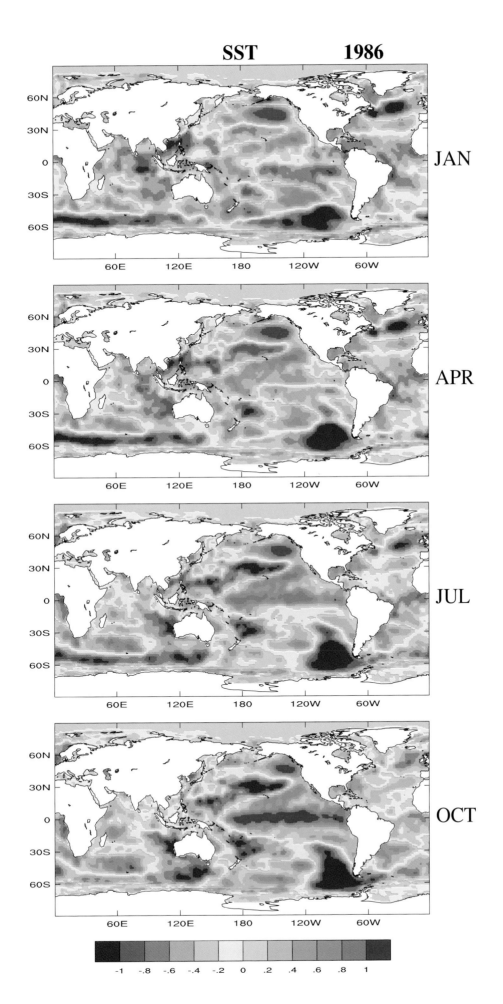

SST   1986

JAN

APR

JUL

OCT

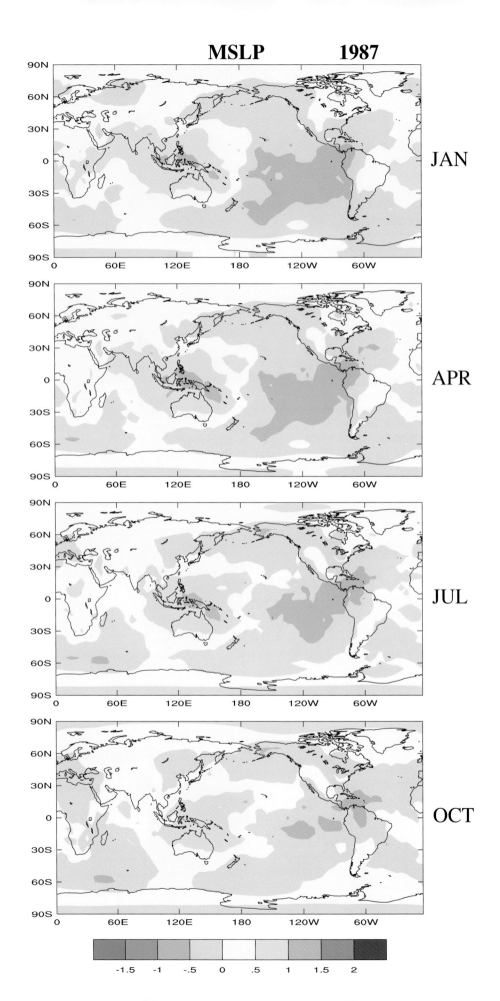

MSLP   1987

JAN

APR

JUL

OCT

-1.5   -1   -.5   0   .5   1   1.5   2

SST   1987

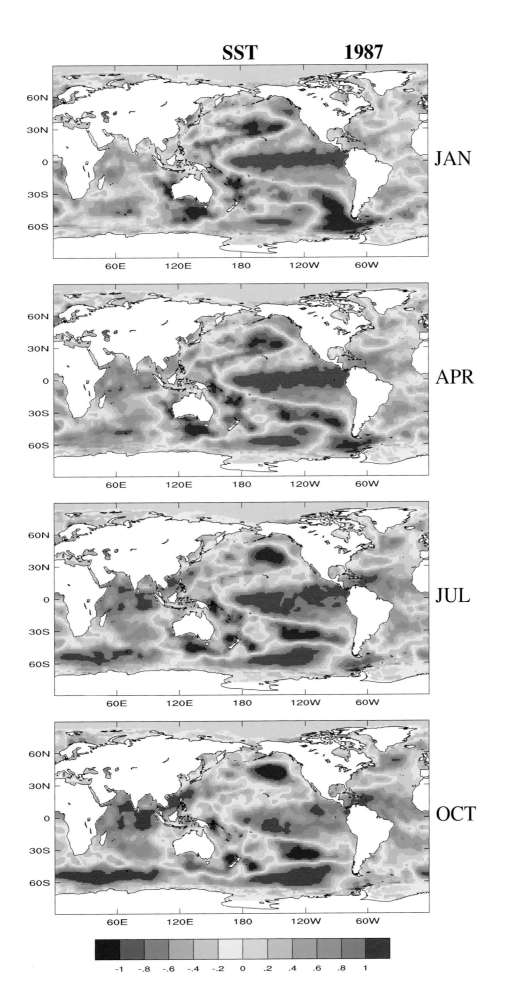

JAN

APR

JUL

OCT

-1 -.8 -.6 -.4 -.2 0 .2 .4 .6 .8 1

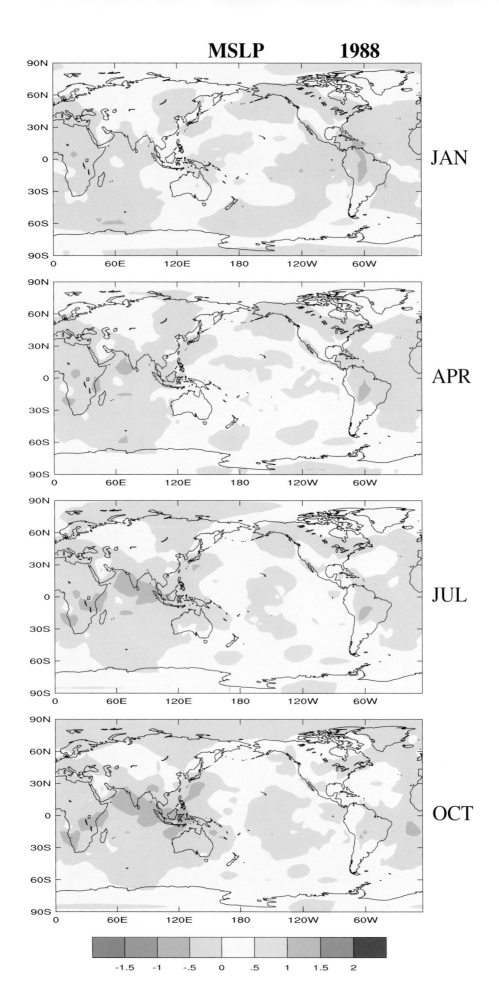

MSLP   1988

JAN

APR

JUL

OCT

-1.5   -1   -.5   0   .5   1   1.5   2

SST   1988

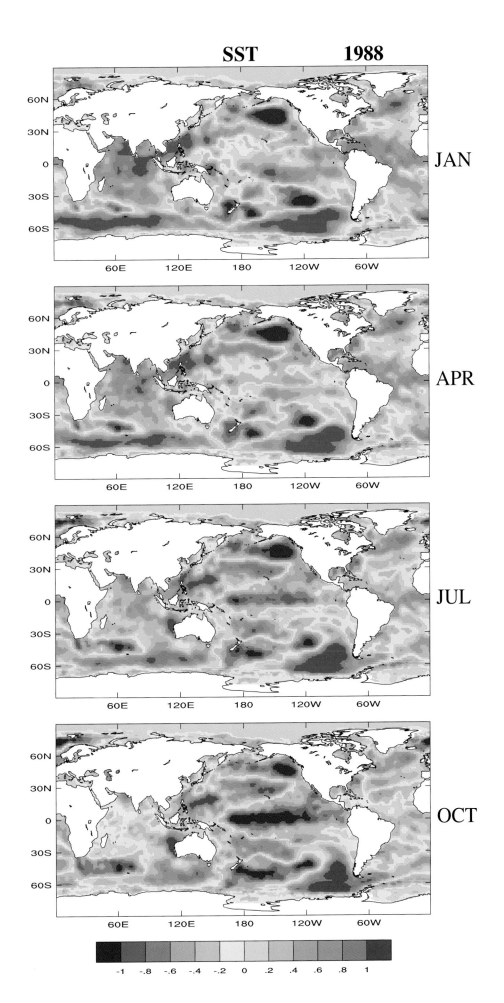

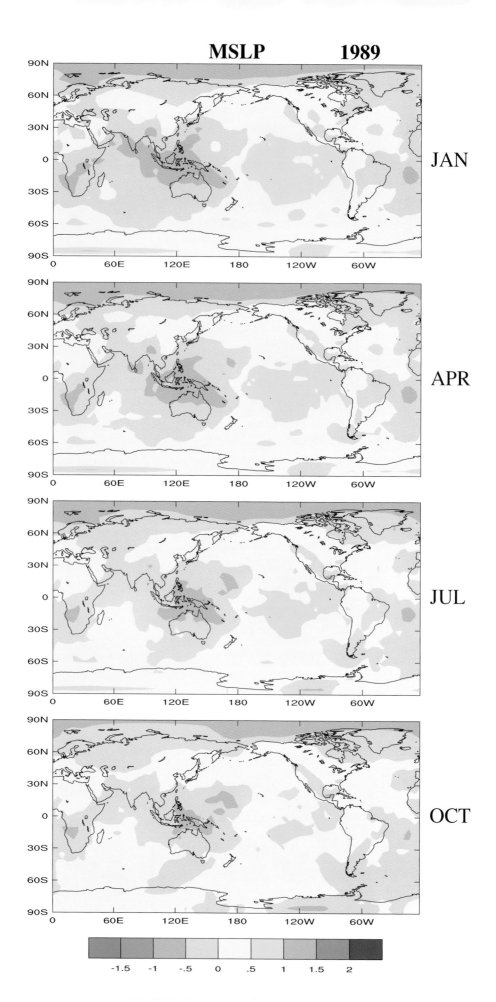

MSLP   **1989**

JAN

APR

JUL

OCT

-1.5   -1   -.5   0   .5   1   1.5   2

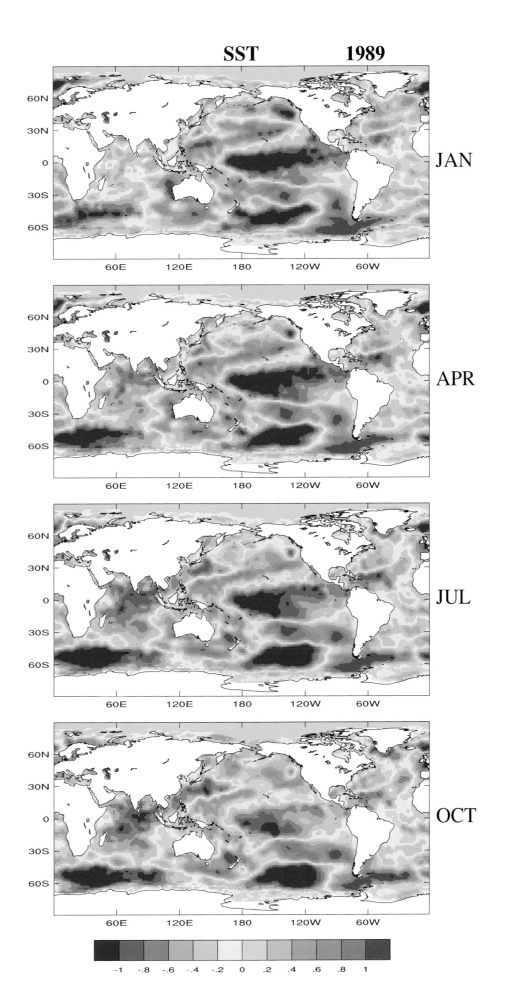

SST   1989

JAN

APR

JUL

OCT

-1 -.8 -.6 -.4 -.2 0 .2 .4 .6 .8 1

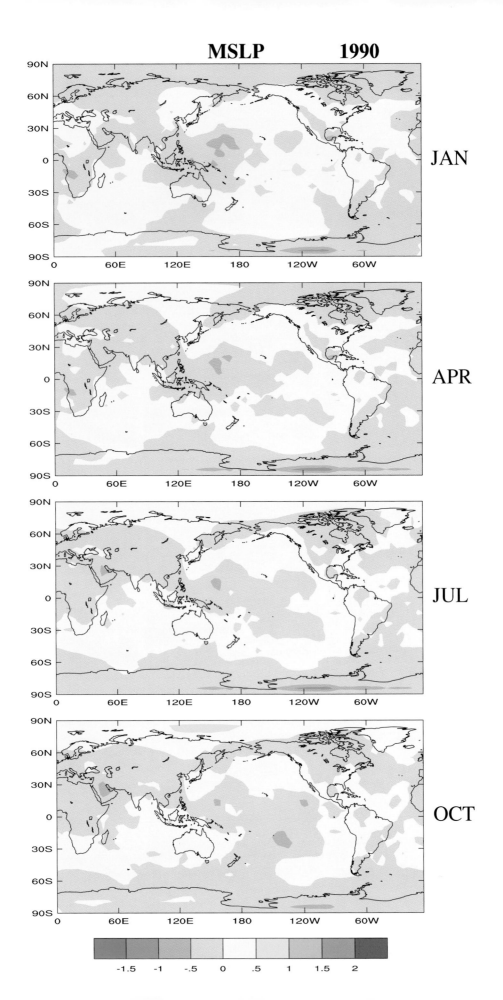

SST   **1990**

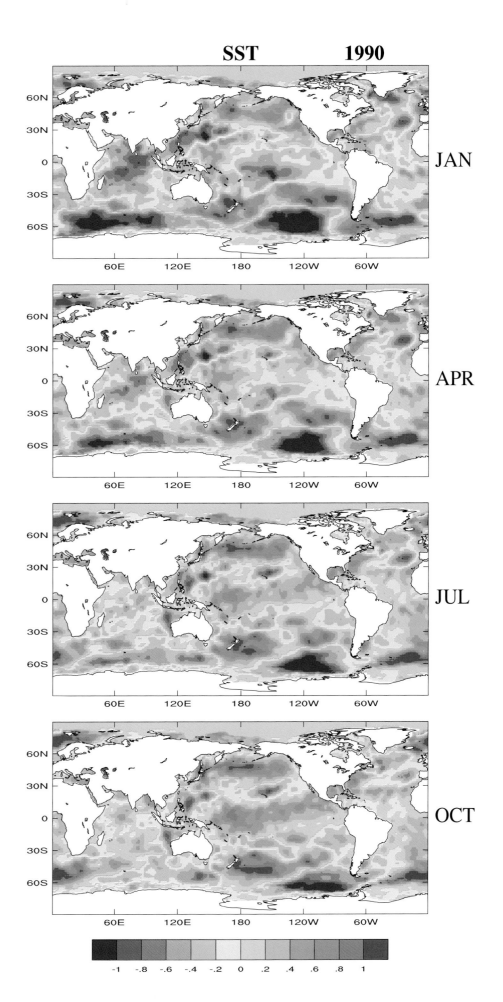

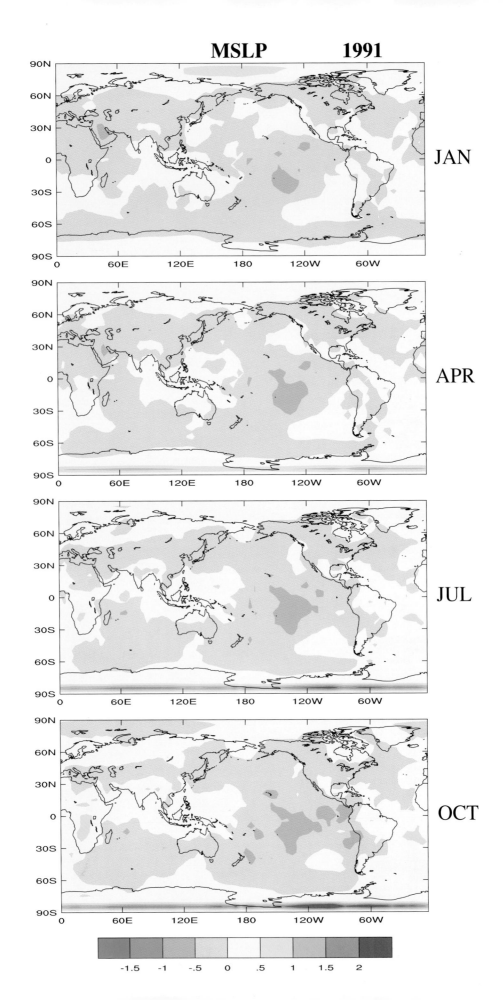

MSLP   **1991**

JAN

APR

JUL

OCT

-1.5   -1   -.5   0   .5   1   1.5   2

SST　　　1991

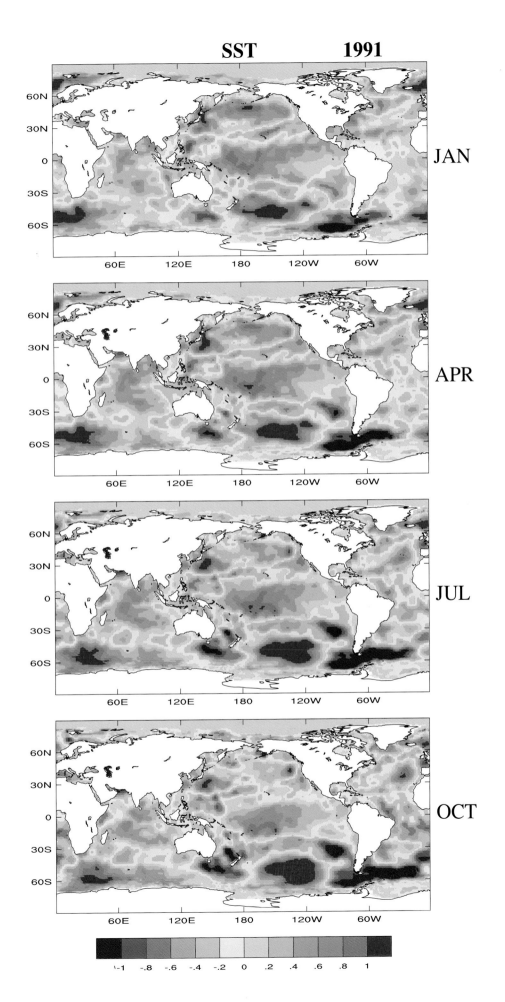

MSLP   **1992**

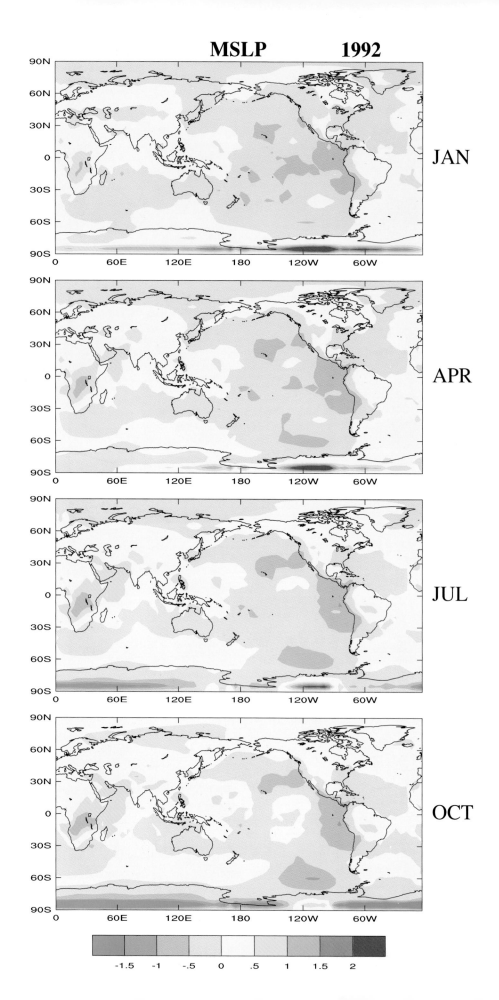

JAN

APR

JUL

OCT

-1.5   -1   -.5   0   .5   1   1.5   2

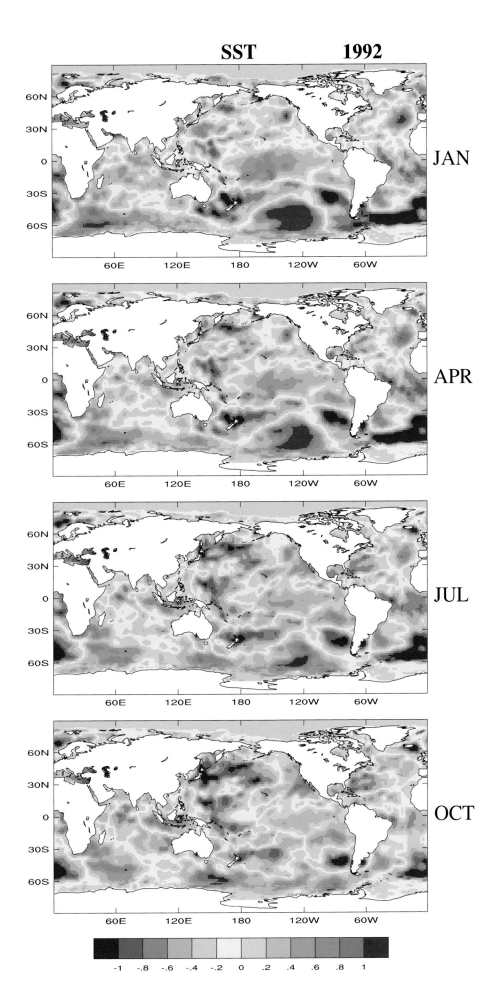

SST   1992

JAN

APR

JUL

OCT

-1   -.8   -.6   -.4   -.2   0   .2   .4   .6   .8   1

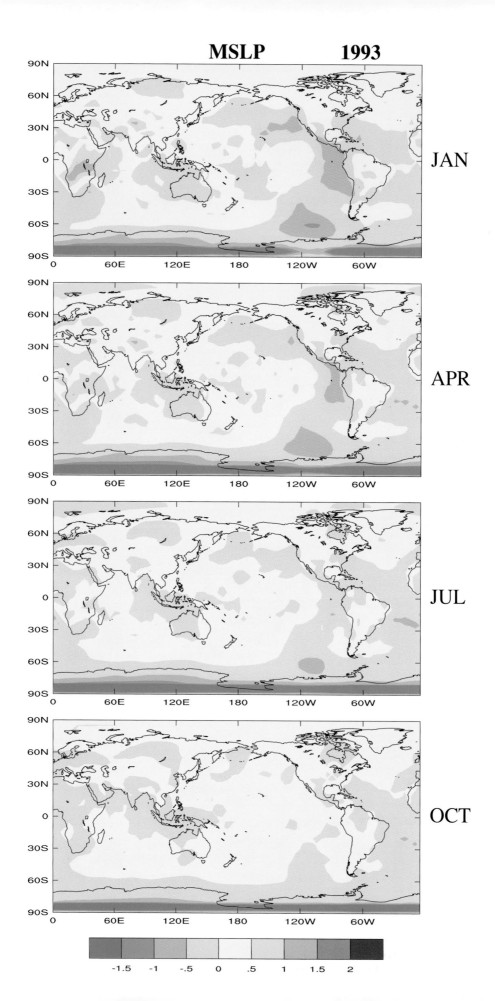

MSLP **1993**

JAN

APR

JUL

OCT

-1.5 -1 -.5 0 .5 1 1.5 2

402

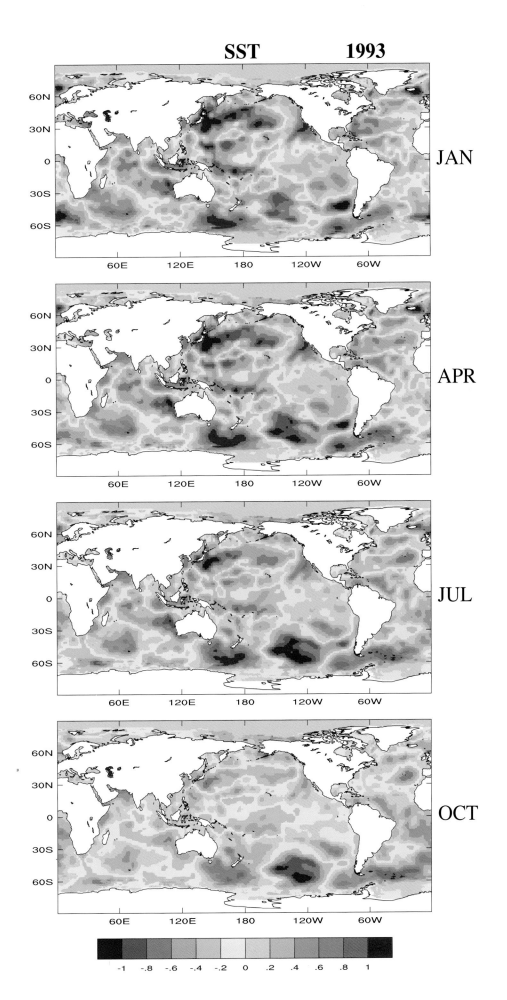

SST   1993

JAN

APR

JUL

OCT

-1 -.8 -.6 -.4 -.2   0   .2   .4   .6   .8   1

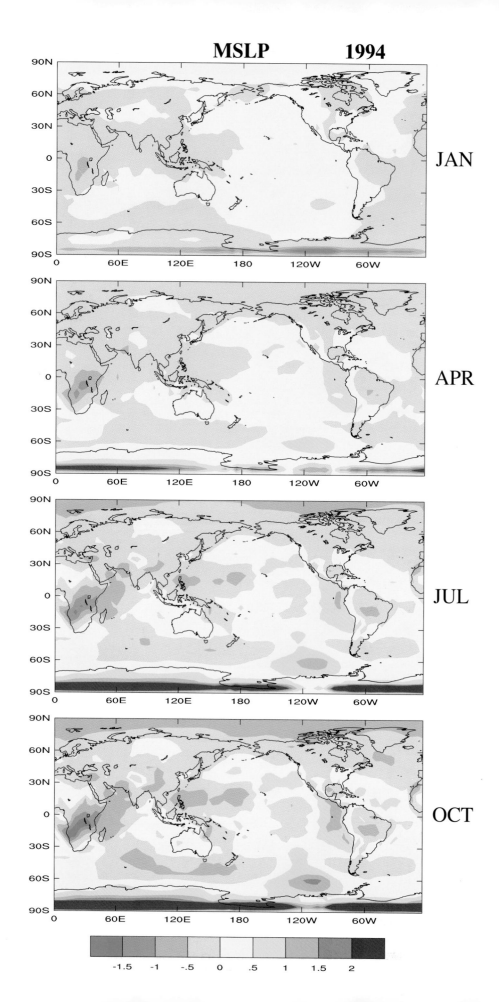

MSLP   1994

JAN

APR

JUL

OCT

-1.5   -1   -.5   0   .5   1   1.5   2

SST   **1994**

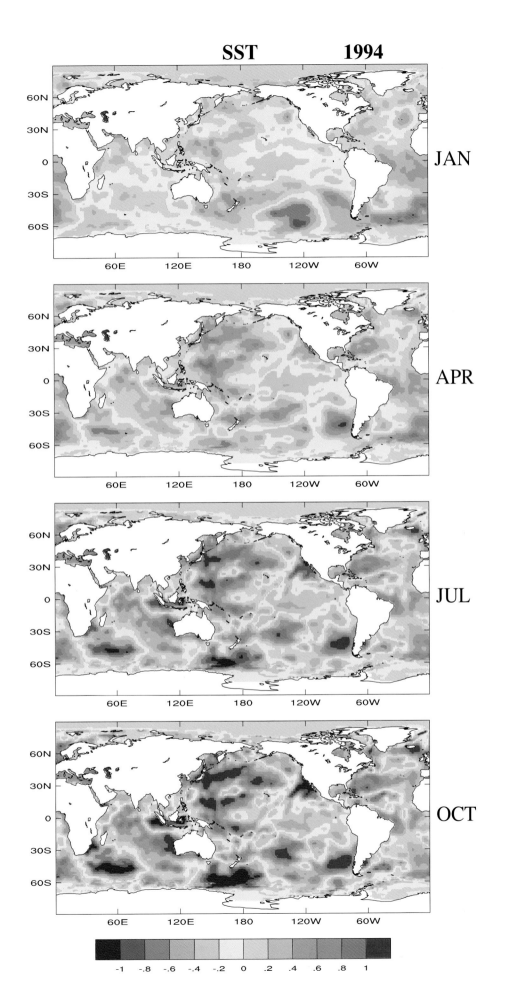

JAN

APR

JUL

OCT

-1   -.8   -.6   -.4   -.2   0   .2   .4   .6   .8   1